Amazonian Ethnobotanical Dictionary

James Alan Duke
Economic Botanist
U.S. Department of Agriculture
Beltsville, Maryland

Rodolfo Vasquez
Assistant Curator
Missouri Botanical Garden
St. Louis, Missouri
Plant Explorer
Herbarium Amazonense
Iquitos, Peru

CRC Press
Boca Raton Ann Arbor London Tokyo

Library of Congress Cataloging-in-Publication Data

Duke, James A., 1929-
Amazonian ethnobotanical dictionary/by James Alan Duke and
Rodolfo Vasquez Martinez.
p. cm.
Includes bibliographical references and index.
ISBN 0-8493-3664-3
1. Ethnobotany—Peru—Dictionaries. 2. Ethnobotany—Amazon River
Valley—Dictionaries. 3. Botany, Economic—Peru—Dictionaries.
4. Botany, Economic—Amazon River Valley—Dictionaries.
5. Medicinal plants—Peru—Dictionaries. 6. Medicinal plants—
Amazon River Valley—Dictionaries. I. Vasquez Martinez, Rodolfo.
II. Title. III. Title: Amazonian ethnobotanical dictionary.
GN564.P4D85 1994
581.6′1′098543—dc20 94-390
CIP

Direct all inquiries to CRC Press, Inc., 2000 Corporate Blvd., N.W., Boca Raton, Florida 33431.

International Standard Book Number 0-8493-3664-3
Library of Congress Card Number 94-390
Printed in the United States of America 2 3 4 5 6 7 8 9 0
Printed on acid-free paper

DEDICATION

To the memory of Alwyn Gentry (January 6, 1945 - August 3, 1993), we gratefully dedicate this Amazonian Ethnobotanical Dictionary (Peru). Al tried so hard to teach so much to Jim Duke, his senior by 16 years, and did teach a great deal to Rodolfo Vasquez, his junior by 15 years. Al was killed in an airplane crash near Guayaquil, Ecuador on August 3, 1993. Jim was in Costa Rica at the time, mentioning Al as one of the greats, to an ecotour in the cloudforest at Monte Verde. Judy duCellier, Duke's right hand woman, was retyping the final draft of this book, and got the heart-breaking news in Beltsville an hour or two before Duke, in Costa Rica. Rodolfo was at the Missouri Botanical Garden, working on the Flora of Peru, and proof-reading this final draft, when Al's plane went down, taking with it, two of the greatest storehouses of rainforest knowledge, Al Gentry and Ted Parker. We suspect that Al probably spotted two species new to Ecuador and one new to science in his last tree-topping minutes. Duke had worked in many of the same forests in Colombia (Choco), Ecuador (Rio Palenque), Panama (Darien) and more recently in Peru (Explorama). Duke enjoyed teaching Al's daughter, Diane, in March of 1993. That was the last time Duke worked with Al on his plot at Explorama Lodge, Peru, home to 300 woody species/hectare (and 600 individuals) 4 inches or more in diameter. As always, Al was such an indefatigable powerhouse that he wore Duke out in the morning, trying to clip specimens from the unknown new forest saplings in his Explorama Plots. That's when Al autographed for Duke a copy of his *Field Guide to the Families and Genera of Woody Plants of Northwest South America* (1993).

Our book has been enhanced by the taxonomic foundation provided by the arduous and intensive perseverance of the late Dr. Gentry. He would not rest until he named an unknown forest giant. It was Al, world authority of the Bignoniaceae, who pointed out how overcollecting was endangering two medicinal Bignoniaceae around Iquitos, the "clavohuasca" and "tahuari". We respectfully but sadly dedicate this small volume to a fallen forest giant, Al Gentry, friend of the forest and teacher to many of us, trying to save the forest that survives him. The Amazonian Center for Environmental Education and Research (ACEER) will dedicate their 250,000-acre forest to his memory. The forest that kept him going like a robot swallowed him up. But his spirit lives on, and will help in the difficult efforts to save the forest of today for the children of tomorrow. Few of us can view any attractive bignoniaceous vine without recalling the ethereal spirit of Al Gentry and the forests he represented. Long live the forest and the spirit of Al Gentry.

TABLE OF CONTENTS

PARADISE LOST

(Could be sung to the tune of John Prine's **Paradise**)

Jim Duke

I praise you John Prine, and I hope you don't mind,
If I mimic your song, to help the forest along.
Even while I am singing, the axeman is swinging,
Choppin' down all that green, to plant corn, rice and bean

♂ Daddy won't you take me to the Primary Forest
By the Amazon River where Paradise lies?
I'm sorry my son, but the forest is gone!
I'll show you some slides, that'll have to suffice!

If you'll not name me, there's something I'll mention
And so folks won't blame me, I'll quote Peter Jenson.
There may be stronger reasons, but I can't think of any,
We're losing the forest "because we're too many"!

♀ Momma won't you take me to the Primary Forest
On the Amazon River where Paradise lies?
I'm sorry my daughter, but I don't think I oughta'
We've waited too long, now the forest is gone!

Oh axeman unkind, you are blowing my mind!
Camu-camu and brazilnut, they can help fill your gut.
But year after year, once the forest is clear,
You'll have less and less food, and you'll run out of wood.

Never thought ecotours, could be one of the cures;
Taking "green" bucks from gringoes, getting mud on their toes.
If the ecotours thrive, Indian cultures survive,
And the children will strive, to keep tradition alive.

No place I'd rather go, than to cruise on the Napo;
Hoping some of my pleas, kinda' help save the trees.
I'd rather you'd find me, sunnin' with the tree huggers
Than back in DC, all arunnin' from muggers!

It's quite element'ry, our praise for Al Gentry,
Whose conserving career really helped at ACEER.
The best botany brain, went down with Al's plane,
And although he is gone, we must still carry on

Jaborandi, papain, curare, and quinine,
The forest's the best, for your medicine chest.
Aware of these goods, you still chop down the woods.
You'd best spare that tree, cause it might help spare thee.

Axeman,
Leave that tree alone!
The life you save
May be your own.
(Wise old owl, 1993)

By Barrie Maguire

INTRODUCTION

This book was conceived and delivered in the Amazonian rainforest, and is about the rainforest, for its inhabitants and devotees, near and far. In addition to helping the curious visitor identify some of the amazing Amazonian phytodiversity, this book may help rainforest inhabitants who cannot afford modern medicine identify plants that their ancestors used for forest maladies. All royalties will revert to ACEER[1] for rainforest conservation.

In 1991, as an ethnobotanical instructor, I participated in my first ecotour. At Explorama Lodge, 3 hours downstream by fast boat from Iquitos, Peru, I taught forest product workshops, ably assisted by Don Segundo Inuma. At the workshop, several interested people got a close look at many of the plants upon which Amazonians depend so heavily, first in the thatched classroom, then out in the field. (During one workshop, one participant was even nipped by a tapir, which also threatened the instructor.) Explorama Lodge is an amazingly comfortable facility nestled in a 250,000-acre rainforest preserve, in an area with 300 woody species per hectare. Good forest guides know the Amazonian names and uses of many plants. Generally, they are pleased to share this information with interested visitors. They don't often know the universal Latin scientific names of the plants.

Ah, serendipity! Sra. Maria Wright had sent me a xerox copy of Rodolfo Vasquez Martinez' Spanish draft, *Plantas Utiles de la Amazonia Peruana*. I had it copied and bound to take to Amazonian Peru, without having even studied it. When Segundo started talking about a plant, I had merely to look up the local name in the index. That led me to the scientific name (key to most published ethnobotanical and phytochemical data). In later travels in Peru, I had the privilege of learning also from Antonio Montero and Lucio Pano. Their Peruvian folklore naturally matched that of Rodolfo and Segundo, confirming the value of the lore presented herein.

How well I remember my first arrival in Panama 30 years ago, after studying the two-dimensional flora of Panama in the herbarium at Missouri Botanical Garden. I was ready to get back on the plane when I saw the flora in three dimensions for the first time. Like many temperate-zone taxonomists, I was overwhelmed by my first real look at the tropical flora. There too, I had a good guide, Afroamerican Narciso Bristan, who was well experienced in the forest. He knew the local names of many plants of Magic Mountain (Cerro Pirre) in Darien, Panama. By the trunk characteristics, blaze, color, latex, aroma, etc., he could even name many of the forest giants. Readily he identified as "cuipo" the dominant species. It was a long time before I could equate that common name with the scientific name, *Cavanillesia platanifolia*. Then there was my Choco Indian confidant, Loro, who taught me Indian ethnobotanicals by their Choco names, "almirajó", "borojó", "cangrejó", "mamejó", etc. In an eight-year period, including an aggregate of three years' residence in Panama, by sending herbarium specimens to specialists all over the world, I

---

[1]ACEER = Amazon Center for Environmental Education and Research. A tax-exempt organization, ACEER goals are to (1) construct and equip a field station to serve as a natural laboratory for integrating education and research; (2) provide environmental and economic benefit to neighboring inhabitants of the reserve; (3) complete the canopy walkway for observation and research; (4) provide a financial basis for the creation of scholarship grants and research stipends; (5) expand the protected reserve area to 400,000 ha (1,000,000 acres); and (6) demonstrate the potential for ecologically responsible tourism as an effective conservation strategy. To attain these goals, the ACEER foundation was created by Explorama Tours, International Expeditions, and concerned individuals. All royalties revert to the rainforest, via ACEER.

constructed my *Isthmian Ethnobotanical Dictionary*, equating the local names of Panama plants with their scientific names, and some of the ethnobotanical lore recorded about these species. I was pleased to see that Vasquez consulted my Isthmian dictionary in writing his dictionary. Whether in forests of Panama or Peru, newcomers will find these dictionaries useful when they have reliable guides. Children of the forest guides are not necessarily learning the tools of the trade. All the more reason to record as much ethnobotanical wisdom as we can, while we can! The shamans, and their knowledge, and their medicinal species are disappearing. There is hope. Occasionally, I see the children appreciating the admiration their parents are receiving from the ecotourists. Perhaps they realize that by learning and carrying on the traditions, they can earn more money than by planting corn or joining the youth in the ghettoes of the asphalt jungles.

Expanding its horizons, following the counsel of Dr. Henry Shands and Dr. Allen Stoner, the USDA encouraged Mr. Eric Rosenquist and me to undertake programs devised to help conserve tropical biodiversity. The USDA sees the link between the health of the remote rain forests and the health of the planet. Rain forests and adjacent clearings in Latin America still house wild relatives of some of North America's most important crops, imported long ago from Latin America. After my visit to Explorama Lodge, I convinced the USDA to sponsor Rodolfo Vasquez's book and its translation into English. So on my second trip to Explorama, I went with an English translation of Rodolfo's dictionary. Al Gentry made dozens of corrections as we motored down the Amazon. Rodolfo met me and we began assembling illustrations of the species most important to the Amerindians.

After the Amerindians discovered America, perhaps 12, maybe 50, millennia before Columbus, all their clothing, food, medicine and shelter — those essential things we call "minor" forest products today — were derived from the forests. Those millennia gave the Indians time to discover and learn empirically the virtues and vices of the thousands of edible and medicinal species in the neotropical forests.

Quinine is one of the more amazing stories in Latin America's pharmacopoeia. The quinine tree, with its dozens of alkaloids, was here before the Indian, long before Columbus, and smallpox and malaria. The history of the continent might have been different had quinine cured smallpox. Instead smallpox decimated the Indians, killing millions, before malaria arrived, perhaps from Africa. The malaria organism was all but controlled by early efforts with quinine. Gradually the malaria organism developed a tolerance for quinine, and we switched to chloroquine and other synthetics and semisynthetics. Gradually the plasmodium developed a tolerance for these as well. WHO-sponsored studies on "qing hao" (*Artemisia annua*) and its derivatives, provided the answer, albeit temporary, to chloroquine-resistant malaria. I predicted that, if natural artemisinin or its semisynthetic derivatives proved out for chloroquine-resistant malaria, ten or twenty years later, the malaria organism would evolve resistance to the "qing hao" compounds. We would then be faced with artemisinin-resistant malaria, and go back to Mother Nature's "Farmacy", the forest, again, hat in hand, seeking a drug for artemisinin-resistant malaria. (I'm told that artemisinin-resistant malaria has already evolved.) This semicircular fable should impress upon us the importance of the forest and biodiversity. If we lose half our species, we cut our odds for finding the new drug in half. Worse! The species most likely to be lost are those least likely to have been studied. New strains of many of our older diseases, measles, tuberculosis, etc., keep cropping up, requiring new medicines. We didn't even know AIDS twenty-five years ago. Each HIV virus is said to be unlike its parent, each generation evolving. Amazonian *Alexa* contains castanospermine, *Abrus* contains glycyrrhizin, *Capsicum* contains caffeic acid, *Momordica* contains momordicin, *Phytolacca* contains phytolaccin, and *Ricinus* contains ricin, to name a few compounds occurring in native or introduced Amazonian species that may possibly help in treating AIDS. Among the thousands of species that have not been analyzed are thousands of unknown chemicals, many evolved to protect the plants from pathogens. These chemicals may help us in our constant struggle with our constantly evolving pathogens. The lower the phytodiversity, the lower our chances of finding new remedies for the newly evolving scourges of mankind. Preservation of biodiversity is self preservation.

NOTES ON AMAZONIAN CRAFTS

(Jim Duke, Celia Larsen, Jan Propst and Rodolfo Vasquez)

Thanks to interactions between the Explorama Guides, the Yagua Indians, and Pamela Bucur de Arévalo, Jim Duke, Francis Gatz, and Peter Jenson, larger ecotours to Explorama Lodge are now greeted by a Sunday afternoon Yagua craft fair. One impressive thing about the craft fair is the importance of the palms to the Yagua. Palms are extremely important among Amerindian cultures throughout tropical America.

At the craft fair, visitors see homes thatched and floored with palms, drink beverages made from fruits of "aguaje" and "ungurahui" palms (the latter loaded with monounsaturated fatty acids like olive oil), eat delicious palm hearts from "chonta" palms, drink liquid endosperm of the coconut or ivory palm, eat the "flan" of the immature ivory palm seed, avoid eating the nutritious "suri" (beetle larvae) nurtured in fallen "aguaje" palms, see carvings made from ivory palms, blowguns reamed with palm, darts fashioned from palm leaf-stalks, see skirts (worn by males and females) fashioned from "aguaje" palm fibers, see hammocks being woven from the "chambira" palm, and fine hats and baskets made from the Panama Hat Palm (not a true palm). Palms provide beverage, clothing, fiber, fishbait, food, handicrafts, housing, meat, medicine, rattan, water, weaponry and wine to the Yagua Indians.

HOUSE CONSTRUCTION

From Costa Rica to Bolivia, most of the raised Indian houses have hard resilient floors made of stilt palm called "pona". Heavier and preferred, if locally procured, is the "huacrapona", *Iriartea deltoidea,* the stilt palm with a "belly" up the trunk, hence the alternative name "barrigon". Natives tell us that it takes more than two men to carry this heavier species' trunk any distance. So, when building off site, the natives prefer the lighter of the two, the "cashapona", *Socratea exorrhiza,* which generally lacks the pot belly.

With either, the hard outer wood of the trunk, above the stilts, is partially but not completely slatted, to make lengths of hardwood flooring, so familiar to those of us who have enjoyed overnighting at Napo. This flooring is a combination floor/springs; whenever anybody makes a midnight visit to the john, everybody rebounds gently with each midnight footstep.

Not only did Rodolfo Vasquez provide the first definitive draft of this book, he provided most of the illustrations. How well I remember one January night when a small group sat around the Napo Lodge, singing "Paradise Lost" (see frontispiece). That January night at the Napo Lodge, guides Aristides Arevalo, Segundo Inuma and Lucio Pano were sitting around with botanists Jim Duke and Rodolfo Vasquez. They were accompanying a small

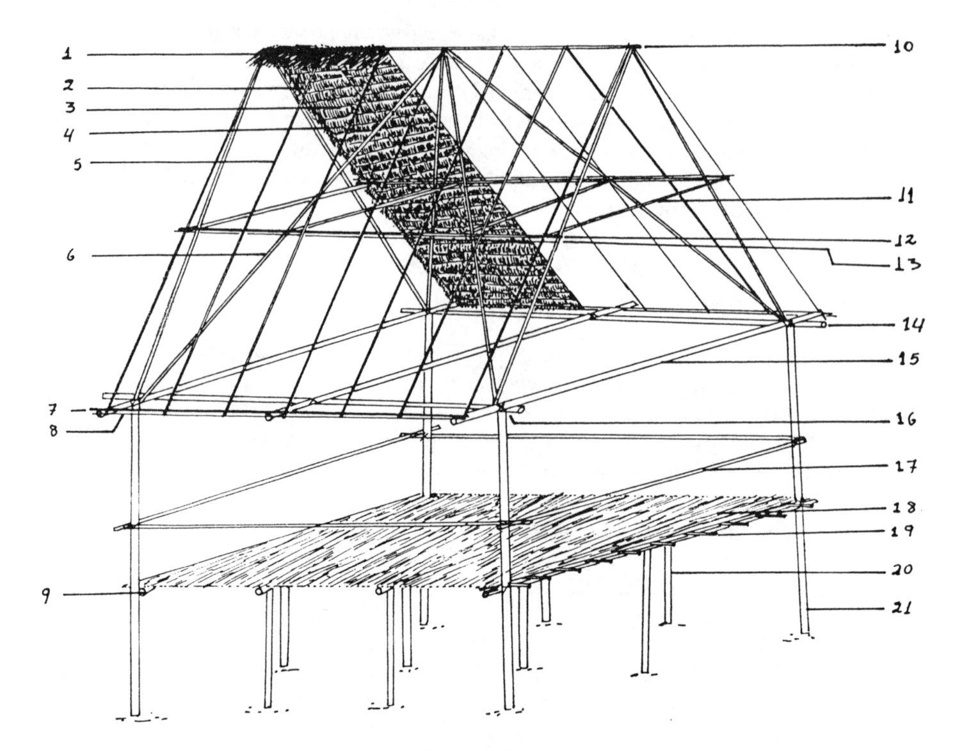

Fig. 1. PLANTS USED IN RURAL HOUSING IN AMAZONIAN PERU
(ExplorNapo Camp on Rio Napo)
Art by Rodolfo Vasquez
Loosley Translated by Jim Duke

	House Part	Plant Used	English Name	Spanish Name
1	CUMBA	Leaves of *Phytelephas microcarpa*	Ivory Palm	Yarina
2	CRISNEJA	Leaves of *Lepidocaryum tessmannii*	Thatch Palm	Irapay
3	RIPA DE LA CRISNEJA	Stem of *Socratea exorrhiza*	Stilt Palm	Pona
4	SOGA PARA CRISNEJAS	Bark of *Trema micrantha*	Bay Cedar	Atadijo
5	CAIBRO	Wood of *Croton palanostigma*	False Balsa	Topilla
6	TIJERALES	Wood of *Croton palanostigma*	Wive's Pole	Señora Vara
7	SOBRE-VIGA	Wood of *Croton palanostigma*	Purma Tree	Purma Caspi
8	ISHPANA	Wood of *Croton palanostigma*	False Balsa	Topilla
9	VIGAS DEL PISO	Wood of *Croton palanostigma*	Wive's Pole	Señora Vara
10	CUMBRERA Y SOBRE-CUMBA	Wood of *Croton palanostigma*	Purma Tree	Purma Caspi
11	VIGUILLAS	Wood of *Croton palanostigma*	False Balsa	Topilla
12	RESISTENCIA	Wood of *Croton palanostigma*	Wive's Pole	Señora Vara
13	SOBRE-VIGUILLA	Wood of *Croton palanostigma*	Purma Tree	Purma Capsi
14	SOLERAS	Wood of *Croton palanostigma*	False Balsa	Topilla
15	VIGAS	Wood of *Croton palanostigma*	Wive's Pole	Señora Vara
16	AMARRAJES	Stems of *Heteropsis* sp.	Aroid	Tamshi
17	BARANDA	Wood of *Oxandra xylopioides*	Belt Tree	Cinta Caspi
18	PISO	Slats of *Iriartea* stem	Belly Palm	Huacra Pona
19	ENRRIPADO	Slats of *Iriartea* stem	Belly Palm	Huacra Pona
20	HORCONES DE PISO	Poles of *Minquartia guianensis*	Column Tree	Huacapú
21	HORCONES DE LA CASA	Poles of *Minquartia guianensis*	Column Tree	Huacapú

Note: Different builders in different places may use different species. These species were noted at ExplorNapo Camp, January 1993.

herbal ecotour group, the "clavohuasqueros", to study useful plants of the Amazon, including the "clavohuasca", *Tynnanthus panurensis,* a reportedly non-gender-specific aphrodisiac. While we can't guarantee that the vine is aphrodisiac, we are sure it contains eugenol, a compound widely used as a dental antiseptic analgesic. Steeped in local "aguardiente" (distilled sugar-cane rum), the "clavohuasca", or its solvent, did make the "clavohuasqueros" a bit more talkative. We soon got up a pool to guess how many different plants were used in the construction of the Napo house. Most of the guides came up with numbers below 12, but one guess put the estimate at 20 different species of plants used in the construction of that attractive rustic edifice. If you have been there, you may have noticed that the whole house is constructed of native materials, without a single nail. It's literally tied together with bark ("atadijo" or *Trema micrantha*) or vines ("huambe" or *Philodendron).*

We still don't have the exact count, but its probably closer to 20 than to 10. A few months after that episode, Vasquez prepared a diagrammatic representation of the house, shown here as Figure 1. The diagram alone shows more than ten species of plants, omitting the cutgrass (*Scleria*) so often hung in the eaves to discourage bats and the balsa, *Ochroma pyramidale,* from which the plaques at Napo are carved.

Whether at Explorama, Napo or ACEER, or on the boats between the camps, you have been kept dry by the favorite thatch palm, called "irapay", *Lepidocaryum tessmannii*, leaves of which are woven around a slat of *Socratea* to make the sheaves that are the favorite roofing material of that part of Peru. As far as we know, the "irapay" is the only stemmed palm in Loreto that consistently has four blades arising at the tip of the leaf stalk as depicted in the diagram (Fig. 2). Eaves of the roof are covered not with "irapay", but with the leaves of the ivory palm they call "yarina", *Phytelephas microcarpa.* So the roofs and floors of many rural Amazonian houses are derived from palms.

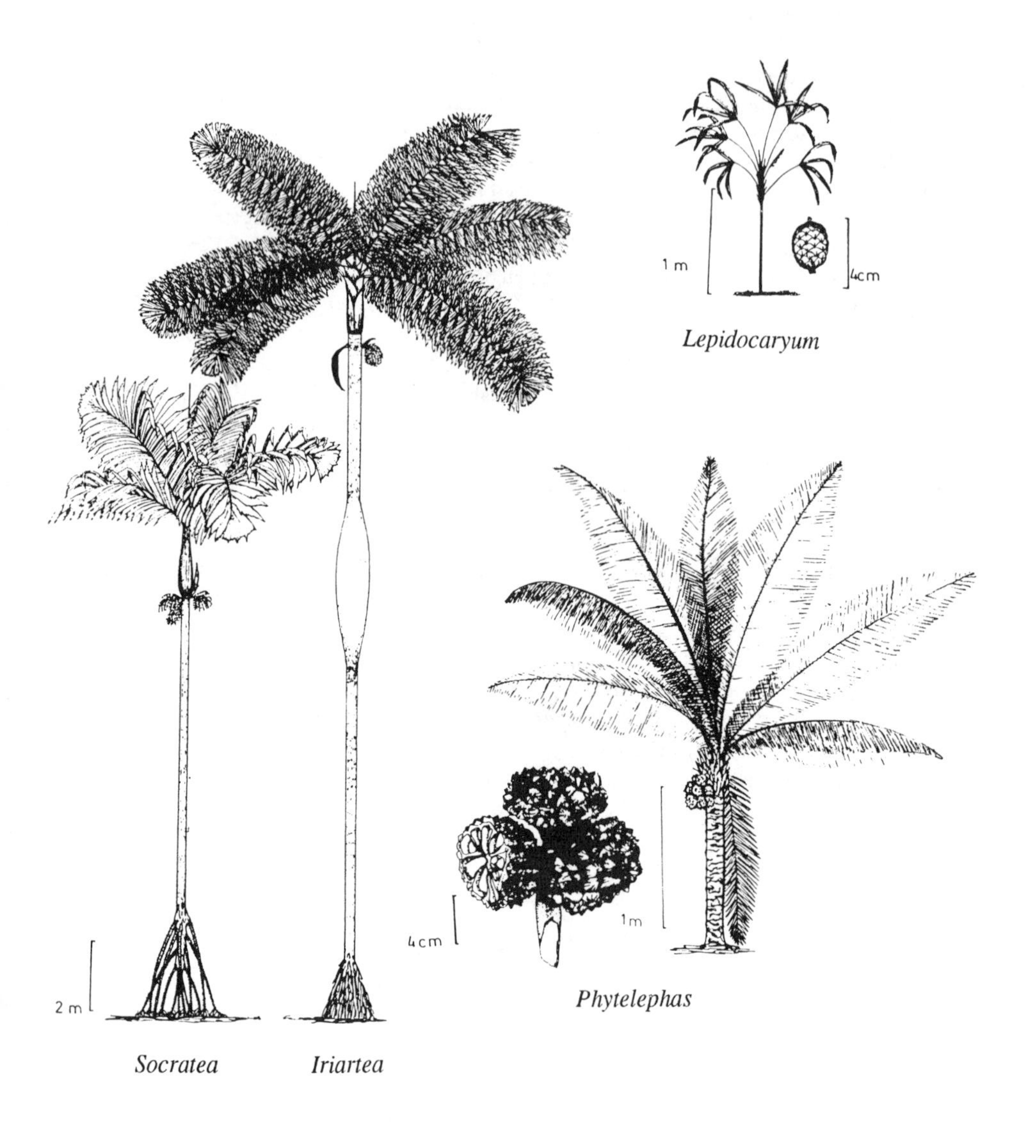

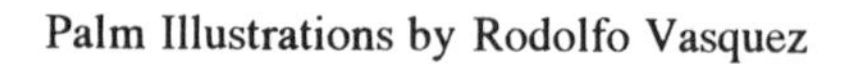

Palm Illustrations by Rodolfo Vasquez

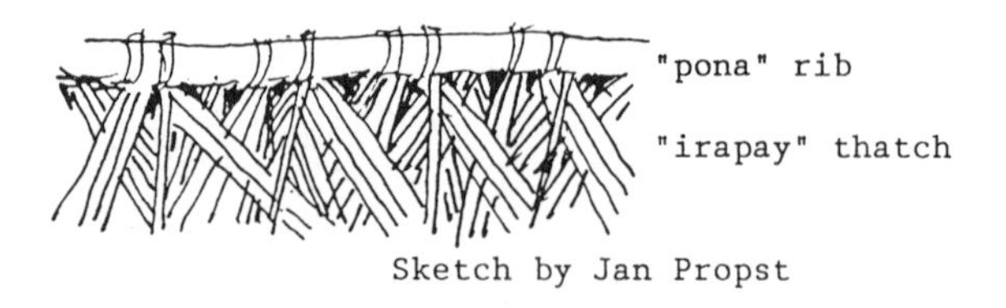

Sketch by Jan Propst

Fig. 2. PALMS

THE AYAHUASCA CEREMONY

In an herbal ecotour in January and five times again following a music therapy ecotour (under Dr. Joe Moreno) to the ACEER camp, Celia Larsen, Rodolfo Vasquez and/or Jim Duke observed as "ayahuasquero" Don Antonio Montero simulated an "ayahuasca" ceremony. He apparently performs medicinal, social and/or spiritual ceremonies on the Napo on special Friday nights. At the ACEER camp, he had his earthen pot on the fire by 2:00 PM. Stems of the "ayahuasca" were pounded and added to the heating water, which would normally be boiled for hours. Antonio had several other essential constituents in his brew. Two big leaves (or 4 small leaves) of "toé", probably *Brugmansia aurea*; a few leaves of "sacha ajo", probably *Mansoa alliacea*; a few leaves of "chiricsanango", probably *Brunfelsia grandiflora*. A vital acoutrement was the "yagé", *Psychotria* sp., which like the "ayahuasca" and "ipecac", has emetic properties. Antonio suggested it was added to make the bitter beverage a bit sweeter. All of these, except the "ayahuasca", are rather common at ACEER, as wild plants or cultivars. But the "ayahuasca", like "clavohuasca", "chuchuhuasi", and "uña-de-gato" are disappearing near civilization.

Chances are good that the *Psychotria,* like *Psychotria viridis,* contains N,N,dimethyltryptamine, which is not active when ingested, unless taken with monoamine-oxidase-inhibitors like the harmaline, harmine, and tetrahydroharmine in "ayahuasca". Add scopoletin from *Brunfelsia* and atropine and scopolamine from *Brugmansia* and you have a pot-pourri of hallucinogenic (but dangerous) synergy. As Antonio warns, there are bad "ayahuasqueros" in Iquitos who have sent some gringo initiates home "basket cases". Other evil "ayahuasqueros" drug their initiates with *Brugmansia* or *Datura* and take their money while they are under the influence. We're not sure how to tell genuine from phoney "ayahuasqueros". Antonio takes up to four small calabash cups (while none of his associates/patients take more than three) of this witches brew, usually starting late at night. First one becomes nauseated, then inebriated, in an hour or so, and all passes within a few hours. During the "highs", Antonio sees beautiful and colorful visions, of long lost or deceased friends, of friends who have moved to large cities in the US, strange animals and spirits of the trees, etc. Most impresive of all his visions, and a real show stopper for ACEER classes, are Antonio's sweeping gestures, as he discusses the thousand-color rainbow that wraps around the "ayahuasquero" like a cosmic whirlpool. As he described it so vividly, Duke lost track of his translation duties, caught up in the vortex.

Under the influence, Antonio divines the cause and cure for his associates' illnesses. Then he may chant or whistle as he purifies them with the rhythmic "shacapa" fan, and as he blows organic tobacco smoke onto selected body points of the patient. (Tobacco smoke is elsewhere blown in the ear to treat earache.) He may use any of several medicinal herbs at this point, or prescribe them for later use, depending on the diagnosis and prognosis of the patient.

Antonio, who stated he was 48 years old in 1993, was initiated into ayahuasca at about age 12, by a serious "ayahuasquero", like himself, who selects his apprentice. He considers himself at once, an "ayahuasquero", a black magician (in league with Lucifer), a healer, an herbalist and a shaman. First experiences with "ayahuasca" can be terrible, for Indians and gringos alike, and they may run into the forest, shouting, terrified by their visions. Antonio's first experience, too, was very frightening and he wanted to back out. But he was chosen. He and his mentors and mentees believed they were destined to be "ayahuasqueros", further believing it difficult to change their destiny (and with Antonio, his league with Lucifer). Like many healers though, he, too, likes to use alien pharmaceuticals for himself and his family, when available. He does not plan for any of his children to become "ayahuasqueros". He picks up a little spending money with his Friday night "ayahuasca" sessions, much like a musician going to town to pick up a few bucks on the weekend. To him, "ayahuasca" is not a drug of abuse, but a source of divination, inspiration, power, telepathy and visions. To the uninitiated it can sometimes be a very dangerous, even life-threating, "bad trip". A few cases have been mental one-way trips. Beware!

Fig. 3. "Ayahuasca" (*Banisteriopsis caapi*)
Source: E.W. Smith, Courtesy R.E. Schultes

THE BLOWGUN (*"Pucuna"*)

Blowguns are fashioned from the "Pucuna caspi" which means "blowgun tree", *Iryanthera tricornis*, a member of the nutmeg family. Amazonian members of the nutmeg family have a pagoda-like branching habit wherein many branches emerge at right angles to the trunk, like a few scattered spokes from the center of a wheel. One long straight branch is selected and cut off to the desired length, halved, and the center reamed out with a reamer made from the hard outer bark of the "pona" palm (*Iriartea* or *Socratea*). With the "pona" palm they "sand" the bore. Then the two halves of the blowgun are glued with a resin or tar-like "brea" (probably *Protium*) and tied tightly with coils of the "huambe", aerial roots of a *Philodendron*, which may also be further affixed with "brea". The spool-like mouthpiece is fashioned from "insira", *Maclura tinctoria*, a relative of our mulberry tree.

Darts are usually made from slivers of the petioles of palm leaf stalks, most often the "inayuga", *Maximiliana venatorum*. At one end of the sliver they affix a wad of kapok or "huimba", *Ceiba samauma*. They dip the other end in one of their curare mixtures, of arrow or dart poisons. The business end of the dart is often sharpened between piranha teeth. Curare recipes vary from individual to individual and tribe to tribe. Rather constant ingredients include members of the moonseed family, Menispermaceae, a group of lianas often endowed with powerful muscle relaxants, including one that is the source of the myorelaxant tubocurarine used in modern medicine, especially in open heart surgery. Probably the most common moonseed ingredient is the "ampi huasca", *Chondrodendron tomentosum*. Skins of poison-dart frogs, common on the forest floor, are often utilized in the mixture. Ants are sometimes added around Iquitos.

Enrique, the Yagua gentleman who has demonstrated his incredible accuracy with the blowgun, capable of hitting a monkey at 30 meters, usually has a kapok bag fashioned from the "chambira" palm, *Astrocaryum chambira*. To this, he usually has affixed a monkey call, fashioned from a hollow monkey or bird bone.

Fig. 4. BLOWGUN CONSTRUCTION
(Illustrations by Jan Propst)

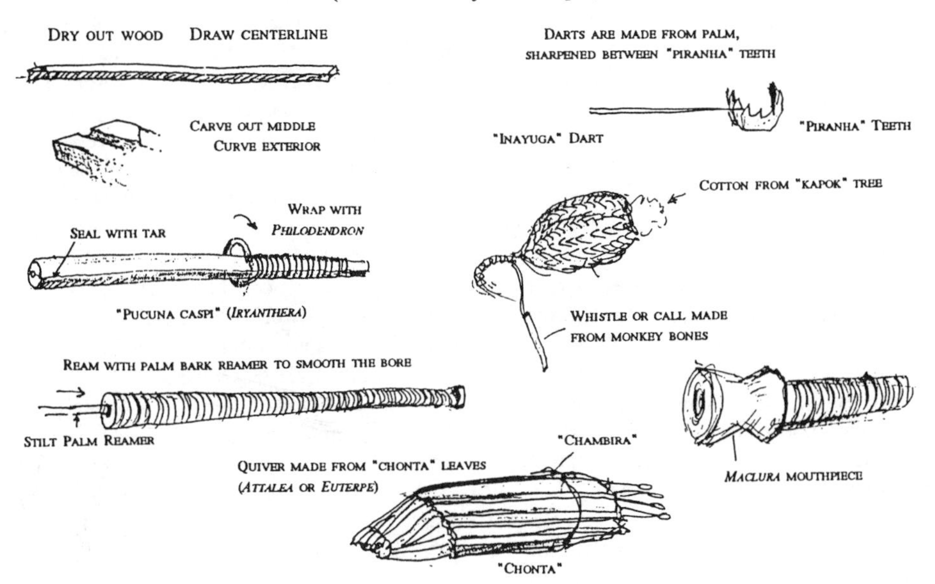

FORMAT AND CAVEATS

Our dictionary is alphabetized by scientific name (in italics), followed by the author. Then comes the family name, followed by local names in quotation marks and occasionally an English name, in bold face. Finally we provide succinct comments on use. Interspersed with the utilitarian notes are three-letter abbreviations, indicating our major sources. POISON and TOXIC have been capitalized as a warning. We do not endorse self-diagnosis and self-medication. If an author said a plant was purgative we used his/her words citing our source. We cannot guarantee the accuracy of our sources, but we have tried to cite them faithfully. The reader is advised to view this as folklore, which may or may not prove out.

ACKNOWLEDGEMENTS

I am indebted to Margarita Darlington for a first translation of the dictionary, which I amplified and reworded. I augmented the draft by incorporating Amazonian data from such books as Schultes and Raffauf's *Healing Forest* and constructing a folk medical index. Robert A. DeFilipps read through the manuscript thoroughly. Most of all, I'm indebted to Rodolfo Vasquez for his pioneering work in Spanish and to Judy duCellier who converted my chicken scratch into the camera-ready copy for this book. We thank Janis Alcorn, Michael Balick, Stephen Beckstrom-Sternberg, Bryan Boom, Rose Broome, William Burger, Robert Bye, Tom Carlson, Jim Castner, Howard Clark, Paul Donahue, Elaine Elizabetsky, Memory Elvin-Lewis, Hardy Eshbaugh, Ingrid Fordham, Al Gentry, Gary Hartshorn, Steven King, Joseph Kirkbride, Celia Larsen, Walter Lewis, Barrie Maguire, Dennis McKenna, Willem Meijer, Dick Mills, Buzz Peavey, Mark Plotkin, Jan Propst, Robert Raffauf, Peter Raven, Jim Reveal, Richard Ryel, R. E. Schultes, E. E. Smith, H. M. Smith, Calvin Sperling, Steve Timme, Joe Tosi, David Williams, Tom Wolfe, and Teresa Wood for their encouragement, help, and/or contributions.

In the Spanish version, Vasquez cited the specialists and herbaria involved with identifications of his specimens, and provided his introduction to the Spanish version, along with all his references. The reader is referred to that version for Vasquez's acknowledgments of the taxonomists, ethnobotanists, botanists and scientists of other disciplines whose works laid the foundation for this Amazonian Dictionary. In adding to Rodolfo's compilation, I eliminated some of Rodolfo's citations, e.g., all herbarium collections and some references. These details can be found in the Spanish version, a copy of which will be deposited at USDA's National Agricultural Library. I used three-letter abbreviations, like RVM, to indicate the sources of information. Important sources include ALG (various papers and personal communications of the late Dean of Amazonian "Diversitologists" and Botanists, Al Gentry), AYA (Ayala), BDS (Branch and daSilva, Rio Ragajos, Brazil, 1983), CAA (Cayon & Aristizabal, 1980), CRC (my own CRC Handbooks), DAD (Duke and duCellier, 1993), DAT (Denevan and Tracy, 1987), DAW (Duke and Wain, 1981), FAO (Food and Agriculture Organization, 1986), FBL (F. Bruce Lamb), FEO (V. de Feo. Fitoterapia 63:417-440), FNF (Father Nature's Farmacy), FOR (Forero), GAB (Garcia-Barriga), GAV (Gentry and Vasquez), GMJ (Grenand, Moretti and Jacquemin), HAC (Hitchcock and Chase), IIC (Leon, 1987), JAD (my own books and personal observations), JBH (J. B. Harborne's Phytochemical Dictionary, 1993), JFM (Julia Morton), JLH (Jonathan Hartwell's Plants Used Against Cancer), HHB (Hager's Handbook), LAE (Lewis and Elvin-Lewis), LAW (Little and Wadsworth), MAC (Macbride's Flora of Peru), MAR (Mors & Rizzini), MJB (Michael Balick's various palm papers and books), MJP (Mark J. Plotkin, 1993), MVA (MM. Vilmorin-Andrieux), NIC

(Nicole Maxwell), PEA (Perez-Arbelaez), PKD (Peggy Kessler Duke), POV (Poveda), RAF (Ramon Ferreyra, 1970), RAR (R. A. Rutter, 1990), RBI (Florula de las Reserva Biologicas de Iquitos), RVM (Rodolfo Vasquez Martinez), SAR (Schultes and Raffauf), SOU (Soukup, 1970), TRA (Tramil), VAG (Vasquez and Gentry), VAM (Valdizan and Maldonado, 1922), VDF (V. de Feo, 1991), and WOI (Wealth of India).

In the January class of 1993 was my daughter Celia Larsen, savoring the Amazon for the first time, and like me, enamored of what she saw. Her biggest contribution was to force me to recognize the gender issue, and add the feminine stanza to "Paradise Lost", and the ♀ sign to flag those herbs more important to women. After all, she said, most of the ecotour groups have more women than men, a ratio not yet reflected among the mostly male guides and instructors. More importantly, most of the ethnobotanical writings on female health issues were by foreign men, interpreting native men in turn interpreting native women. Clearly, as the Lewises indicate, the gender issue has not received the attention it deserves.

This book would have been impossible without all the scientists and Amerindians who preceeded us and their contribution to our efforts to save the forest through rational and sustained utilization and conservation of the forest species. With sincere but inadequate gratitude, the book is specifically dedicated to Antonio, Loro, Lucio, Narciso, Rodolfo, and Segundo, and their ancestors who augmented our empirical knowledge of forest foods and medicines. And foremost to Al Gentry, who lost his life, trying to tie it all together taxonomically.

JIM DUKE

Al Gentry and *Tabebuia*

VASQUEZ'S PROLOGUE

(To *Plantas Utiles de la Amazonia Peruana*)

Hoping to help satisfy the thirst for knowledge welling up in all scientists, I set out several years ago to catalogue for posterity the *Useful Plants of Amazonian Peru*. I wanted to consolidate under one cover the scattered information about ethnobotany of the area. I hoped to provide information about local uses and colloquial names of wild and cultivated plants. To city dwellers, the forest is remote and vague, perhaps something to be feared. With this book, I hope to show that the Amazonian forest, properly utilized, can be the friend of (wo)mankind, by no means its enemy. By enumerating all the diverse uses to which the forest is put, I hope to enhance chances for conservation of this irreplaceable resource, which has provided native Americans with sustenance for some 12,000 years.

In this initial effort, I hopefully covered most uses of the more important species. Much remains to be done. We treat here only some 20% of the flora of the area. In compiling this document, I hope to show that Peruvians are interested in studying Peruvian plants and their ethnobotany, and to encourage collaboration among the many concerned agencies, not only to study, but to save, for the future of Peru and the World at large, this magnificent Amazonian forest.

Sincerest gratitude is expressed to Dr. Al Gentry, Curator at Missouri Botanical Garden, to Dr. F. Ayala, Director of the Herbarium Amazonense, and to Mr. Nestor Jaramillo for their counsel, help, and guidance and for their contributions to Amazonian ethnobotany. Special gratitude is due the rural people and the native Americans of the Amazon region, as well as the herbalists at the Iquitos market, all of whom were invaluable sources of information.

RODOLFO VASQUEZ MARTINEZ
as loosely translated by Jim Duke
February 1992

- A -

Abarema laeta (Poepp.) L.Rico. Fabaceae. "Shimbillo". Used for malaria around Iquitos (RVM).

♀ *Abelmoschus moschatus* Medik. Malvaceae. "Almizclillo", "**Musk**". Cultivated ornamental at Yanomono. Seeds used to make necklaces for children suffering bronchitis and cough. Source of musk for the perfume industry, also used to scent cosmetics. Seeds used as tonic and antispasmodic (SOU). "Créoles" and "Wayãpis" use the plant to protect against poisonous snakebite. The stimulating and antispasmodic attributes of the seeds are well known. Seeds also used as antipyretic, analgesic, and anti-inflammatory. The "Palikur" use a decoction of leaves and green capsules in baths to ease childbirth GRE. Boiling leaves 15 minutes yields an emmenagogue tea (SAR). Leaf decoction alleviates menstrual cramps and metrorrhagia (SAR). Crushed seed with chicken fat are given for bronchitis (SAR). Leaves boiled with crushed cashew leaves taken for diarrhea (SAR). (Fig. 5)

♀ *Abuta grandifolia* (Mart.) Sandwith. Menispermaceae. "Abuta", "Motelo sanango", "Trompetero sacha". The decoction of the stems and roots mixed with wild bee honey is used to treat sterile women. Root decoction used for post-menstrual hemorrhages, the alcoholic maceration, for rheumatism. Macerated leaves, bark and roots, mixed with rum, are used by the "Créoles" as aphrodisiac. "Wayãpi" use the decoction of the bark and stem as a dental analgesic. Root decoction used as a cardiotonic and antianemic (SOU). Antimalarial (DAW). "Sionas" use leaf decoction for fever. Ecuadorian "Ketchwas" use the leaf decoction for conjunctivitis and snakebite SAR. Ecuadorians use the root tea for difficult delivery and for nervous or weak children with colic SAR. Contains palmatine and other berberine derivatives (GMJ). (Fig. 6)

Abuta imene (Mart.) Eichl. Menispermaceae. "Trompetero sacha". Natives use it as fish POISON (SOU). Amazonian Brazilians regard the root as diuretic, emetic, resolvent (for bruises), tonic and toxic (SAR).

♀ *Abuta rufescens* Aubl. Menispermaceae. "Abuta". Used to make "curaré". Considered antimalarial, emmenagogue, tonic. Good for kidney stones, sterility, vaginitis, and weak vision (RAR). This plant contains alkaloids which relax the muscular system. Contains imenine, homoschatoline, imerubrine, imeluteine, rufescine and norrufescine (AYA, RVM).

Abuta sandwithiana Krukoff & Barneby. Menispermaceae. "Trompetero sacha". "Wayapi" use the decoction of the bark and stem as a dental analgesic. Root contains protoberberine derivatives well known for their analgesic properties (GMJ).

Abuta solimosensis Krukoff & Barneby. Menispermaceae. "Abuta macho". Bark decoction used for anemia, alcoholic maceration for rheumatism (RVM).

Acalypha macrostachya Jacq. Euphorbiaceae. "Cimora leon", "**Lion's tail**". Chopped leaves used, externally, as a decongestant (FEO).

Aciotis fragilis (Rich.) Cogn. Melastomataceae. "Kéécui". "Boras" use the leaves to bathe children with high fever (DAT).

Aciotis purpurascens (Aubl.) Triana. Melastomataceae. "Kiiwahe". Fruit edible (RAR). Used by "Boras" to reduce fever (DAT).

Acroceras zizanioides (HBK) Dandy. Poaceae. "Grama", "**Grama grass**". Forage.

Acrocomia erioacantha Barb. Rodr. Arecaceae. "Gru-gru". Cultivated ornamental.

Adenaria floribunda HKB. Lythraceae. "Gurima-cy", "Pega perro", "Puca varilla", "Rumo caspi". Kept in the fields to provide shade for pigs. The small branches are commonly used as mosquito repellent, i.e. as "tambinas" (implements to repel mosquitoes) (RVM). Pucallpa natives use the shoot tincture topically for rheumatism.

♀ *Adiantum capillus-veneris* L. Adiantaceae. "Culantrillo", "Cebolla de venus", "Shopumbillo", "**Maidenhair fern**". Cultivated ornamental. Fronds diaphoretic, emollient, pectoral; to treat certain urinary disorders. Once used to treat cough. Now used as mosquito repellent. Emmenagogue, expectorant, and emollient when made into infusion or syrup; also used as aperitive and diuretic. A 10% infusion mixed with honey is expectorant, for rheumatism, and colds, heartburn, and sour stomach (RVM). Considered diuretic, pectoral, sudorific (FEO); decoction used for alopecia, gallstones, icteria (FEO). (Fig. 7)

Adiantum humile Kunze. Adiantaceae. "Culantrillo". Used by the "Boras" for snakebite (DAT).

Adiantum spp. Adiantaceae. "Culantrillo". Ornamental.

Adiantum tomentosum Kl. Adiantaceae. "Culantrillo". Ornamental. Used ceremonially at community balls because aromatic when dry. Cooked shoots used as an emetic; also used to treat stomachaches. Rubbed on the body to alleviate redbug bites. Crushed leaves sniffed to alleviate respiratory infections; hot infusions used to cure bronchitis (RVM).

Aechmea spp. Bromeliaceae. "Sacha piña", "Bromilia", "**Wild pineapple**". Occasionally planted as ornamentals.

Agave americana L. Agavaceae. "Maguey", "Mexico", "**Century plant**". Leaf decoction, considered decongestant, depurative, and vulnerary, used for jaundice and syphilis (FEO).

Aiphanes caryotifolia (Kunth.) Wendl. Areaceaea. "Quindio". Cultivated. Fruit edible (RVM). Extremely rich in vitamin A, 16,000 IU/100 g fresh weight basis = 44 mg beta-carotene/100 g (MJB). Seed contains 37% fat, of which 63% is lauric acid.

♀ *Alchornea castaneifolia* (Willd.) Juss. Euphorbiaceae. "Iporoni", ""Iporuro", "Ipururo", "Ipurosa", "Macochihua". Alcoholic bark maceration used to treat rheumatism, arthritis, colds, and muscle pains after a long fishing day. The "Candochi-Shapra" and the "Shipibos" used the bark and roots to treat rheumatism. Iquitos herbalists recommend it for rheumatism (RVM). Pucallpa citizens take the leaf decoction orally for cough and rheumatism (VDF). "Tikunas" take one tablespoon bark decoction before meals for diarrhea (SAR). Around Piura, the leaves are used to increase fertility of females where

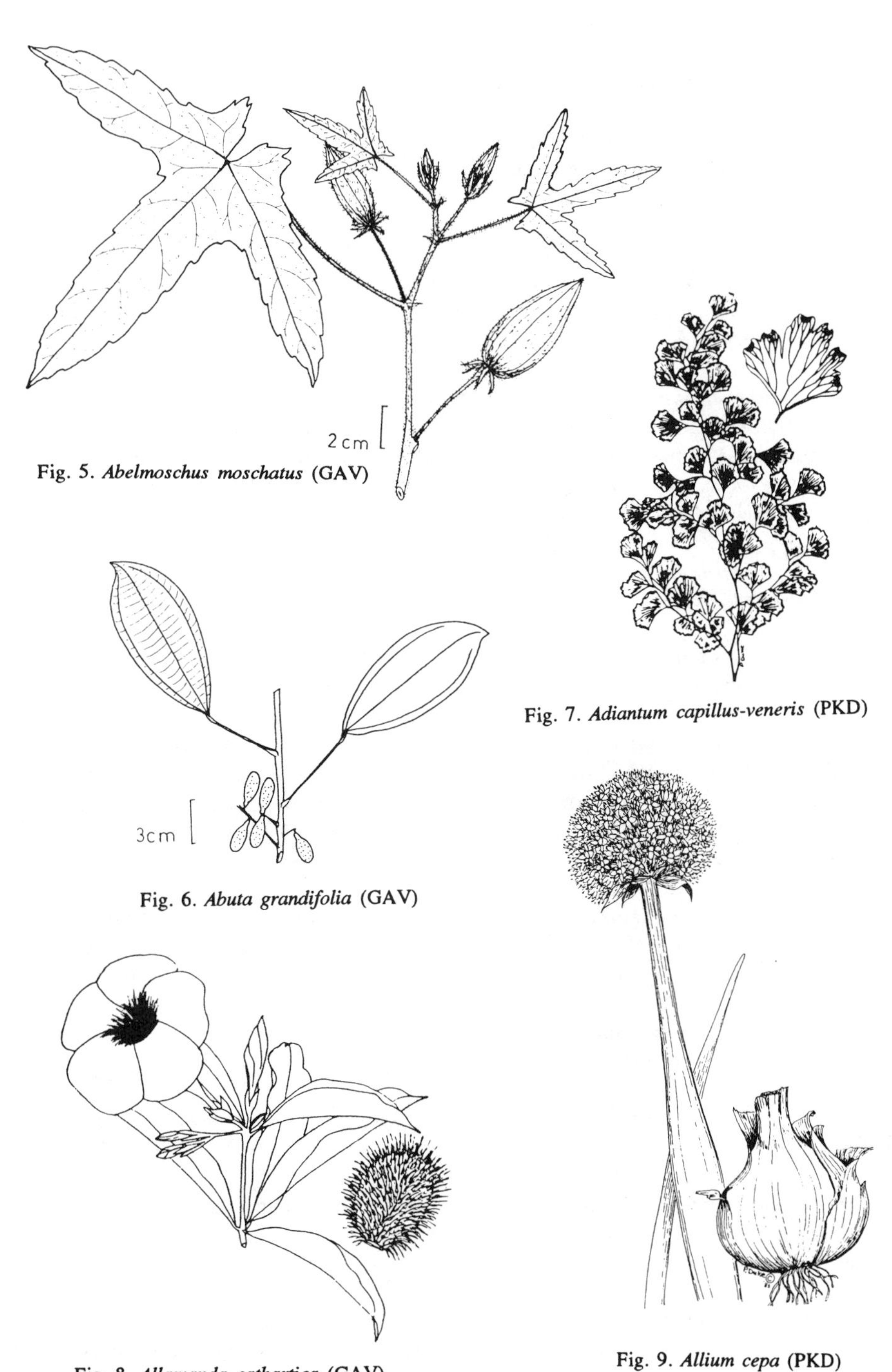

Fig. 5. *Abelmoschus moschatus* (GAV)

Fig. 6. *Abuta grandifolia* (GAV)

Fig. 7. *Adiantum capillus-veneris* (PKD)

Fig. 8. *Allamanda cathartica* (GAV)

Fig. 9. *Allium cepa* (PKD)

the male is relatively impotent (FEO). Rutter stresses that it is aphrodisiac and geriatric for males (RAR). Sometimes found in the famous "Rompe Calzon" aphrodisiac.

Alchornea discolor Endl. Euphorbiaceae. "Palometa huayo". Fishermen like to fish in the shade of this tree (RVM).

Alchornea triplinervia (Spreng.) Muell. Arg. Euphorbiaceae. "Zancudo caspi", "Mojarra caspi". Wood used for temporary construction, possibly for paper pulp (RVM). "Witotos" use the leaves for diarrhea (SAR).

♀ *Alibertia edulis* (L.Rich.) A.Rich. ex DC. Rubiaceae. "Huitillo". Fruit edible. "Cuna" put the bark in cold water to make a lactagogue beverage. "Kayapó" cultivate it for use in food and hunting (RVM). Herniated Brazilians steep the lower part of their bodies in the leaf decoction (BDS).

Allamanda cathartica L. Apocynaceae. "Campanilla de Oro", "Canaria", **"Golden bell"**. Cultivated. As an ornamental. The latex is toxic and caustic. "Wayãpi" use the bark to reduce fever by preparing a "solar tea", and rubbing it over a patient's body. "Palikur" use leaves for anxiety, washing their heads with the decoction (RVM). Cathartic, emetic and POISON (DAW). Leaves contain ursolic-acid (HHB). (Fig. 8)

Allium cepa L. Liliaceae. "Cebolla", **"Onion"**. Surely an acquired remedy, the Rio Tapajos natives eat raw onions to lower cholesterol (BDS). (Fig. 9)

Allium sativum L. Liliaceae. "Ajo", **"Garlic"**. Chopped cloves applied to mange (FEO). Brazilians make a tea of the cloves for cold (grippe). (Fig. 10)

♀ *Allium fistulosum* L. Liliaceae. "Cebolla peruana", **"Peruvian onions"**. Cultivated herb, used to poultice first degree burns. Mothers-to-be drink the liquid of the cooked crushed pulp to hasten childbirth. For upset stomach they wrap the pulp in banana leaves, warm over hot coals, and apply on the patient's stomach as hot as comfortable (RVM). Used elsewhere for cancer (uterine), dysentery, fever and headache DAW.

Aloe vera L. Liliaceae. "Savila", **"Aloe"**. Juice applied topically to inflammation and toothache. Leaf decoction used as an antidote to poisoning and as a purgative (FEO). Brazilians use the jelly for burns and sores (BDS) and to prevent alopecia. (Fig. 11)

Alpinia nutans Rosc. Zingiberaceae. "Alpinia". Cultivated ornamental. Brazilians use the flower tea as a sedative (BDS).

Alseis peruviana Standl. Rubiaceae. "Misho quiro", "Mucla de Gato", "Pino regional", **"Cat tooth"**. Wood for lumber.

Alternanthera halimifolia (Lam) Standl. Amaranthaceae. "Ojo de pollo", **"Chicken's eye"**, "Picurillo", "Sanguinaria". Well known for diabetes (RVM). In the Sierra, the chopped plant is used for muscular sprains (FEO).

♀ *Amaranthus spinosus* L. Amaranthaceae. "Ataco", **"Spiny Pigweed"**. Whole plant used as a laxative and antieczemic. Decoction used in washes to reduce high fever, and in poultices to treat swollen sores (RVM). The decoction of the shoots is considered

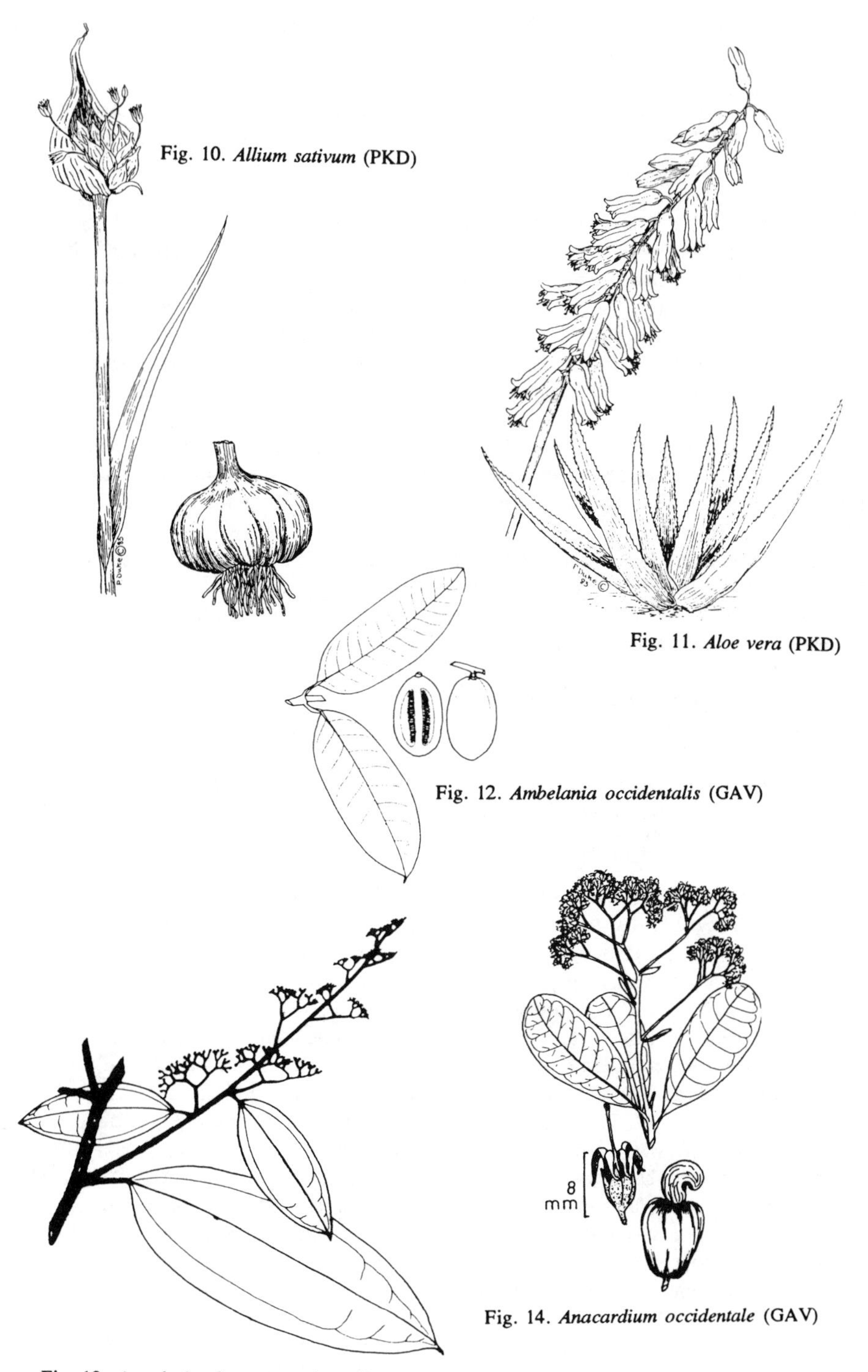

Fig. 10. *Allium sativum* (PKD)

Fig. 11. *Aloe vera* (PKD)

Fig. 12. *Ambelania occidentalis* (GAV)

Fig. 13. *Ampelozizyphus amazonicus* (GAV)

Fig. 14. *Anacardium occidentale* (GAV)

antirheumatic, antiseptic (urinary, throat, pharynx), and emmenagogue; the root infusion is considered astringent (FEO).

Ambelania occidentalis Zarucchi Apocynaceae. "Cuchara caspi". Edible fruit (RVM). "Witotos" coat their extremities with latex of closely related *Mucoa duckei* to protect against gnats. "Karijonas" use *Mucoa* for fever (SAR). (Fig. 12)

♀ *Ambrosia peruviana* Willd. Asteraceae. "Altamisa", "Marco", "Marquito", **"Peruvian ragweed"**. Used with other plants in baths for magic or religious rituals (RVM). Considered astringent, antirheumatic, and tonic (DAW). Shoot decoction, considered antirheumatic, antispasmodic, digestive, tonic, and vermifuge, is used for dysmenorrhea (FEO). Juice of the plant used by the "Incas" to preserve corpses. In popular medicine, the juice is used for rheumatism and late menstrual periods. Root decoction used for neuralgia and hysteria. Floral infusion used as vermifuge (SOU).

Amburana cearensis (Fr. Allen) A.C.Smith. Fabaceae. "Ishpingo". Hard light wood for furniture and decorative plaques (RVM). Folk remedy for respiratory ails (DAW). Containing coumarin, seeds are used to make perfumes and soaps (SOU).

Ampelozizyphus amazonicus Ducke. Rhamnaceae. "Saracuramira". "Boras" use it for insect bites (DAT). Roots used as a depurative; leaves are caustic; new shoots mixed with water produce a foamy beverage which tastes like beer (RVM). Root extract shows antimalarial activity (RVM). (Fig. 13)

Amphitecna latifolia (Mill.) Gentry. Bignoniaceae. "Peche sheti". Around Pucallpa, bark used as topical antiinflammatory in rheumatism (VDF).

Anacardium giganteum Hancock ex Engler. Anacardiaceae. "Sacha cashu", **"Wild cashew"**. Wood for lumber, with possibilities for plywood. The peduncles are edible fresh or in beverages (RVM).

♀ *Anacardium occidentale* L. Anacardiaceae. "Cacho", "Cashu", "Marañón", **"Cashew"**. Cultivated. Roasted seeds are edible, but oil from the fruit (cardol) is a strong vesicant. The swollen peduncles of the fruit are edible fresh, in drinks or ices. Juice from green fruits used to treat hemoptysis. The leaf infusion is used to treat diarrhea. Oil used for warts; good for teeth. Wine obtained from fruit is a good antidysenteric. Seeds used as worm medicine to kill bot-fly larvae (MJP). A gum like gum-arabic is extracted from sap; fruit juice used as a permanent marker for clothing (SOU). Bark decoction used to treat diarrhea (RVM). Important products are: the cashew nuts, and the cashew nut shell liquid, also called cardol. Containing phenol it is an important raw material for the plastic and resin industry (RVM). From the tender shoots they make expectorants (RVM). Fruit juice used for warts. The "Wayãpi" use the bark as a remedy for infants (GMJ). "Tikunas" use the "apple" juice for flu. (Fruit juice contains three antitumor compounds JAFC 41:1012. 1993.) Bark decoction, taken each month during the menses, is said to be contraceptive (SAR). Brazilians use it as a douche for vaginal secretions (BDS), or as an astringent to stop bleeding after tooth extraction. (Fig. 14)

Anadenanthera peregrina (L.) Benth. Fabaceae. "Cohoba", "Yopo". Brazilians wash cuts and bruises with the bark decoction (BDS). Hallucinogenic snuff; used as antidiarrheal, antivenereal, hemostatic, and stimulant (RAR). Brazilians believe that

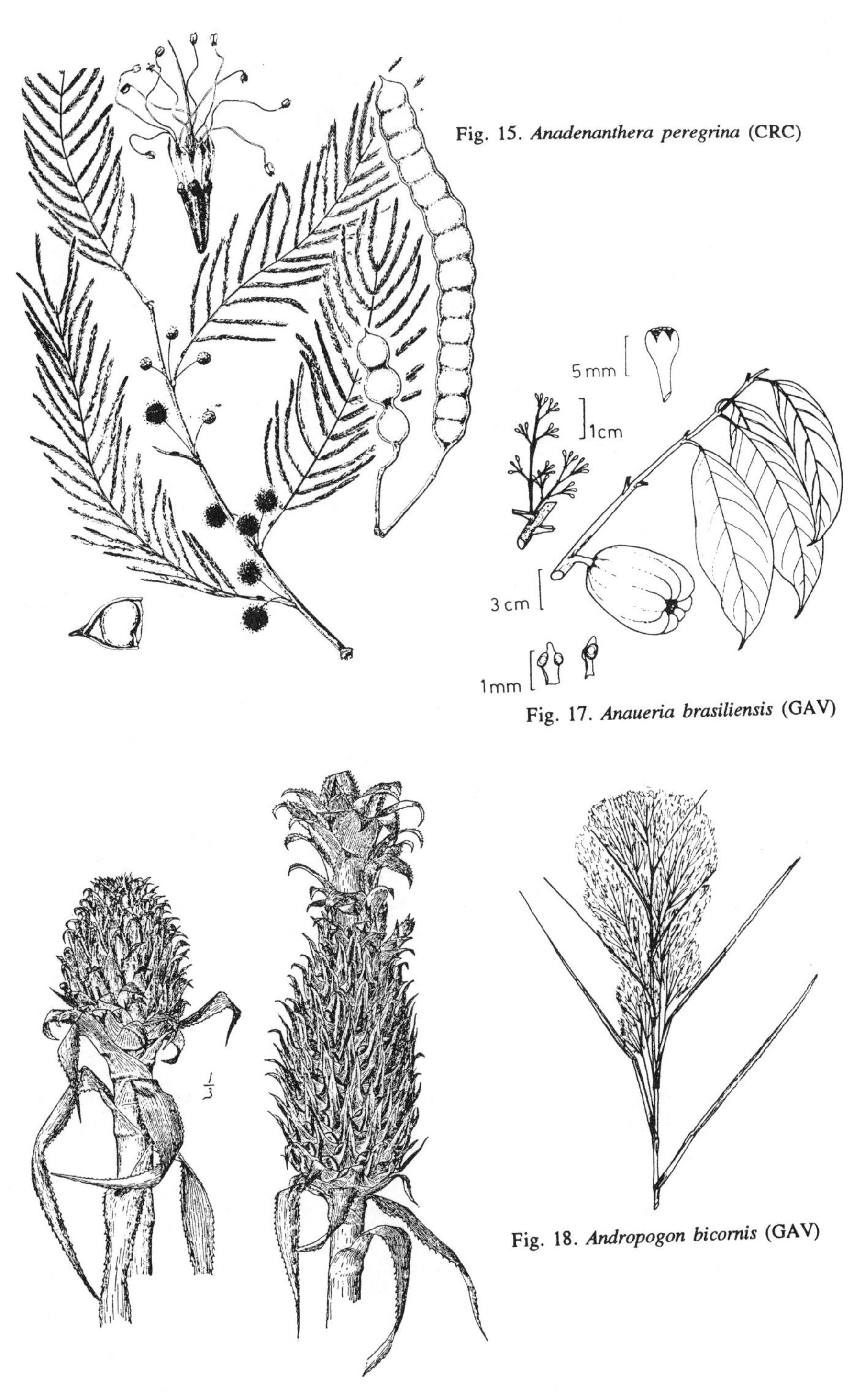

Fig. 15. *Anadenanthera peregrina* (CRC)

Fig. 17. *Anaueria brasiliensis* (GAV)

Fig. 18. *Andropogon bicornis* (GAV)

Fig. 16. *Ananas comosus* (CRC)

burning the bark with dried cow feces will keep insects, snakes, and other beasts away (BDS). Contains N,N,dimethyltryptamine. (Fig. 15)

♀ *Ananas comosus* L. Bromeliaceae. "Piña", "Piña negra", "Huacamayo piña", "Gebero piña", "Garrafón piña", "Lagarto piña", "Jambo piña", **"Pineapple"**. Cultivated. Edible fruit. A refreshing drink is made from the pericarp decoction, which is also added to "chicha" to improve its taste. Preserves made with the fruit. The juice is astringent and anthelmintic. In the Philippines the fiber yields a very fine white thread (SOU). "Tikunas" grate the green fuits in water and take in the first or second month of pregancy as abortifacient. Amazonian Brazilians take the fruit for dyspeptic flatulence (SAR). In Piura, practicing food "farmacy", the fruit is ingested for blenorrhagia, kidney stones, rheumatism, and worms (FEO). (Fig. 16)

Anaueria brasiliensis Kosterm. Lauraceae. "Añushi moena". For lumber; boats, canoes. While still young, the poles are used for beams. Also used as a hiding tree for hunting rodents. The roasted seeds are edible (RVM). (Fig. 17)

Anaxagorea brachycarpa R.E.Fr. Annonaceae. "Espintana". Wood used for beams; bark tincture antirheumatic (RVM).

Anaxagorea brevipes Benth. Annonaceae. "Carahuasca", "Espintana". In the construction of fences and rustic balconies. The bark is used as a primitive rope.

Andropogon bicornis L. Poaceae. "Cola de caballo", "Rabo de zorro", **"Horse's tail"**. Dry flowers used to stuff toys. "Bora" use it to plug their nose and ears (DAT). Valuable for forage and paper production; also used in making mats and brooms. Roots used as diuretic and sudorific. Brazilians use the rhizomes as tourniquets around snakebites (RVM). Used for beriberi and hepatitis (RAR). (Fig. 18)

Aniba perutilis Hemsley. Lauraceae. "Moena negra". Wood for lumber.

Aniba puchuri-minor (Mart.) Mez. Lauraceae. "Moena amarilla". Wood for lumber and for building canoes. Carminative, digestive, laxative, tonic, for dysentery, leukorrhea (RAR).

Aniba roseodora Ducke. Lauraceae. "Palo de rosa", **"Rosewood"**. Wood used for lumber and extraction of linalool, valued in perfumery. Also contains terpineol.

Annona cherimolia Mill. Annonaceae. "Chirimolia", **"Custard apple"**. Fruit edible. Chopped leaves applied to the nape of the neck for headache; leaf decoction drunk for dysentery; crushed seeds used to kill parasites (FEO). (Fig. 19)

Annona exellens R.E.Fr. Annonaceae. "Anona". Edible fruit.

Annona hypoglauca Mart. Annonaceae. "Guanábana sacha", **"Wild soursop"**. Edible fruit.

Annona montana Macf. Annonaceae. "Anona", "Guanábana", **"Mountain soursop"**. Edible fruit. Rio Tapajos Brazilians make solar tea of the leaves with those of bitter orange to make bath water for flu (BDS). (Fig. 20)

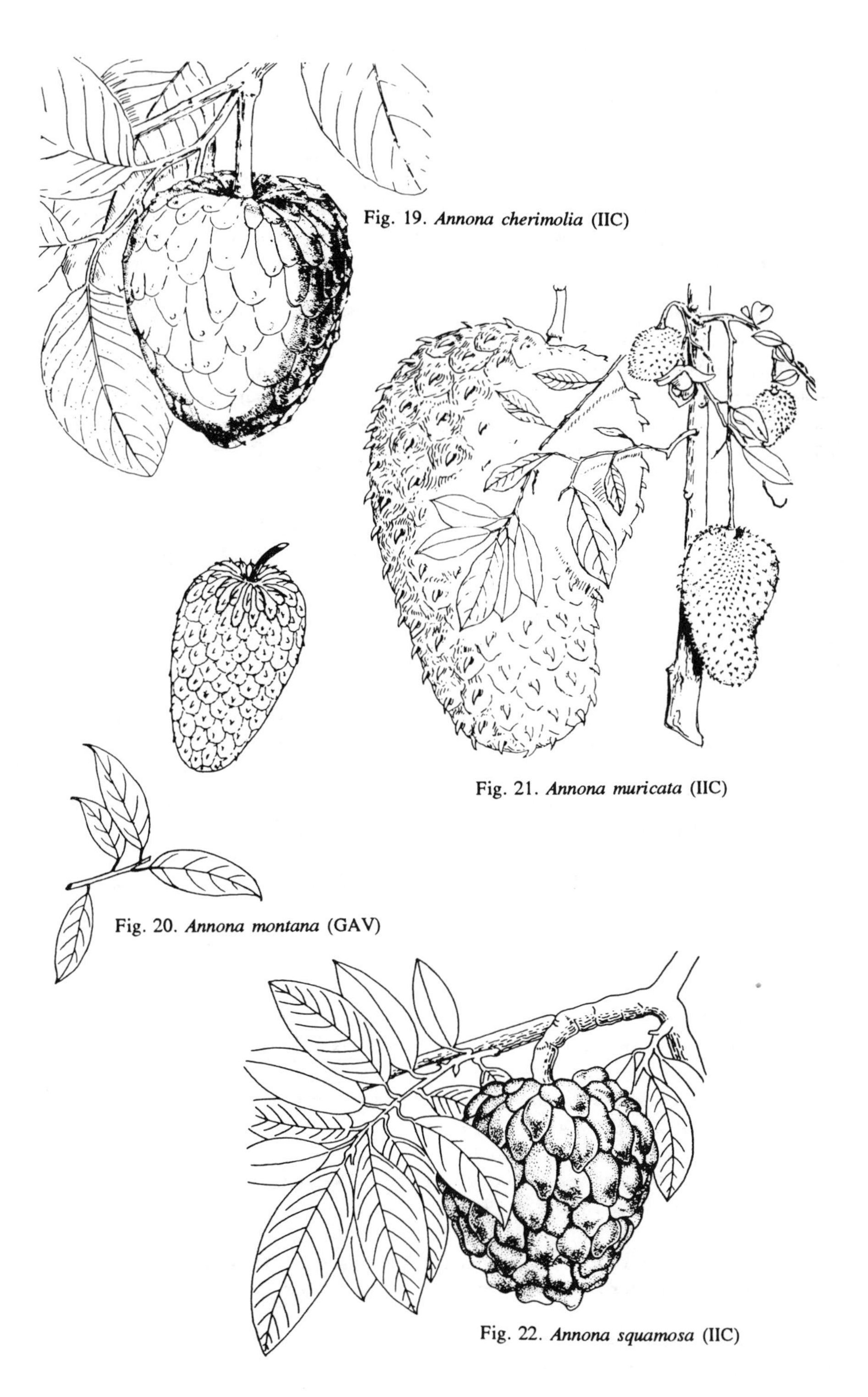

Fig. 19. *Annona cherimolia* (IIC)

Fig. 21. *Annona muricata* (IIC)

Fig. 20. *Annona montana* (GAV)

Fig. 22. *Annona squamosa* (IIC)

Annona muricata L. Annonaceae. "Guanábana", "Chirimoya", "Chirimoya brasilera", "**Soursop**". Cultivated. Fruit edible fresh or in ice creams. Leaf decoction used for catarrh in Piura; crushed seed to kill parasites (FEO). Colonists from Risaralda use the plant for rachitic children. Bark, roots and leaves are used in teas for diabetes; also used as a sedative and antispasmodic (RVM). "Créoles" use the decoction of the leaves and bark as a sedative, yet heart tonic. They use *A. montana* the same way (GMJ). Tapajos natives use the leaf tea for the liver (BDS). Elsewhere used for chills, colds, diarrhea, dysentery, dyspepsia, fever, flu, gallbladder attacks, hypertension, insomnia, kidneys, nervousness, palpitations, pediculosis, ringworm, sores and internal ulcers (DAW). (Fig. 21)

Annona squamosa L. Annonaceae. "Anona", "**Sweetsop**". Cultivated. Fruit edible (RVM). Elsewhere used for abortion, bruises, carbuncles, chancre, cold, diarrhea, dyspepsia, fever, puerperium, rheumatism, spasm, syphilis, tumors, ulcers and venereal disease; considered astringent, insecticide, pectoral, pediculicide, purgative, soporific, tonic and vermifuge (DAW). Brazilians use the leaves in cough syrup (BDS). Like so many Annonaceae with seeds used to control insects and lice, this contains pesticidal acetogenins (JAD). (Fig. 22)

Anomospermum chloranthum Diels ssp. *confusum* Krukoff & Barneby. Menispermaceae. "Abuta Sacha". "Wayãpi" consider this a dangerous POISON (GMJ).

Anomospermum reticulatum (Mart.) Eichl. Menispermaceae. "Tikuna" boil the flowers in soapy water to make a shampoo. They once used the bark in making arrow POISON (SAR).

Anthodiscus klugii Standl. Caryocaraceae. "Tamara". Bark maceration used internally for gastritis; mashed fruits applied to wounds (VDF).

Anthodiscus peruanus Baillon. Caryocaraceae. "Tahuarí". Wood for lumber, dormers and heavy construction. Widely used as a fish POISON (SAR).

Anthodiscus pilosus Ducke. Caryocaraceae. "Tahuarí amarillo", "botón caspi". Wood for lumber, dormers and heavy construction (RVM). "Maku" rub fresh fruit rind on sprains. "Tanimuka" consider the leaves insecticidal or insectifugal (SAR). In Piura, the fruit decoction is drunk for rheumatism and tumors (FEO). (Fig. 23)

Anthurium crassinervium (Jacq.) Schott. Araceae. "Rona setha rao". Powdered root used for bugbites and inflammation around Pucallpa (VDF).

Anthurium fosteri Croat sp. nov. ined. Araceae. "Jergón quiro". The "Boras" burn the leaves to obtain salt (DAT).

Anthurium pentaphyllum G. Don. Araceae. "Nea niti rao". Leaf decoction for arthritic rheumatism around Pucallpa (VDF).

Aparisthmium cordatum (Juss.) Baill. Euphorbiaceae. "Yana vara" "Ukshaquiro". Wood occasionly used for temporary construction or firewood.

Apeiba aspera Aubl. ssp. *membranacea* (Spruce) Meijer & Setser. Tiliaceae. "Llausaquiro", "Maquizapa ñaccha". Bark used as rope, fruits in handicrafts.

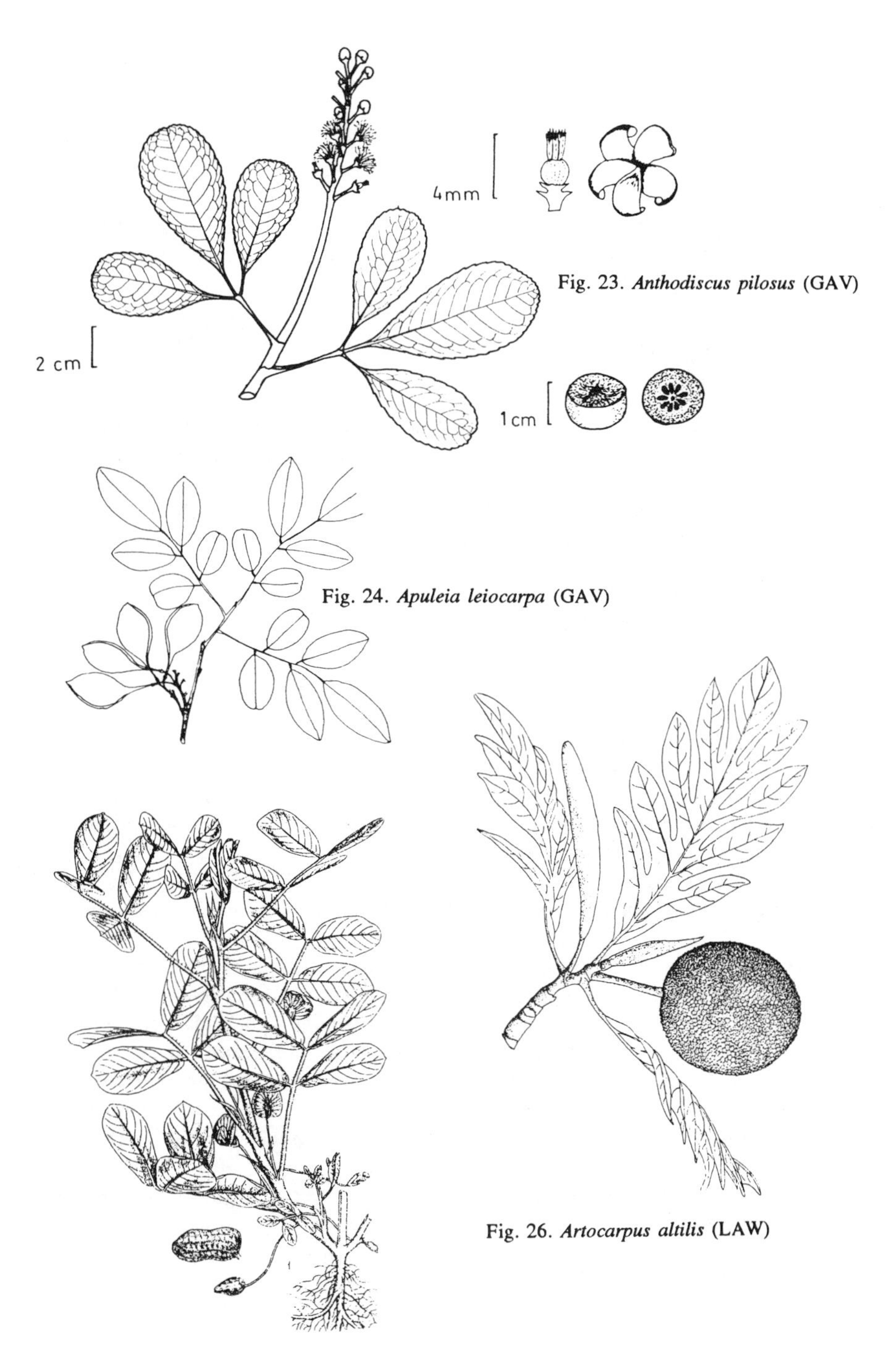

Fig. 23. *Anthodiscus pilosus* (GAV)

Fig. 24. *Apuleia leiocarpa* (GAV)

Fig. 26. *Artocarpus altilis* (LAW)

Fig. 25. *Arachis hypogaea* (DAD)

Apeiba tibourbou Aubl. Tiliaceae. "Maquizapa ñaccha blanco", "**Monkey comb**". The bark is used as rope. Bark decoction used as vermifuge (SOU). Considered antirheumatic and antispasmodic (DAW). Febrifuge (RAR).

Apodanthera smilacifolia Cogn. Cucurbitaceae. "Shanin Rao". Topical anitinflammatory around Pucallpa (VDF).

Apuleia leiocarpa (Vog.) MacBr. Fabaceae. "Ana caspi", "Guacamayo" (RAR). Wood used in heavy contruction, dormers, jam posts on bridges, boat construction, canoes, keel plates, mills. (Fig. 24)

Arachis hypogaea L. Fabaceae. "Maní", **Peanuts**". Cultivated. Nourishing seeds used as food. (Fig. 25)

Aristolochia spp. Aristolochiaceae. "Pisi rao". Decoctions or macerations of the leaves used for gastritis and toothache. Considered TOXIC around Pucallpa (VDF).

♀ *Arrabidaea brachypoda* (DC.) Bur. Bignoniaceae. "Balsa huasca". "Karaja" give the macerated roots orally to expectant mothers (RVM).

Arrabidaea candicans (Ric.) DC. Bignoniaceae. "Osen rao". Macerated leaves used in febrifugal baths around Pucallpa (VDF).

Arrabidaea chica (HBK) Verlot. Bignoniaceae. "Puca panga". Fresh leaves used in decoction alone or mixed with the fruits of *Renealmia alpinia* to dye fibers of *Astrocaryum chambira* or to make tattoos. This dye is also used to treat skin infections and herpes (RAR). Leaves also used as antiinflammatory. "Chami" from Risaralda extract the red tint to dye baskets (RVM). "Tikuna" use leaf infusion for conjunctivitis (SAR). Achual "Jivaros" chew the leaves with clay to blacken the teeth (SAR). Tapajos residents use leaf tea for anemia, blood disorders, inflammation.

Arthrostylidium sp. Poaceae. "Carricillo". The split stems are used in basketry.

Artocarpus altilis (Park.) Fosb. Moraceae. "Pandisho", "Pan del árbol". Cultivated. Green fruit eaten cooked or roasted. Latex used to treat hernias and extract worms from the skin (RVM). "Créoles" use leaf decoction to treat high blood pressure (GMJ). Anodyne, effective against hypertension, used for boils, burns, diabetes, gout, oliguria, rheumatism, sores and tumors (DAW). Mature seeds of seed-bearing varieties are eaten roasted, though they may cause indigestion and flatulence. Used in production of the regional drink called "mazato". "Wayãpi" use the latex as antirheumatic (GMJ). (Fig. 26)

Artocarpus heterophyllus Lam. Moraceae. "Jaca", "Jackfruit". Cultivated. Weighing up to 15 kilos, fruits are used to feed pigs. Some local people eat them also, because of the very sweet pulp. Considered astringent, demulcent, laxative, refrigerant, and tonic; used for alcoholism, carbuncles, caries, leprosy, puerperium, smallpox, sores, stomach problems, toothaches and tumors (DAW). (Fig. 27)

Asclepias curassavica L. Asclepiadaceae. "Benzenyuco", "Flor de Seda", "Flor de muerto", "Pucasisa", "**Bloodflower**", "**Tropical milkweed**". Generally considered POISONOUS (SOU). In Colombia the latex is used to extract decayed teeth (RVM). Inflorecences are hemostatic, and used to treat diarrhea. The dry roots are ground and used as emetic; the latex is anthelmintic and used for toothache. It is very toxic and may cause paralisis. Brazilians poultice the crushed leaves onto wounds. Latex is used for rat POISON. The "Cuna" use the roots to treat diarrhea (RVM). Around Pucallpa, they bathe in the leaf decoction for persistent fever (VDF). "Palikur" make drops for eye infections out of the shoot decoction (GMJ). "Siona-Secoya" use the latex as a vermifuge. "Tikuna" use it for toothache (SAR). (Fig. 28)

Aspidosperma excelsum Benth. Apocynaceae. "Canalete", "Remo caspi", **"Paddle tree"**. In house construction as beams, decks, and columns. Antiseptic (SAR).

Aspidosperma marcgravianum Woods. Apocynaceae. "Naranjo", "Quillobordón". Wood for lumber, dormers (RVM). Antiseptic (SAR).

Aspidosperma nitidum Benth. Apocynaceae. "Pinsha caspi", "Quillobordon", "Remo caspi". Wood used in house construction (RVM). Latex used in leprosy (SAR). Used for malaria in Brazil (DAW), for fever and lung ailments (RAR).

Aspidosperma schultesi Woods. Apocynaceae. "Quillobordón". Wood for lumber, dormers (RVM). "Makuna" paint the latex between the toes, perhaps as a fungicide. Around Rio Piraparana, roots are crushed with mud and smeared on the house as a termite repellent (SAR). Vaupes Indians use the plants for fever (SAR).

Asplenium cuneatum Lem. Aspleniaceae. "Shapumba". Sometimes planted as an ornamental.

Asplenium serratum L. Aspleniaceae. "Lengua de vaca", **"Bird's nest fern"**, **"Cow's tongue"**. Rhizome decoction for liver diseases (SOU).

Asplundia spp. Cyclanthaceae. "Puspo tamshi". Many species are used as ropes, and in basketry.

Astrocaryum chambira Burret. Arecaceae. "Chambira", **"Fiber palm"**. The fruits have edible seeds. The shells from the dried seeds are used to make baskets to carry cotton. From the leaf buds they extract fibers used in the manufacturing of hammocks, handbags, and other handicrafts where they need thread and ropes (RVM). (Fig. 29)

Astrocaryum huicungo Dummer ex Burr. Arecaceae. "Huicungo". Fruits edible; source of fiber and construction materials.

Astrocaryum jauari Mart. Arecaceae. "Huiririma". The seed is edible; from the leaf buds they extract fiber.

Astrocaryum macrocalyx Mart. Arecaceae. "Huicungo". Fruit edible; the leaves are used sometimes to build temporary roofs after the thorns have been removed.

Astrocaryum murumuru Mart. Arecaceae. "Huicungo", "Huiririmi". Seeds yield edible oil. Aphrodisiac (RAR). Fruit with 3 times more beta-carotene than carrot (MJB) (oil contains 313,000 IU). Oil composed of 48.9% lauric-, 21.6% myristic-, 13.2% oleic-, 6.4% palmitic-, 4.4% capric-, 2.5% linoleic-, 1.7% stearic-, and 1.3% caprylic-acids (MJB). (Fig. 30)

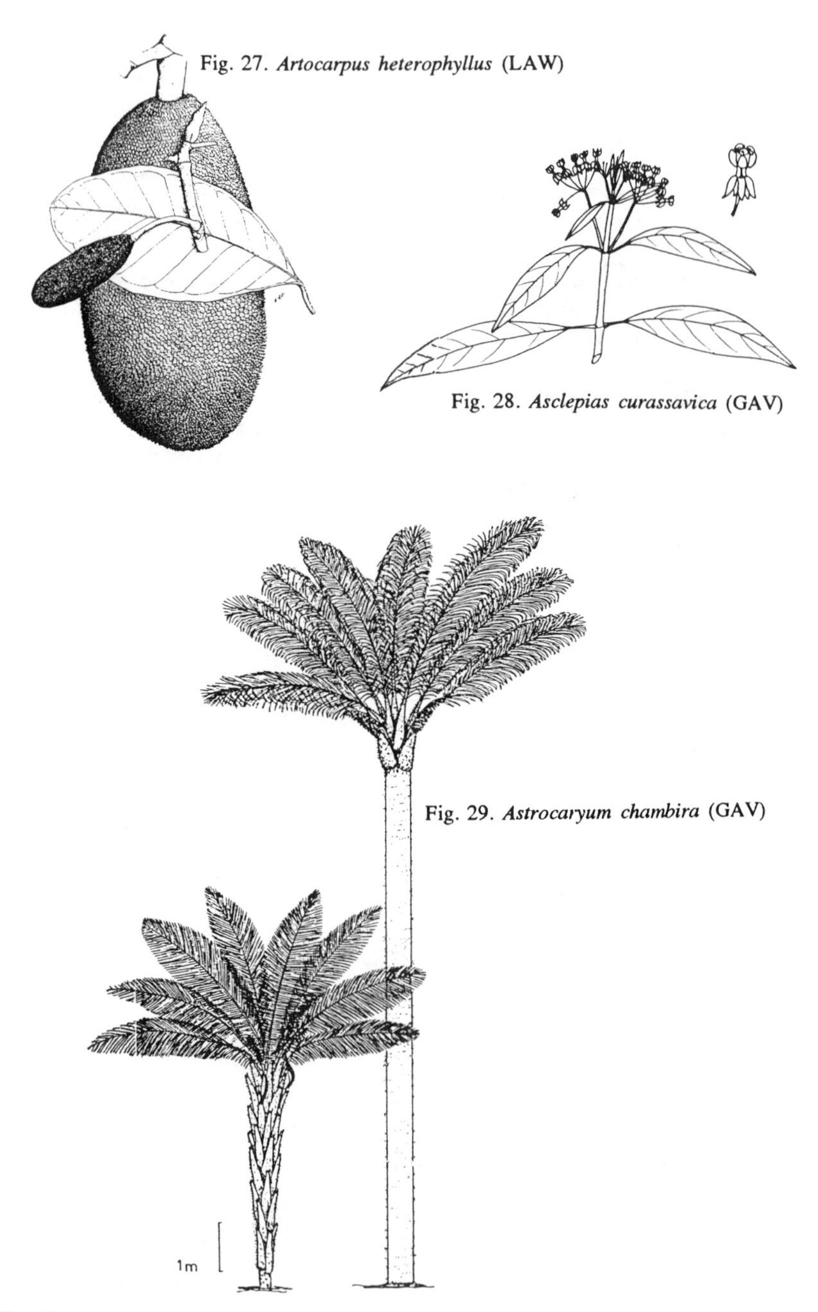

Fig. 27. *Artocarpus heterophyllus* (LAW)

Fig. 28. *Asclepias curassavica* (GAV)

Fig. 29. *Astrocaryum chambira* (GAV)

Fig. 30. *Astrocaryum murumuru* (GAV)

Attalea tesmannii Burret. Arecaceae. "Chonta", "Conta". The fruit has edible seeds (RVM).

Averrhoa bilimbi L. Oxalidaceae. "Limón chino", **"Chinese lemon"**. Cultivated as a fruit tree.

Averrhoa carambola (Sw.) Beauv. Oxalidaceae. "Carambola", **"Star Fruit"**. Cultivated fruit tree.

Axonopus compressus (Sw.) Beauv. Poaceae. "Cola de Caballo", "Grama", "Zacate amargo", "Nudillo". Lawn for sport fields and yards; for forage. Used for whooping cough (RAR). (Fig. 31)

Axonopus scoparius (Flugge) Hitchc. Poaceae. "Maicillo", "Imicayo". Forage.

Fig. 31. *Axonopus compressus* (HAC)

- B -

Bactris acanthocarpoides Barb.-Rodr. Arecaceae. "Ñeja". Fruits edible.

Bactris actinoneura Drude. Arecaceae. "Ñejilla". Fruits edible.

Bactris amoena Burret. Arecaceae. "Ñejilla". Fruits edible.

Bactris brongniartii Mart. Arecaceae. "Ñejilla". Fruits edible, leaves used for thatch (RAR).

Bactris concinna Mart. Arecaceae. "Ñejilla". Fruits edible. Wood used for carving (RAR).

Bactris gasipaes HBK. Arecaceae. ""Pejibaye", "Pijuayo", "Chontaduro", "**Peach palm**". Cultivated. Fruit and seeds edible. From cooked fruit is prepared the regional drink called "mazato"; oil is also extracted. First shoots are edible fresh or in salads, as substitute for palm hearts (SOU). (Fig. 32)

Bactris simplicifrons Mart. Arecaceae. "Cúwarahííba". Fruits edible, used by the "Bora" as soporific (DAT).

Bambusa multiplex Raeush. Poaceae. "Marona". Cultivated. As an ornamental.

Bambusa superba (Huber) McClure. Poaceae. "Marona", "Chingana". The canes are used to build fences, to make flutes; the internode of the stalk is used to store small things.

Banara guianensis Aubl. Flacourtiaceae. "Raya caspi", "Sacha rifari". The "Créoles" use the decoction of five leaves in 1/3rd liter water, drinking a cup a day in order to keep a healthy liver (GRE).

Banisteriopsis caapi (Spruce ex Griseb.) Morton. Malpighiaceae. "Ayahuasca", "**Soul vine**", "**Spirit vine**", "Yagé". In some cases semicultivated by witches, shamans and ayahuasqueros. Stems used by native farmers and city folk as a purge. The "ayahuasca" is an hallucinogenic drink, much used in the old days in rituals. Now it is used for medicinal purposes and divination. It is said not only to cure all kinds of sickness, but to help in diagnosis, divination, and telepathy. It is also laxative and emetic. To prepare the "purge" it is recommended that the stem collector (the brujo) abstain from sex for at least a week before cutting the stems, and in the day of the gathering he should go without eating. This should be done either Tuesdays or Fridays in the morning. They have to cut and grind the stems and boil them until the liquid becomes dark; then pass through a sieve; once cool, it is ready to be taken. They generally use one species, sometimes mixed with other species of *Banisteriopsis*, as *B. longialata*, and occasionally with other plants, such as: *Psychotria viridis*, *P. carthaginensis*, *Nicotiana tabacum*, *Brugmansia suaveolens*, *Malouetia tamarquina*, etc. *Tabernaemontana* sp., *Brunfelsia*, sp., *Datura suaveolens*, *Iochroma fuchsioides*, *Juanulloa*, cactus, ferns, etc. Contains alkaloids such as harmaline, tetrahydroharmine, harmol, harminic-acid methyl-ester, harminic acid, acetyl-norharmine, N-norharmine, N-oxyharmine, harmalinic acid, ketotetrahydronorharmine (RVM). (Refer back to Fig. 3 in Introduction.)

Banisteriopsis elegans (Triana & Planch.) Sandwith. Malpighiaceae. "Yagé". "Bora" treat oral sores in children with the leaves (DAT).

Banisteriopsis longialata (Ndz.) B. Gates. Malpighiaceae. "Ayahuasca", "Yagé". Sometimes used to prepare "ayahuasca". Contains high quantities of tryptamine derivatives: DMT, MMT, 5-Meo DMT, bufotenine, and traces of beta-carboline (RVM).

Banisteriopsis muricata Cav. Malpighiaceae. "Ayahuasca sacha", "Sarcelo". Used by natives of Tocache to treat dogs (RVM).

Batesia floribunda Spruce ex Benth. Fabaceae. "Huayruro colorado". The wood is mainly used for carving and decorative plaques. The seeds are used in necklaces and other handicrafts (RVM).

Batocarpus amazonicus (Ducke) Fosberg. Moraceae. "Tulpay", "Mashonaste", "Najahe". Wood for inferior lumber. It is used to build keel plates on boats and canoes. Fruit edible.

Batocarpus costaricensis Standl. Moraceae. "Sacha tulpay". To make keel plates for boats and canoes.

Bauhinia glabra Jacq. Fabaceae. "Motelo huasca". Stem infusion used by the Huallaga to treat pulmonary diseases (RVM).

Bauhinia guianensis Aubl. Fabaceae. "Escalera de mono", "**Monkey ladder**". Used as an ornamental and for handicrafts (RVM). Amazonian "Tukuna" use stem for kidney diseases (SAR). "Taiwanos" consider the seed diuretic (SAR). The root is boiled with "sarabatuco" to treat ameba in Amazonian Brazil (BDS). Considered ichthyotoxic (RAR). (Fig. 33)

Bauhinia tarapotensis Benth. in Mart. Leguminosae. "Pata de vaca", "**Cow hoof**". Semi-cultivated as an ornamental.

Begonia palmata DC. Begoniaceae. "Begonia". Cultivated ornamental.

Begonia semiovata Liebm. Begoniaceae. "Pava chaqui". The stem is chewed to quench the "hangover" thirst. "Waunana" grind the stems, cook them and drink the tea for intestinal worms (RVM).

Begonia semperflorens Link. & Otto. Begoniaceae. "Begonia". Cultivated ornamental.

Begonia spruceana A. DC. Begoniaceae. "Begonia". Leaf decoction used internally for gastritis around Pucallpa (VDF).

♀ *Bellucia grossularioides* (L.) Trian. Melastomataceae. "Níspero". Antiseptic fungicide, the inner bark steeped in rum and used for massage in case of excess vaginal excretions (BDS). Eat fruit for worms (BDS).

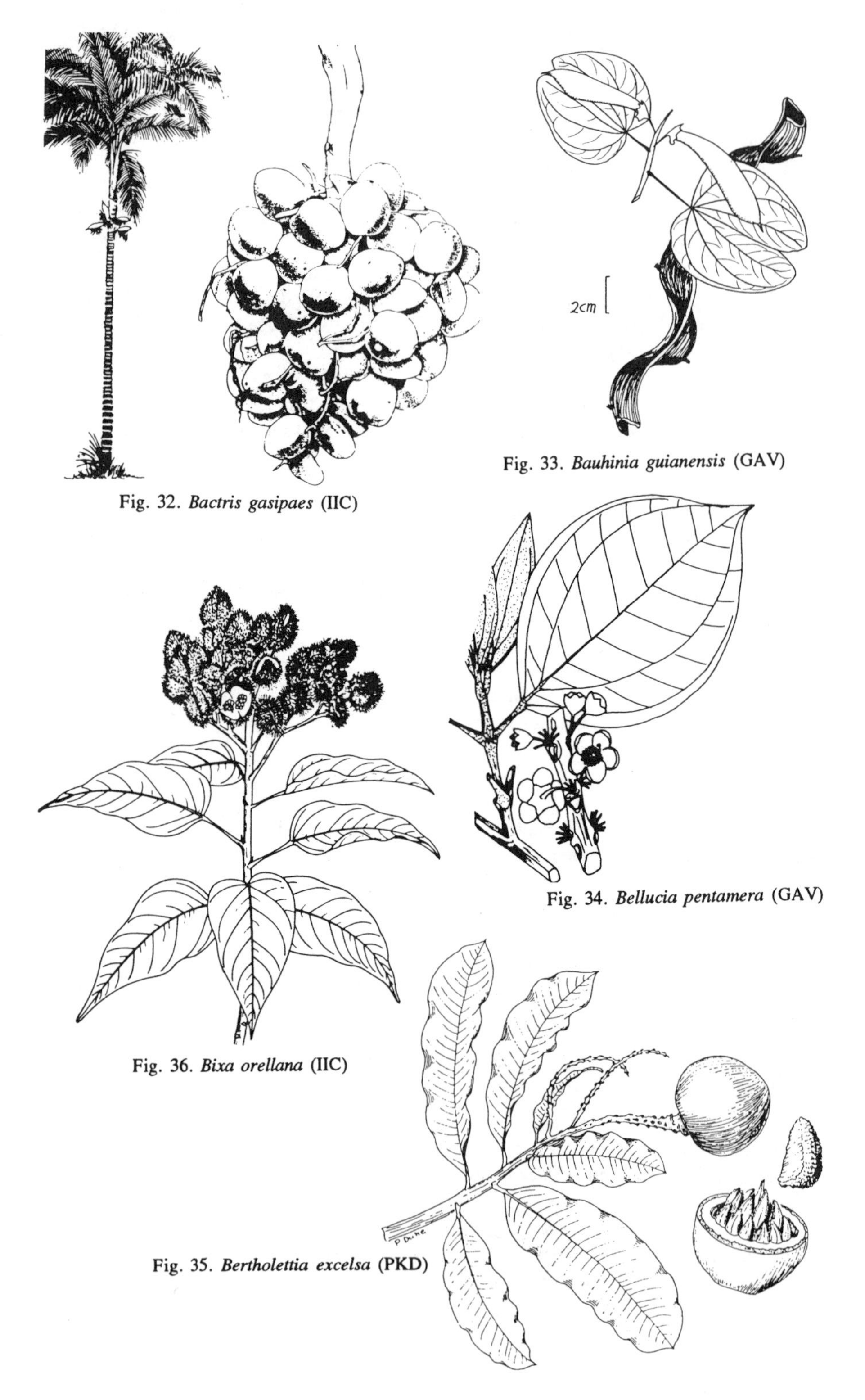

Fig. 32. *Bactris gasipaes* (IIC)

Fig. 33. *Bauhinia guianensis* (GAV)

Fig. 34. *Bellucia pentamera* (GAV)

Fig. 35. *Bertholettia excelsa* (PKD)

Fig. 36. *Bixa orellana* (IIC)

Bellucia pentamera Naud. Melastomataceae. "Sacha níspero". Fruits edible. Rasped stem used to dye gourds (SAR). "Andokes" use fresh fruits as anthelmintics (SAR). (Fig. 34)

Bertholletia excelsa HBK. Lecythidaceae. "Castaña". Cultivated near Iquitos. Good wood for lumber, used in construction. The seeds are edible. Oil extracted from the seeds to manufacture soaps (RVM). Considered insect repellent (DAW). An average brazilnut contains more than the US RDA for selenium, overdoses of which can cause hair loss, garlic breath and other problems. Selenium is an important antioxidant (JAD). Amazonian Brazilians use the bark in liver ailments (SAR). Fruit husk tea used for stomachache (BDS). (Fig. 35)

Besleria aggregata (Poepp.) Hanst. Gesneriaceae. "Puca sisa". Ornamental.

♀ *Bidens pilosa* L. Asteraceae. "Amor seco" "Cadillo", "Chilca", "Isha sheta rao", "Pacunga", "Pirco", **"Dried love"**. Chewing or gargling may help angina and sores in the mouth; infusions used as emmenagogue, antidysenteric, and to alleviate chills. Decoction mixed with lemon juice for angina, sore throat, water retention, hepatitis, dropsy. Sometimes mixed with aguardiente and milk (SOU). In Piura, the root decoction is used for alcoholic hepatitis and worms (FEO). Around Pucallpa, the leaf is balled up and applied to toothache. Leaves also used for headache VDF. In Brazil it is used as a diuretic and to treat jaundice. In the Philippines, flowers, mixed with cooked rice, are fermented to make an alcoholic beverage. In Tonga the infusion of the flowers is used to treat upset stomach in food poisoning. The "Exumas" grind sun-dried leaves and mix them with olive oil to make poultices for sores and lacerations. "Cuna" mix the crushed leaves with water to treat headaches (RVM). Used for aftosa, angina, diabetes, dysentery, dysmenorrhea, edema, hepatitis, jaundice, laryngitis, worms (RAR).

♀ *Bixa orellana* L. Bixaceae. "Achote", "Achiote amarillo". Cultivated. Natives mainly use it for food coloring and to decorate their bodies. There are experimental plots for the extraction of bixin. In Piura, the shoot decoction is considered antidysenteric, antiseptic, antivenereal, aphrodisiac, astringent, and febrifugal (FEO). The foliage is used to treat skin problems and hepatitis; also used as aphrodisiac, antidysenteric, and antipyretic. Considered good for the digestive system, and for treatment of liver disease. Very effective as a gargle for tonsilitis (RVM). "Chinatecas" poultice leaves on cuts to avoid scars (RVM). People from Cojedes use the flower infusion as purge and to avoid phlegm in newborn babies. "Kayapo" massage stomachs of women in labor with the leaves. "Waunana" use to dye demijohns and baskets. Bark yields a gum similar to gum arabic. Fiber used as cordage. "Kayapo" use to tint to the body (RVM). Dye said to be an antidote for HCN (SAR). Seeds believed to be expectorant, the roots, digestive (SAR), antitussive (BDS). Around Explorama, fresh leaf stalks, devoid of blades, are inserted into a glass of water; the mucilage that forms is applied in conjunctivitis. (Fig. 36)

Blepharodon nitidum (Vell.) Macb. Asclepiadaceae. "Mútsújkeu". Used by the "Boras" to tie the beams of their houses (DAT).

Bocconia pearcei Hutchinson. Papaveraceae. "Yanali". Latex used as escharotic, shoots to treat "aire" (FEO).

Boerhavia caribaea Jacq. Nyctaginaceae. "Pega-Pega", "Pega josa". Root decoction antispasmodic, choleretic (FEO).

Boerhavia paniculata Rich. Nyctaginaceae. "Solidonio". Diuretic, antibilious, used for the gall bladder and liver. Brazilians, e.g., use the root tea for hepatitis (BDS, RAR).

♀ *Bonafousia longituba* Mgf. Apocynaceae. "Coca sanango", "Nane repote", Ruro de paloma", Sanango macho". Latex used for toothache, bark as contraceptive around Pucallpa (VDF).

Bonafousia sananho (R. & P.) Mgf. Apocynaceae. "Lagarto micunan", "Sanango". Powdered root taken for rheumatism around Pucallpa, avoiding salt and grease (VDF).

Bonafousia tetrastachya Mgf. Apocynaceae. "Capeshini", "Coca sanango", "Osha rao". Powdered root taken for rheumatism around Pucallpa (VDF).

Bonafousia undulata (Vahl) A.DC. Apocynaceae. "Sanango". "Palikur" use as an analgesic for headaches (GMJ).

Bothriospora corymbosa (Benth.) Hook. Rubiaceae. "Afasi caspi", "yacu quinilla". The wood is used for construction of rural houses. Indians consider the wood POISONOUS (SOU). Said to contain antiparasitic cephaeline and emetine, the leaves are ingested to expunge intestinal parasites (SAR).

Bougainvillea spectabilis Willd. Nyctaginaceae. "Papelillo". Cultivated. Ornamental, used to treat scrofula and pneumonia (SOU).

Brachiaria mutica (Forssk.) Stapf. Poaceae. "Grama", "Nudillo". Forage.

Brachiaria purpurascens (Rad) Henr. Poaceae. "Grama". Forage.

Brosimum acutifolium ssp. *obovatum* (Ducke). C.C.Berg. Moraceae. "Tamamuri", "Congona" (FEO). Trunk and bark used as antirheumatic. Used by the "Palikur" and "Wayãpi" as a hallucinogen and in ritual initiations. Bark used in baths to reduce fever. Also used by the "Wayãpi" to protect against witchcraft (GMJ). Tincture used in Piura for headache and rheumatism (FEO). Depurative, tonic, vermifuge, to stimulate the appetite, for rheumatism and syphilis (RAR).

Brosimum alicastrum Sw. ssp. *bolivariense* (Pittier) C.C.Berg. Moraceae. "Machinga". Heavy wood used for dormers, beams, and parquets. Tree used by the "Campas" as a blind for hunting tapir and peccary. (Fig. 37)

Brosimum aubletii P.&E. Moraceae. "Tamamuri", "Machinga", "**Snakewood**". Fruits edible. Wood excellent (RAR).

Brosimum guianense (Aubl.) Huber. Moraceae. "Huayra caspi", "Misho chaqui", "Palo brujo". The wood is used in construction, and the latex to treat abscesses.

Brosimum lactescens (S.Moore) C.C.Berg. Moraceae. "Machinga", "Tamamuri". Wood used for beams, posts, for keel plates on boats. Fruit edible.

Brosimum parinarioides Ducke ssp. *amplicona* (Ducke) C.C.Berg. Moraceae. "Chimicua", "Machinga". The wood is used for keel plates on boats, it is also used for beams, posts, and forked poles. The latex is used to mix with latex of higher quality in order to increase the volume. The latex is used as cicatrizant, as a pectoral, as well as tonic to treat weakness; in poultice it is used for bruises (RVM).

Brosimum potabile Ducke. Moraceae. "Machinga". For lumber. Latex a medicinal tonic (RVM).

Brosimum rubescens Taubert. Moraceae. "Palisangre". Wood used for handicrafts, forked poles, posts, dormers, and to make decorative plaques. "Yaguas" use the wood to carve a mythological representation called "mayantu". Heartwood mixed with aguardiente to fortify the body after child birth. In Peru wood used to build the gigantic cross in 1985, to honor the visit of the pope Juan Pablo II (RVM).

Brosimum utile (HBK) Pittier ssp. *longifolium* (Ducke) C.C. Berg. Moraceae. "Machinga", "Chingonga". The wood is used to make plaques, in lumbers, in rustic carpentry and indoor building. The latex is used to adulterate better quality latices. Wood also used for pulp and plaques. Some Colombian natives drink the latex with sugar water as a tonic (RVM, SAR). Bark used for asthma and other lung ailments (SAR). Brazilians drink a little latex for fever (BDS).

♀ *Brownea ariza* Benth. Fabaceae. "Rosa del monte" (Wild rose), "Palo cruz". Loreto's finest wood is used to make canes, rulers, parquets, and small objects. Considered hemostatic (RVM). "Siona" use flowers in emetic hemostat teas for excessive menstruation (SAR).

Brownea disepala Little. Fabaceae. "Trueno shimbillo". Used in making tool handles.

Brugmansia aurea Lagerhein. Solanaceae. "Toé", "Maricahua", "Floripondio", **"Angels trumpet"**. Cultivated. Ornamental used as an hallucinogen, for telepathy and divinations. Some people smoke the leaves and the flowers in small quantities, as a substitute for marijuana. Brujos make a purge for dogs to make them good hunting dogs (EXP). Leaf decoction externally used for dermatitis and orchitis; chopped leaves antispasmodic, decongestant (FEO). The main alkaloid in *Brugmansia* is scopolamine, also found are: norscopolamine, atropine, meteloidine, noratropine, 3alpha,6beta-ditigloyloxytropane-7beta-ol, tropine, 3alpha-tigloyloxytropane RVM. (Fig. 38)

Brugmansia insignia (Barb.-Rodr.) Lockwood ex R.E.Schultes. Solanaceae. "Toa-toé", "Toé", "Maricahua" Cultivated. Rio Pastaza natives tie the leaves (transdermal "scope" JAD) or simply apply the juice onto aches and pains (SAR). Loreto natives use the leaf infusion as a calmant and sedative (SAR).

Brugmansia suaveolens (H.&B. ex Willd.) Berchtold & Presl. Solanaceae. "Toé", "Floripondio", "Maricahua". Cultivated. Similar applications as *B. aurea* (RVM). "Kofáns" believe the white flowered forms are more toxic than the pink-flowered forms (SAR). (Fig. 39)

Brugmansia versicolor Lagerhein. Solanaceae. "Toé", "Floripondio". Cultivated. Similar applications as *B. aurea*. Used as an ornamental like *Brugmansia* x *candida* and *Brugmansia* x *insignis*.

Brunfelsia chiricaspi Plowman. Solanaceae. "Chiricaspi", "Chiric sanango". Used as an additive in the preparation of hallucinogenic beverages (RVM). The hallucination has serious side effects, among them: chills, cold sweats, heavy tongue, itchiness, nausea, stomachache, temporary insanity, tingling, and vomiting (SAR). Used by the Indians for fever (SAR).

Brunfelsia grandiflora D.Don ssp. *schultesii* Plowman. Solanaceae. "Chiric sanango", "Chuchuwasha", "Moca pari". Sometimes cultivated as an ornamental or medicinal plant. Curanderos add it to ayahuasca or to prepare special initiations. Also used for bad luck. Around Pucallpa, the leaf decoction is used internally for arthritis and rheumatism (VDF). Root infusion with aguardiente for rheumatism and venereal diseases. "Boras" take the root decoction for chills (RVM). Amazonians take the plant, regarded as diaphoretic and diuretic, for fever, rheumatism, snakebite, syphilis and yellow fever (SAR). The main phytochemical is scopoletin, which induces psychopharmacologic effects. Also: tartaric acid, lactic acid, quinic acid, etc. (RVM).

Brunfelsia miro Monach. Solanaceae. "Chiric sanango". Sometimes planted as ornamental. The alcoholic root maceration used as an antirheumatic.

Buchenavia cf *fanshawei* Exell & Maquire. Combretaceae. "Yacushapana". The hard and heavy wood is used for dormers, jam posts for bridges; sometimes it is used for keel plates on boats.

Buchenavia viridiflora Ducke. Combretaceae. "Lagartillo de altura". Wood is used for forked poles, posts, beams, parquets, decorative plaques, and also in handicrafts (RVM).

Bunchosia armeniaca (Cav) DC. Malpighiaceae. "Ciruelo", "Indano". Cultivated. Ornamental with edible fruit.

Bursera graveolens (HBK.) Trel. Burseraceae. "Carana", "Palo santo". Twigs chewed for toothache; boiled as depurative, febrifuge, and sedative. Resin massaged in for headache and rheumatism (FEO). Used in amulets against shamanism (RAR).

Byrsonima chrysophylla HBK. Malpighiaceae. "Indano". Brazilians use the bark in cough syrups (BDS).

Byrsonima coriacea (Sw) DC. var. *spicata* (Cav.) Ndz. Malpighiaceae. "Indano colorado". The astringent bark is used for dysentery. Fruit edible.

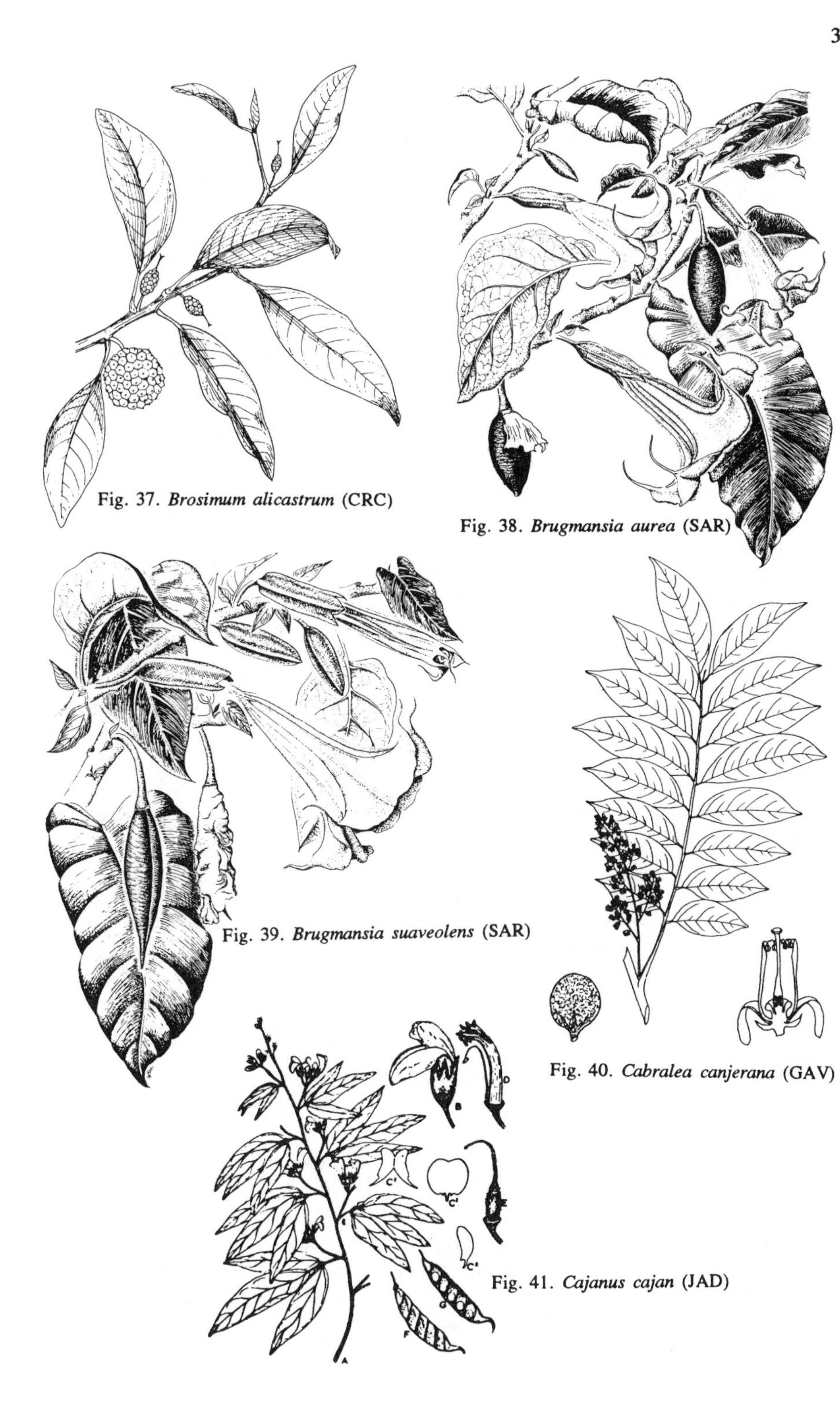

Fig. 37. *Brosimum alicastrum* (CRC)

Fig. 38. *Brugmansia aurea* (SAR)

Fig. 39. *Brugmansia suaveolens* (SAR)

Fig. 40. *Cabralea canjerana* (GAV)

Fig. 41. *Cajanus cajan* (JAD)

- C -

Cabralea canjerana (Vel.) Mart. Meliaceae. "Cedro macho". Wood excellent for carpentry, used like *Cedrela odorata*. Bark is used to reduce fever. The decoction of some meliaceous fruits is used as insecticide, hazardous for domestic animals (SOU). (Fig. 40)

Caesalpinia ferrea Mart. Fabaceae. "Cedro", "**Ironwood**". Brazilians drink water containing grated fruits for bruises (BDS, RAR).

♀ *Caesalpinia pulcherrima* (L.) Sw. Fabaceae. "Angel sisa", "**Pride of Barbados**". Cultivated ornamental. Leaves cathartic, bark used as febrifuge and abortive. Flowers used in purges and antipyretics. Leaves used to intoxicate fish (SOU).

♀ *Caesalpinia spinosa* (Mol.) Kuntze. Fabaceae. "Talla", "Tara". Leaf infusion used for stomatitis, branches for abortion, fever, constipation (FEO).

♀ *Cajanus cajan* (L.) Millsp. Fabaceae. "Puspo poroto", "**Pigeon pea**". Cultivated. The seeds are edible. The flowers and foliage are used to stabilize the menstrual period, and for dysentery. Used elsewhere as a diuretic, astringent, antidysenteric, detersive, febrifuge, laxative, and vulnerary (RAR, RVM). "Créoles" recommend the leaf tea for lung infections. The seed infusion is diuretic (GMJ). Brazilians take the seed tea for blood ailments and inflammation (BDS), the leaf tea for cough. (Fig. 41)

Caladium bicolor (Ait.) Vent. Araceae. "Corazón de Jesús", "**Jesus's heart**", "Oreja de perro", "**Dog's ear**". Cultivated ornamental. "Wayãpi", superstitious about this plant, use it for love potions and amulets (GMJ). Febrifuge (RAR).

Caladium picturatum (Lind.) Koch & Bouche. Araceae. "Corazón sangriento", "**Bleeding heart**". Cultivated ornamental.

Calathea allouia (Aubl). Lindl. Marantaceae. "Dale dale", "**Sweet cornroot**". Cultivated. The tuberous roots are used as a starchy food. "Chocó" and "Cuna" wrap vegetables and meat in the leaves. Tonic; used against scrofula (RAR). (Fig. 42)

Calathea gigas Gagn. Marantaceae. "Wira bijao del bajo". Leaves are used by fishermen to line baskets, helping keep fish fresh. Leaves used as thatch (RVM).

Calathea inocephala (Kuntze) H.Kennedy & D.Nichols. Marantaceae. "Wira bijao". Leaves used to wrap food.

Calathea insignis Peters. Marantaceae. "Bijao chancaquero". Mainly used to wrap brown sugar (RVM).

Calathea lutea (Aubl.) G.F.W.Meyer. Marantaceae. "Bijao", "Cauassu". Sometimes semicultivated. Leaves are mainly used as wrapping, to cook tamales, etc. The "Chinatocas" of Oaxaca, use the leaves to wrap tamales, and also to make tortillas. Cooked roots are used by the "Cuna" to control nausea and diarrhea. Resinous wax extracted with solvents similar to wax from *Calathea gigas* Gagn. (Fig. 43)

Calathea microcephala (P.E.) Koern. Marantaceae. "Motelilla enana". Ornamental.

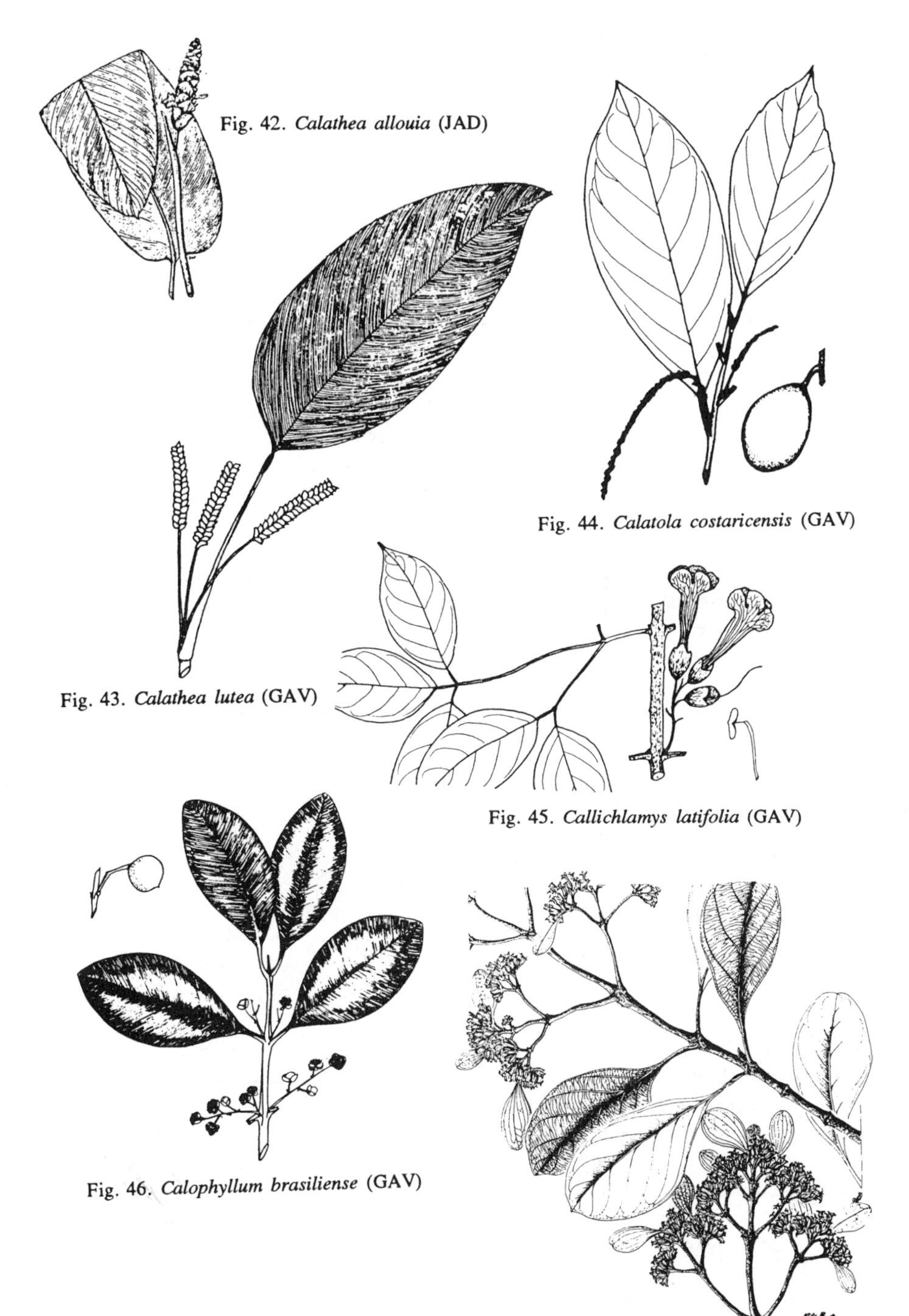

Fig. 42. *Calathea allouia* (JAD)

Fig. 43. *Calathea lutea* (GAV)

Fig. 44. *Calatola costaricensis* (GAV)

Fig. 45. *Callichlamys latifolia* (GAV)

Fig. 46. *Calophyllum brasiliense* (GAV)

Calathea roseo-picta (Lindl.) Reg. Marantaceae. "Motelilla". Ornamental, medicinal for the "Boras" (DAT).

Calatola costaricensis Standley. Icacinaceae. "Pió", "Nashum". "Achuales" along the Rio Huasaga chew the tender leaves to coat their teeth and protect against cavities. The blackish color will disappear ca 15 days after use. This old custom is vanishing because natives are embarrased when caucasians call them "huanganas". "Achueles" from Pastaza and "Mayna" from Corrientes also use the blackish coat to protect their teeth (RVM). The "Candoshi-shapra" also chew the leaves to protect their teeth. (Fig. 44)

Calliandra angustifolia Spruce. Fabaceae. "Bobinsana", "Bubinsana". "Quinilla blanca". Cultivated ornamental (RVM). Around Pucallpa, bark decoction taken internally for dyspnea and rheumatism (VDF).

Callichlamys latifolia (Rich.) Schumann. Bignoniaceae. "Manapeuí". "Wayãpi" consider this the best medicine for leishmaniasis. Grated stem bark poulticed onto the sores or made into a decoction applied to the sores (GMJ). (Fig. 45)

Calophyllum brasiliense Cam. Clusiaceae. "Alfaro", "Lagarto caspi". Fruit edible (RAR). Excellent wood for naval construction; keel plates for boats, canoes. Used to make plaques and for lumber. Part of this wood is exported to the USA and Europe. In Central America the slow-burning, long-lasting wood has a great value as a fuel. Elsewhere used for dormers, and on the railroads. "Karajá" drink the bark infusion for diarrhea, and build canoes with the wood (RVM). The bark, mixed with *Coutarea hexandra* Schumann, is used by the "Palikur" in infusions for diabetes and worms (GMJ). Considered antiherpetic, antirheumatic (RAR). Wood contains the antiinflammatories dehydrocycloguanan and jacareubin (JBH). (Fig. 46)

Calopogonium caeruleum Hemsl. Fabaceae. "Kudzu". Cultivated as forage and green manure.

Calycophyllum acreanum Rubiaceae. "Capirona". For firewood and charcoal.

Calycophyllum obovatum (Ducke) Ducke. Rubiaceae. "Capirona". For firewood and charcoal. (Fig. 47)

♀ *Calycophyllum spruceanum* (Benth.) Hook. Rubiaceae. "Capirona". The wood, used for contruction, is a favorite for firewood and charcoal. Natives boil 1 kg of bark in 10 liters of water to obtain 4 liters of medicine from which they drink 150 ml 3 times a day for 3 consecutive months for diabetes (RVM). Peruvians use the bark against "sarna negra", an arachnid that lives under the skin. Powdered bark is applied to mycoses (SAR). Considered contraceptive, emollient, vulnerary. (Fig. 48)

Campomanesia lineatifolia R.& P. Myrtaceae. "Palillo". Cultivated. Fruits edible. Natives soak the wood chips in water to make a beverage to treat hemoptysis (SOU). Good timber. Antiemetic, hemostatic, for skin infections (RAR). (Fig. 49)

Campsiandra angustifolia Spruce ex Benth. Fabaceae. "Huacapurana". Seed edible cooked. Wood burned in rubber processing. Bark infusion mixed with aguardiente. Used as an antirheumatic and antidiarrheic (RVM). Used for malaria (RAR). (Fig. 50)

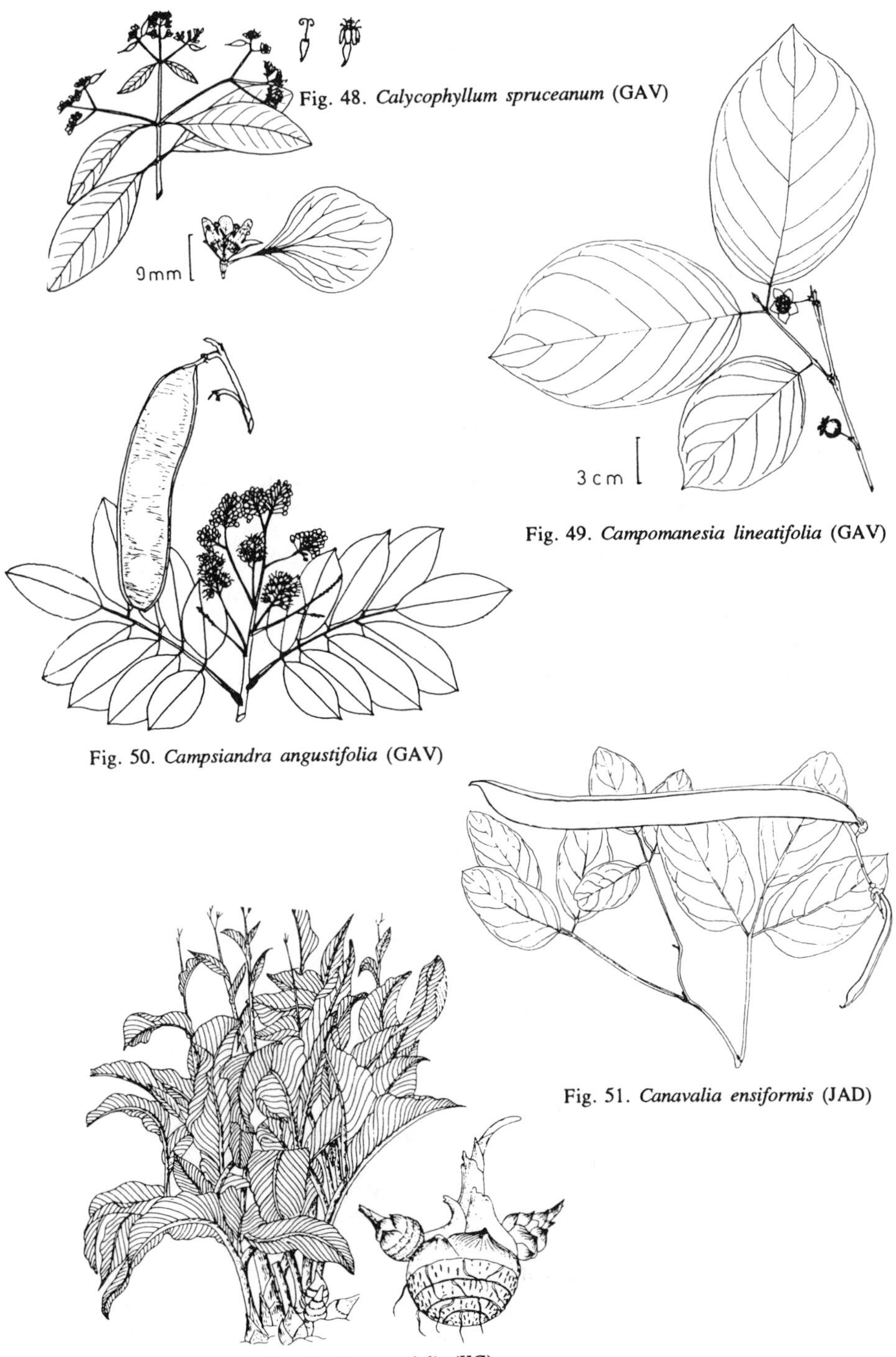

Fig. 48. *Calycophyllum spruceanum* (GAV)

Fig. 49. *Campomanesia lineatifolia* (GAV)

Fig. 50. *Campsiandra angustifolia* (GAV)

Fig. 51. *Canavalia ensiformis* (JAD)

Fig. 52. *Canna edulis* (IIC)

Campsiandra comosa Benth. Fabaceae. "Huacapurana". Bark tincture in aguardiente taken for malaria around Iquitos (SAR). "Witoto" use powdered bark as a vulnerary to treat wounds (SAR). Fruits steeped in vinegar and salt are used for oral infections in Amazonian Brazil (SAR).

Canavalia ensiformis (L.) DC. Fabaceae. "Nescafé". Cultivated. The toasted ground seed are used as a substitute for coffee (RVM). Considered anodyne, digestive, mitogenic, POISON, yet tonic (DAW). (Fig. 51)

Canna edulis Ker. Cannaceae. "Achira", **"Canna lily"**. Cultivated. Ornamental with edible corms (RVM). Considered diuretic and emollient (DAW). (Fig. 52)

Canna indica L. Cannaceae. "Achira". Cultivated ornamental. The rhizome contains starch, used in soups for children and the sickly. Root decoction diuretic. In Brazil, leaves are used to clean ulcers, and stems to treat rheumatism. The roots, with mild aroma, are used for fever and dropsy. A docoction of the roots and leaves is used to treat gonorrhea and as a diuretic. "Cuna" cook 4 pieces of stem; after the infusion has cooled, they drink a cup and bathe in the remaining water to regain energy RVM.

Canna paniculata R.&P. Cannaceae. "Achira cimarrona". Cultivated ornamental. Rhizome eaten by humans and pigs (RVM).

Cannabis sativa L. Moraceae. "Marihuana". Cultivated. Narcotic, hallucinogen (RVM), often a drug of abuse. Useful in glaucoma, multiple sclerosis and the nausea of chemotherapy (JAD). (Fig. 53)

Caperonia castaneifolia (L.) St. Hil. Euphorbiaceae. "Yacu ishanga". Leaves poulticed onto swellings and tumors (SOU).

Caperonia palustris (L.) St. Hil. Euphorbiaceae. "Yacu ishanga". Same as previous species.

Capirona decorticans Spruce. Rubiaceae. "Capirona de altura". For firewood and house construction (RVM). Used for acarosis, psoriasis (RAR).

Capparis sola Macbr. Capparidaceae. "Intuto caspi". An important ingredient in "Java" "curaré" (SOU).

Capsicum annuum L. Solanaceae. "Pimiento", "Pucunucho", **"Sweet pepper"**. Cultivated. Natives believe that to become a good blowgun shooter, one must chew and eat slowly a half dozen fruit before breakfast for 8 days. Studies report that this species is hallucinogenic, but they don't use it for this purpose. Curanderos use it in a maceration mixed with aguardiente to give as a purge for dogs to make them good hunting dogs. This species and *C. frutescens* are present in this maceration and also *Nicotiana tabacum, Brunfelsia grandiflora* ssp. *schultesii* and *Brugmansia* spp. (RVM). "Jivaro" apply the fruit directly to toothache (SAR). In Piura, the fruit infusion is considered antipyretic, tonic, and vasoregulatory; the decoction used as a gargle for sore throat or pharyngitis; the tincture is applied to bugbites, mange, hemorrhoids, and rheumatism (FEO). (Fig. 54)

Capsicum conicum Mey. Solanaceae. "Carolito",* "Coralito"*. Cultivated. Considered one of the strongest chillies.

Fig. 53. *Cannabis sativa* (CRC)

Fig. 54. *Capsicum annuum* (MVA)

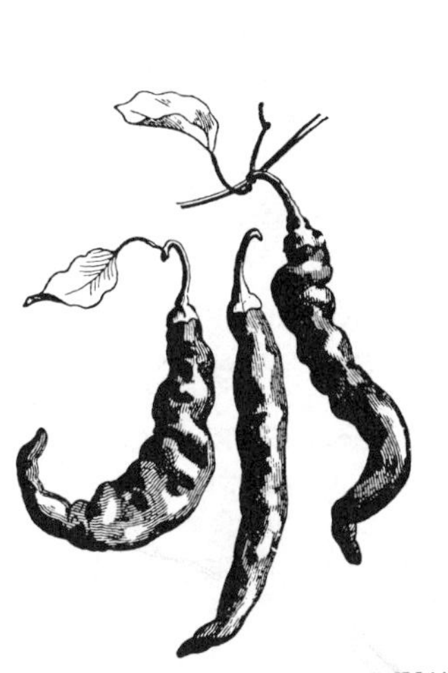

Fig. 55. *Capsicum frutescens* (MVA)

Fig. 56. *Carapa guianensis* (GAV)

Capsicum frutescens L. Solanaceae. "Pimiento", "Charapilla", **"Hot pepper"**. Cultivated. Spice, believed to be hallucinogenic. "Créoles" use for throat diseases of pigs. "Wayãpi" use it in making "curaré" (GMJ). Rio Apaporisa natives take the fruits for flatulence, and use small quantities of powdered fruit when breathing is difficult. Used also for scorpion stings, toothache, hemorrhoids, fever, flu, rheumatism (RAR). The active hot ingredient, capsaicin, is used to reduce the pain of cluster headaches, diabetic neuropathy and herpes zoster (JAD). (Fig. 55)

Caraipa densifolia ssp. *densifolia* Mart. Clusiaceae. "Brea caspi". Wood used in house construction, beams, poles, columns, decks, etc. Leaves reportedly aphrodisiac; stem latex used for herpes, dermatitis, eczema, itch, and impetigo. The product "óleo de tamakouré", extracted from this plant, is used in various skin diseases, rheumatism, corneous ulcers, and pediculosis (RVM).

Caraipa grandifolia ssp. *grandifolia.* Clusiaceae. "Aceite caspi". Wood used in house construction; beams, decks, and columns. The alcoholic infusion is used to treat local parasitosis, dermatitis, itches, fungus, and pruritus (RVM). Amazonian Brazilians apply the sap to herpes, itch and mange (SAR).

Caraipa jaramilloi Vasquez. Clusiaceae. "Aceite caspi". Wood used in house construction.

Caraipa tereticaulis Tul. Clusiaceae. "Brea caspi", "Aceite caspi blanco". Wood used for construction.

Caraipa utilis Vasquez. Clusiaceae. "Aceite caspi negro". Wood used in house construction; decks, beams, posts. This is the main wood used in house construction around Iquitos.

Caraipa valioi Paula. Clusiaceae. "Aceite caspi". Wood used in house construction.

Carapa guianensis Aubl. Meliaceae. "Andiroba", "Requia", **"Brazilian mahogany"**. An excellent wood for carpentry, comparable with the wood from *Cedrela odorata* and *Swietenia macrophylla.* The bitter bark infusion is believed febrifuge and vermifuge (SAR), also a tonic. Perhaps useful in herpes (RAR). Infusion used to wash dermatoses and sores (SAR). Seeds yield an oil, with the consistency of lard, used to coat wood to protect it from insects (SOU). Brazilians sell seed oil as antiinflammatory and antiarthritic (RVM). Also used in the soap industry. Fruit oil ingested for cough in Brazil (BDS). The "Wayãpi", the "Palikur", and the "Créoles" use it to remove ticks from their heads, also for *Schongastia guianensis*, which gets in the skin. Native Americans trust the oil as an emollient and antiinflammatory for skin rash (GMJ). (Fig. 56)

Carica microcarpa Jacq. Caricaceae. "Papailla". Fruit edible RVM.

♀ *Carica papaya* L. Caricaceae. **"Papaya", "Pawpaw"**. Cultivated. Green fruit eaten cooked; ripe, eaten fresh or in juices. A dozen seed are swallowed as a vermifuge. For constipation, eat half a papaya. Rutter mentions use of papaya for acarosis, enteritis, and tachycardia (RAR). "Chocó" mix the latex with honey as vermifuge. Leaf infusion cardiotonic. "Cuna" use cooked roots for indigestion. Tikuna eat grated immature fruit with 2-6 aspirin, inducing abortion in about two days (SAR). In Piura, the leaf tea is considered

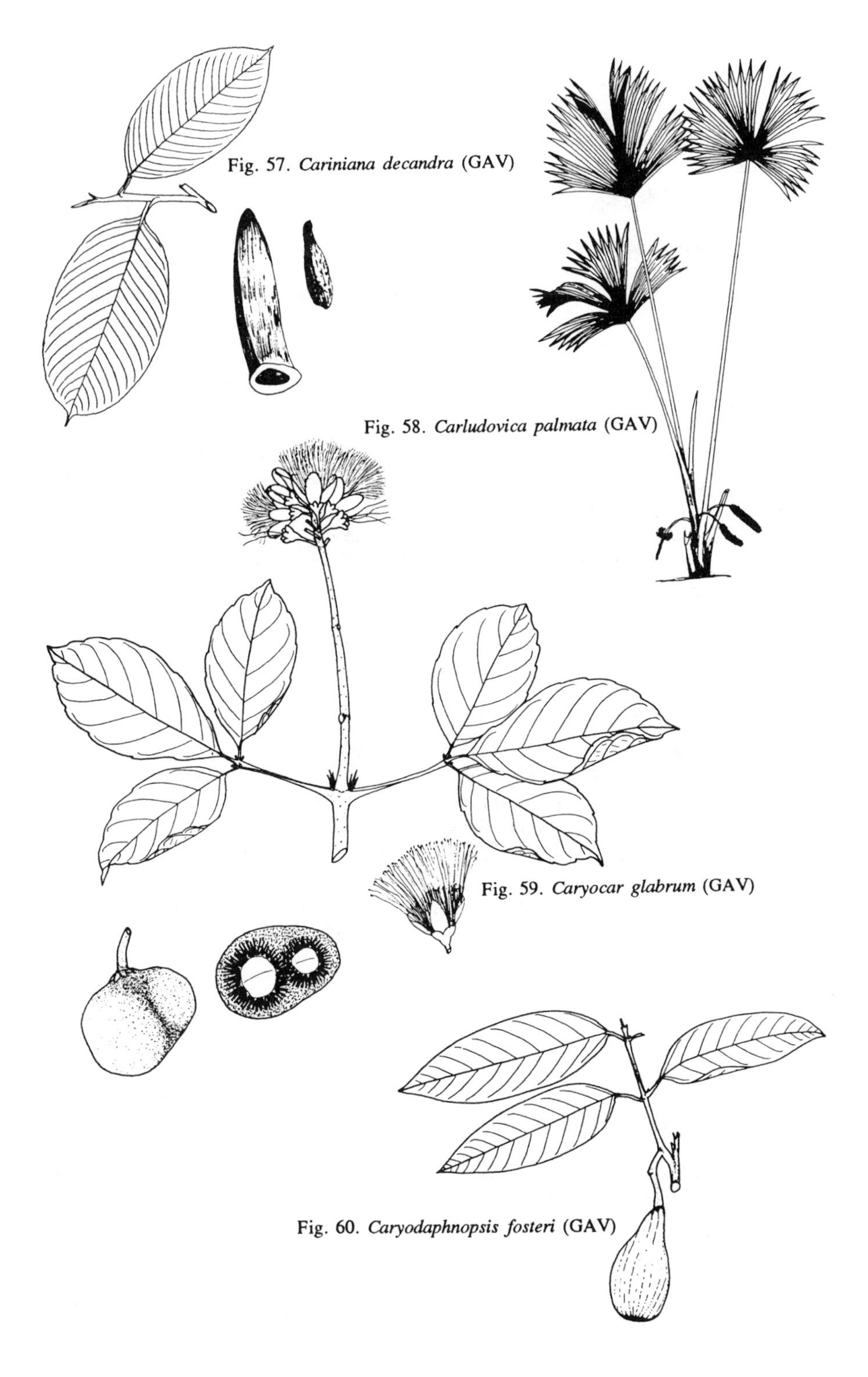

Fig. 57. *Cariniana decandra* (GAV)

Fig. 58. *Carludovica palmata* (GAV)

Fig. 59. *Caryocar glabrum* (GAV)

Fig. 60. *Caryodaphnopsis fosteri* (GAV)

digestive and hypotensive; chopped fruits are used as antiseptic (FEO). Brazilians make flower tea for heart and liver (BDS). Knowing that meat tenderizer (based on papaya's papain) had been used for sea nettle stings, JAD applied papaya juice to the rash Don Segundo induced by flagellating the wrist with stinging nettle. JAD had a reaction. Chymopapain has been used to dissolve herniated disks, but 1 in 4,000 people exposed to this treatment die of anaphylactic shock. Recent news has suggested that too much papaya might induce prostate cancer (JAD).

Cariniana decandra Ducke. Lecythidaceae. "Cachimbo", "Papelillo caspi", "Tahuarí". Good wood for lumber; parquets, and construction in general. Bark commercialized in Iquitos as substitute for "Tahuarí", *Tabebuia* spp., used to treat many sickness and diseases, even cancer. Bark used as room divider in rustic houses. (Fig. 57)

Cariniana multiflora Ducke. Lecythidaceae. "Papelillo caspi". The wood is used for parquets.

Carludovica palmata R.&P. Cyclanthaceae. "Bombonaje", "**Panama hat palm**". Petioles used in making darts for the blowguns. Buds yield a fiber used to make Panama hats, fans, baskets, curtains, cartons for cigarettes. "Chocó" occasionally eat the buds which taste like asparagus. Cooked roots are used to treat sore bruises. Among the "Bora" and "Witoto", the stump of one species is burned, and the ashes used to coat hallucinogenic pellets made from *Virola* bark (SAR). Used for stingray stings (RAR). (Fig. 58)

Carpotroche longifolia Poepp. & Endl. Flacourtiaceae. "Champa huayo", "Cacahuillo". Seed aril is edible.

♀ *Caryocar glabrum* (Aubl.) Pers. ssp. *glabrum*. Caryocaraceae. "Almendro colorado". Wood for lumber; dormers, bridges. Thorny fruit with edible nut. Mesocarp of green fruit crushed and mixed with water as a fish POISON. Mesocarp and endocarp contain saponin. "Kubeos" and "Tukanos" ingest seed for dysmenorrhea (SAR). Some "Sionas" bind the inner bark around the arm to improve their hunting aim, only to find that, after about a half hour, blisters form which leave a scar (SAR). (Fig. 59)

Caryocar microcarpum Ducke. Caryocaraceae. "Almendro blanco", "Almendro colorado" . Seed edible. Bark used for mycoses and filaria, sometimes mixed with *Elephantopus scaber* and *Lagenaria siceraria* (GMJ).

Caryodaphnopsis fosteri Van der Werff. Lauraceae. "Achuni moena". Wood for lumber, used for carpentry, construction in general, and canoes. (Fig. 60)

Caryodaphnopsis inaequalis (A.C.Smith) Van der Werff & Richter. Lauraceae. "Moena". For cleaned lumber.

Caryodendron orinocense Karst. Euphorbiaceae. "Inchi", "Meto huayo". Raw seed are mildly laxative. Seeds, edible cooked or roasted, are rich in polyunsaturated fatty acids (up to 72% linoleic), making them useful in preventing atherosclerosis. Crude oil laxative, used for skin and lung disease. (Fig. 61)

Casearia fasciculata (R.&.P.) Sleum. Flacourtiaceae. "Tamarillo". Fruit edible (VAG).

Casearia pitumba Sleumer. Flacourtiaceae. "Péécojúhe". The leaves mixed with water are used as soothing baths, and to color the hair (DAT).

Casearia praecox Griseb. Flacourtiaceae. "Nea bero". Leaf wads applied to toothache (VDF).

Casearia sylvestris SW. Flacourtiaceae. "Ucho caspi". Brazil's "Karajá" drink the bark maceration for diarrhea (RVM).

Cassia alata L. Fabaceae. "Retama", **"Ringworm senna"**. Cultivated. Decoction used to treat herpes ("ihui"), and infected wounds. The strong decoction is used to kill chiggers (SOU). Colombian Amazonians use the flower decoction as a purge. Juice from macerated flowers used to treat pellagra sores (RVM). "Wayãpi" use the leaf decoction for worms and skin diseases (GMJ).

Cassia bacillaris L. Fabaceae. "Mataro chico". Flower infusion used to wash the swollen muscles caused by trauma. Green fruit crushed and poulticed onto infected wounds. Tip of the fruit cut and squeezed onto small wounds (RVM).

Cassia bicapsularis L. Fabaceae. "Alcapaquilla", "Alcaparillo". Cubans use the leaves as a purge, that may causes stomachache. Branches are used in making baskets (SOU).

Cassia latifolia (G.F.W.Meyer) Irwin & Barneby. Fabaceae. "Mataro huasca". Same as *C. bacillaris*.

Cassia loretensis Killip & Macbr. Fabaceae. "Mataro grande". Same uses as *C. bacillaris*.

Cassia macrophylla Kunth. Fabaceae. "Mataro grande". Same as *C. bacillaris*. "Kofán" use for earache and headache (SAR).

Cassia occidentalis L. Fabaceae. "Achupa poroto", "Hedionda", "Retama", "Ayaporoto", **"Coffee senna"**. Roots are diuretic; root decoction a refreshing drink for fever (SOU). Seeds made into a coffee-like beverage for asthmatics; flower infusion for bronchitis considered abortive, anithepatitic, purgative (RAR).

Cassia quinquangulata (Rich.) Barneby. Fabaceae. "Mataro chico". Same as the *C. bacillaris*. Crushed leaves steeped in cold water to make a refreshing bath (MJP). Roots contain quinquangulin and rubrofusarin, which may have activity against leukemia, in vitro (JBH).

Cassia reticulata Willd. Fabaceae. "Retama". Sometimes planted as an ornamental. Flower infusion used for liver diseases, acid indigestion, upset stomach and kidney inflammation. Leaves and flowers contain antibiotics such as rhein (cassic acid), which is antibacterial against gram-positive and acid-resistant bacteria. The antibiotics reduce swellings of hepatic and renal sickness. Also used to treat venereal and skin diseases (LAE, RVM). Leaves used in baths for gastritis and ulcers (VDF). Used around Explorama for ringworm (JAD). "Boras" burn the leaves to repel sandfly *Lutzomyia* sp. "Manta blanca", vector of leishmaniasis. Used as a purge by the "Chocó". The Piria "Cuna" in the town of Piria (Panama) use it for stomachaches. Infusion of leaves and

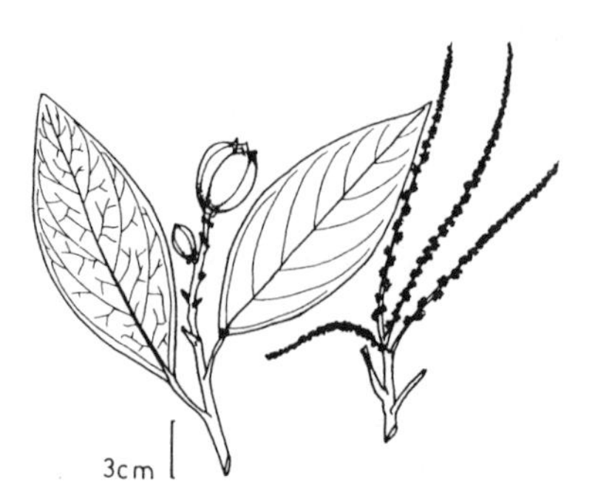

Fig. 61. *Caryodendron orinocense* (GAV)

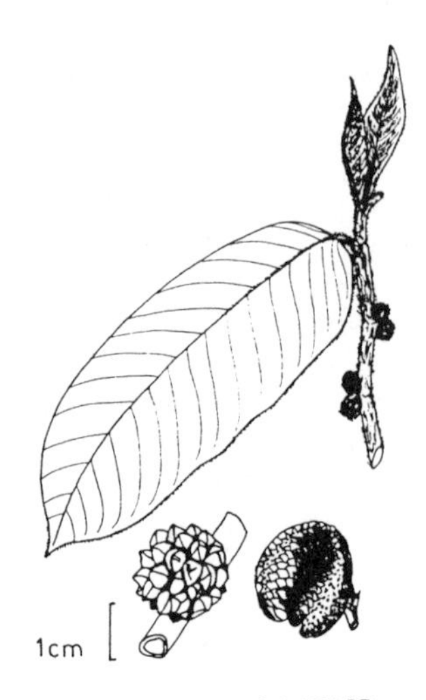

Fig. 62. *Castilla ulei* (GAV)

Fig. 63. *Cedrela odorata* (GAV)

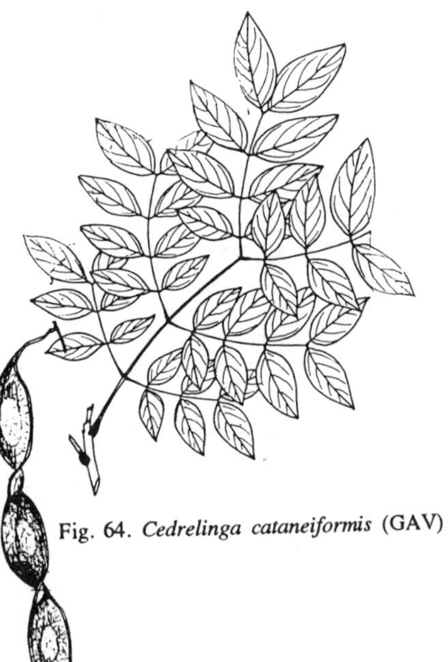

Fig. 64. *Cedrelinga cataneiformis* (GAV)

flowers used by the "Waunana" for stomachaches (RVM). "Witotos" use the roots in a febrifugal tea (SAR). "Tukanos" use leaves as insect repellents in clothes and hammocks (SAR). "Achuanos" value for fungal infections (SAR). Sometimes used for cardiac edema (NIC).

Cassia ruiziana (Vog.) Irwin & Barneby. Fabaceae. "Mataro". Similar to *C. bacillaris*. "Kofáns" use bark branches for earache and headache (SAR).

Cassia tora L. Fabaceae. "Aya poroto", "Dormidera", "**Sicklepod**". Mashed leaves applied to itch (BDS). The antitick folklore is exaggerated, "Mash up leaves, remove juice... Give it to any animal and the ticks will jump off." (BDS).

Castilla ulei Warb. Moraceae. "Caucho". Fruit edible (VAG). Latex yields a gum used in the plastic industries. Vaupes Indians apply the latex to wounds, as a protective coating (SAR). (Fig. 62)

Casuarina equisetifolia L. Casuarinaceae. "Casuarina". Cultivated ornamental tree.

Catharanthus roseus L. Apocynaceae. "Isabelita". Cultivated. Valued in flower gardens, it contains TOXIC compounds. The roots and leaves are used as vermifuges and hemostatics (RVM). Widely used as a dangerous folk remedy for diabetes. Famous as an antileukemic, with vinblastine and vincristine, now used more than 30 years in treating leukemias and other types of cancer (JAD).

Catoblastus drudei Cook & Doyle. Arecaceae. "Ponilla", "Taótaco". "Boras" use stems to smoke fish and to trap them using fish poisons (RVM).

Cattleya spp. Orchidaceae. "Orquidia". Orchids of this genus and *Laelia* are the most valued ornamental orchids, followed by others such as: *Odontoglossum, Miltonia, Cymbidium, Oncidium, Epidendrum,* and *Maxillaria;* other genera from the Old World as *Vanda, Phaleonopsis,* and *Dendrobium,* are rarely cultivated.

Cavanillesia umbellata. Bombacaceae. "Lupuna bruja", "Puca lupuna", "Pretino", "Árbol del tambor". Seeds edible. Seed oil used for medicine and cooking (SOU). Some natives believe witches use it to do harm, spreading clothing of a victim around the tree.

Cecropia distachya Huber. Cecropiaceae. "Cetico". For paper pulp.

Cecropia ficifolia Warb. Cecropiaceae. "Cetico". For paper pulp.

Cecropia membranacea Trecul. Cecropiaceae. "Cetico". "Shiari". Bark is used in building boats and canoes; the internodes are used in handicrafts, and to make blowguns (RVM).

Cecropia sciadophylla Mart. Cecropiaceae. "Cetico", "Purma cetico". Stems used in rafts to transport heavy wood; also used for paper pulp, and matchsticks. Roots provide drinking water. "Boras" use ashes from burned leaves to flavor coca and/or make it strong. From pounded bark is made canvas for painting regional scenes (RVM). "Tirio" drip the sap from crushed plants into sore eyes. Also used for bleeding gums and heart problems (MJP).

Cedrela angustifolia Sesse & Moç. ex A.DC. Meliaceae. "Cedro blanco", "**White cedar**", "Cedro", "**Cedar**". Wood for lumber, decorative plaques.

Cedrela fissilis Vell. Meliaceae. "Cedro blanco", "**White cedar**". Wood for lumber, decorative plaques (RVM). Used as abortifacient and urinary astringent (RAR).

Cedrela odorata L. Meliaceae. "Cedro rojo", "Cedro", "**Spanish Cedar**". One of the finest woods used in the Amazon; also used to make ornamental plaques. The astringent bark infusion is used for diarrhea and urinary problems (RVM). Leaves and bark steeped in bath water for body aches and colds (BDS). Also used for gangrene and orchitis (RAR). (Fig. 63)

Cedrelinga cataneiformis (Ducke) Ducke. Fabaceae. "Tornillo", "Huayra caspi", "Cedro masha". Wood for lumber; used in carpentry, construction in general. Grows on poorer soils than *Cedrela*, not requiring much care (ALG). (Fig. 64)

Ceiba pentandra (L.) Gaertn. Bombacaceae. "Ceiba", "**Kapok**", "Lupuna", "Lupuna blanca". Wood mainly used for plywood exports. Because it grows along the rivers with easy access, it has been overexploited to the point that it is disappearing. In the old days, trees served as guideposts for river navigators. "Wayãpi" associate this tree with jungle spirits. Bark decoction used in baths for fever (GMJ). Branch decoction diuretic and emetic (FEO). The cotton is used with blowguns (JAD).

Ceiba samauma (Mart.) K.Schum. Bombacaceae. "Ceiba", "Huimba", "**Kapok**". Used to make buoys to transport heavy wood, and also to make rafts. Cotton used by natives on darts (RVM).

Celosia argentea var. *cristata* (L.) Kuntze. Amaranthaceae. "Gallo cresta", "**Cock's comb**". Cultivated ornamental.

Celtis iguanea (Jacq.) Sarg. Ulmaceae. "Meloncito blanco", "Palo Blanco", "**Hackberry**". Fruit edible; also ground and cooked for dysentery (SOU). "Wayãpi" use it in making "curaré" (GMJ).

Centropogon cornutus (L.) Druce. Campanulaceae. "Gallo-cresta-rango". Leaves and fruits boiled and eaten (RAF).

Cephaelis ipecacuanha (Brot.) A.Rich. Rubiaceae. "Ipecac" Colombians chew the root as an insect repellent and amebicide. Crude extracts still find their way into millions or prescriptions a year in the US. Found in many US medicine chests to cause vomiting in children who have swallowed poison. Poison control centers should be consulted, though, as vomitting is counterindicated with some poisons. Emetine has elsewhere proved out against ameba, bilharzia, cancer, and guinea worms (JAD).

Ceratopteris pterioides (Hook.) Hieron. Parkeriaceae. "Lechuga de agua", "**Water lettuce**". Used as an ornamental, planted in ponds.

Cespedesia spathulata (R.&P.) Planch. Ochnaceae. "Afasi caspi", "Caballo shupa". "Witotos" and "Boras" use leaves to line baskets, and plug roof leaks. The bare trunk used in construction; bark used to make different objects for ritual purposes. They

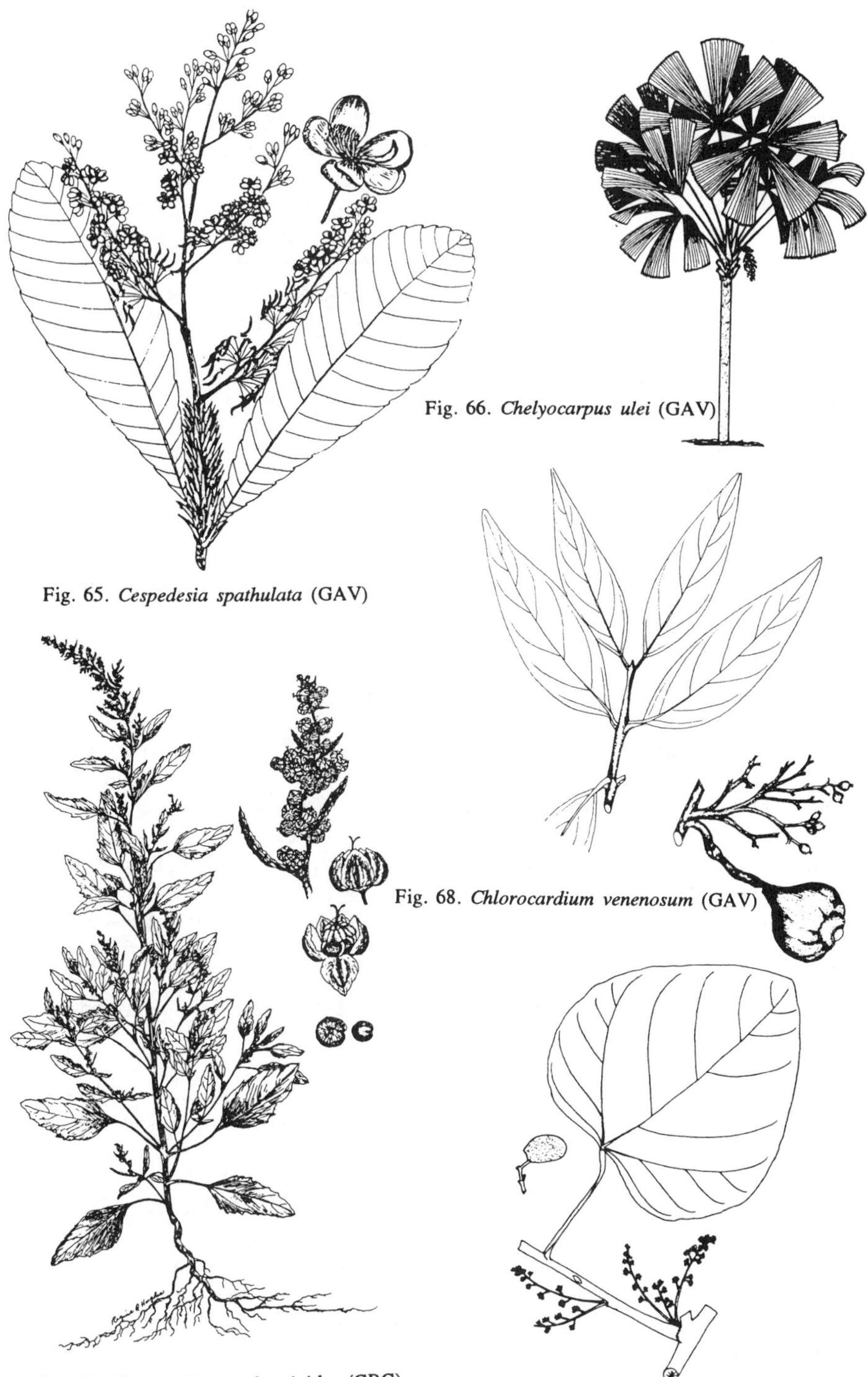

Fig. 65. *Cespedesia spathulata* (GAV)

Fig. 66. *Chelyocarpus ulei* (GAV)

Fig. 67. *Chenopodium ambrosioides* (CRC)

Fig. 68. *Chlorocardium venenosum* (GAV)

Fig. 69. *Chondrodendron tomentosum* (GAV)

make a big pot for the cahuana (a gelatinous starch drink) and a funnel used to filter vegetal salts (RVM). (Fig. 65)

Cestrum auriculatum L'Herit. Solanaceae. "Hierba santa". Leaf decoction applied externally for headache and hemorrhoids, the infusion taken internally for fever and rhuematism (FEO).

Cestrum hediondinum Dun. Solanaceae. "Hierba santa", **"Holy weed"**, "Hierba de la Virgen", **"Virgin's weed"**. Cultivated. Sometimes as an ornamental, and also used in magic-rituals for cures and protection against bad luck. Mainly used in baths, but also rubbed on affected areas. For dandruff, colds, gastralgia, sarampion (RAR).

Cestrum megalophyllum Dun. Solanaceae. "Mitiraey", "Nibi saya", "Yanagara negra". Around Pucallpa, used for headache and toothache (VDF).

Chelyocarpus ulei Dammer. Arecaceae. "Sacha bombanaje", "Ushpa aguaje". Leaves used for roofing and umbrellas (RVM). (Fig. 66)

♀ *Chenopodium ambrosioides* L. Amaranthaceae. "Cashua", "Paico", **"Wormseed"**. Cultivated. Used as an anthelmintic, mainly for *Ascaris* and *Oxyurus* (effective ingredient ascaridole). Plant infusions used for stomachache, cholera and tumors. Around Pucallpa, the maceration is applied topically for arthritis (VDF). "Créoles" use it as a vermifuge for children, cold medicine for adults. "Wayãpi" use decoction for upset stomach; internal hemorrages caused by falls. Tonic used to treat fever, flu, laryngitis, infant dermatosis. "Tikunas" consider the leaf decoction purgative and vermifuge (SAR). Root and leaf decoction, taken each month during menstruation, believed contraceptive (SAR). A cupful is believed to induce labor (SAR). In Piura, the leaf decoction is believed carminative, decongestant, depurative, insecticidal, and vermifuge; used for cramps, gout, hemorrhoids, and hysteria (FEO). With the resurgence of tuberculosis, try beating the leaf juice with the yolk of an egg. "Good for the lungs in general and cures tuberculosis" (BDS). (Fig. 67)

Chimarrhis glabriflora Ducke. Rubiaceae. "Pablo manchana", "Pampa remocaspi", "Yerno prueba". Wood is used for rural construction, firewood, and living fences (RVM).

Chimarrhis williamsii Standl. Rubiaceae. "Pablo manchana", "Yerno prueba", "Palo perro". Like the preceeding.

Chlorocardium venenosum (Koster. & Pink) Rohw., Richt. & v.d.Werff. Lauraceae. "Moena". Timber used in boats. Used in some curaré recipes, contains rodiasine and dimethylrodiasine (RVM). (Fig. 68)

Chondrodendron tomentosum R.&P. Menispermaceae. "Ampihuasca", **"Curaré"**. Some natives, crush and cook the roots and stems, adding other plants and venomous animals, mixing until it becomes a light syrup; they call this decoction "ampi", or "curaré", which they use on the tip of their arrows and darts. The active ingredient in "curaré" is D-tubocurarine, actually used in medicine. Brazilians consider the root diuretic, emmenagogue, and febrifuge (SAR), using it internally for madness and dropsy, externally for bruises. Used for edema, fever, kidney stones, and orchitis (RAR). (Fig. 69)

Chorisia insignis HBK. Bombacaceae. "Lupuna". Wood used for plywood.

Chrysophyllum argenteum Jacq. ssp. *ferrugineum* (Ruiz & Pavon) Penn. Sapotaceae. "Masaranduvilla". Fruits edible.

Chrysophyllum bombycinum Penn. Sapotaceae. "Balatillo", "Caimitillo-hoja grande". Fruits edible.

Chrysophyllum caimito L. Sapotaceae. "Caimito". Cultivated. Fruits edible. "Créoles" prepare a hypoglycemic decoction (GMJ). "Tikunas" spread the latex on infected gums; others use for abscesses and sores (RAR); "Yukunas" use it for fungal crotch infections (SAR). (Fig. 70)

Chrysophyllum nanaosense Penn. Sapotaceae. "Quinilla". Fruit edible (VAG).

Chrysophyllum peruvianum Pennington. Sapotaceae. "Caimitillo". Wood for construction. Edible fruit.

Chrysophyllum prieurii A.DC. Sapotaceae. "Quinilla colorada". Parquets, posts, forked poles, and handicrafts.

Chrysophyllum sanguinolentum (Pierre) Baehni. Sapotaceae. "Balata Sapotina". For house construction; from the latex they extract gum "balata". Latex used in coating wounds (SAR).

Chrysophyllum ulei Krause. Sapotaceae. "Sacha caimito". For house construction.

Chrysophyllum venezuelanense (Pierre) Pennington. Sapotaceae. "Caimitillo", "Magaranduva". For house construction. Edible fruit.

Cinnamomum maynense (Nees) Koster. Lauraceae. "Moena". Wood for timber.

Cissampelos andromorphe DC. Menispermaceae. "Reshno toscan". Around Pucallpa, leaf macerations used to wash the head for fever and headache (VDF).

Cissampelos pareira L. Menispermaceae. "Imchich masha", "Barbasco". "Palikur" use the leaf poultice as an analgesic (GMJ). Seeds used for snakebite; diuretic, expectorant, febrifuge, piscicide, POISON, for venereal disease (RAR). Contains tetrandrine, which is analgesic, antiinflammatory, and febrifuge.

Cissus erosa L. Vitaceae. "Navarria". Leaf decoction used in washes for arthritis (VDF).

Cissus sicyoides L. Vitaceae. "Ampato huasca", "Paja de culebra", "Sapo huasca", "Yedra", **"Toad vine"**. Leaf decoction used around Pucallpa for rheumatic pains (VEF). Leaves used as hypotensive. One of the four species most used as a pain reliever, and for sickness similar to epilepsy; pharmacologically, by inducing convulsions, it was found that the aqueous extract, after removing the fat, is most effective (RVM). Nicole Maxwell describes poulticing crushed leaves on a sprained ankle (NIC). On the Papajos, the leaf tea is boiled for anemia (BDS). For flu and hemorrhoids (RAR).

Citrullus lanatus (Thunb.) Matsumara and Nakai. Cucurbitaceae. "Sandia". Fruit edible. Seeds parched and eaten or used to make coffee. Seeds a good source of phenylalanine, an antisickling agent (DAD).

♀ *Citrus aurantiifolia* (Christm.) Swingle. Rutaceae. "Naranja agria", "**Lime**". Cultivated. Juice mixed with crushed *Neurolaena* and *Petiveria* for measles. Root decoction taken a cup-a-day during the menses, said to act as a contraceptive. (Recently, Explorama guides said the juice of one lime, taken by both partners just before intercourse, will prevent conception.) Juice taken by the "Tikuna" with one aspirin for fever, with three aspirin for abortifacient (SAR). Peels used as antidandruff, antispasmodic, decongestant, and sedative. Tonic flowers used for cramps and enteritis (FEO). Various citrus species are high in limonene, ascorbic acid, and beta-carotene, all noted to prevent cancer (JAD).

Citrus limon (L.) Burm. Rutaceae. "Limón", "**Lemon**", "Limón ácido", "**Acid lemon**". Cultivated. Fruit edible. Leaf infusion digestive.

Citrus medica L. Rutaceae. "Limón cidra". Cultivated. Fruit edible.

Citrus paradisi Macfadyn. Rutaceae. "Toronja", "**Grapefruit**", "Pomelo". Cultivated. Fruit edible.

Citrus peruvianus R.&P. Rutaceae. "Limón sutil". Cultivated. Fruit edible.

Citrus sinensis (L.) Osbeck. Rutaceae. "Naranja", "**Orange**", "Naranja dulce". Cultivated. Fruit edible.

Clarisia biflora R.&P. Moraceae. "Capinuri de altura". Wood used for dormers.

Clarisia racemosa R.&P. Moraceae. "Mashonaste", "Tulpay", "Guariuba". Wood for lumber; dormers, jam posts for bridges, and occasionally as keel plates for boats. The "Campas" use it as a blind when hunting (RVM). (Fig. 71)

Clavija jelskii Szyszl. Theophrastaceae. "Lucma sacha". Planted as an ornamental.

Cleome spinosa L. Capparidaceae. "Tuaruubia". Plant used to treat stomach problems (RVM). Elsewhere regarded as a fish POISON and insect and tick repellent, and used for earache (DAW).

Clerodendron thompsonae Balf. Verbenaceae. "Brinco de dama", "Corazón sangriento". Cultivated ornamental (RVM). Febrifuge, tonic (RAR).

Clibadium asperum (Aubl.) DC. Asteraceae. "Huaco", "**Barbasco**". "Secoyas" mixed crushed leaves with crushed fruits of *Bactris* as POISONOUS fishbait (SAR).

Clibadium surinamense L. Asteraceae. "Waca", "Barbasco". Cultivated. Crushed leaves mixed with water are thrown in ponds to stupefy fish; not TOXIC for humans and domestic animals, it makes fish dizzy, enabling fishermen to catch them (RVM).

Clibadium sylvestre (Aubl.) Baillon. Asteraceae. "Wacamasha". Cultivated. "Kubeos" and "Tikunas" use as piscicide (SAR).

♀ *Clidemia hirta* (L.) D.Don. Melastomataceae. "Mullaca", "Mullaca morada". Fruit edible. Used by "Créoles" as a cicatrizant. Leaf decoction used for dysentery and cramps. Leaf maceration used as antiseptic for vaginal hemorrhages. "Palikur" take one spoonful decoction a day for menstrual hemorrhages (GMJ).

Clitoria arborea Hoff. ex Benth. var. *arborea*. Fabaceae. "María buena", "**Butterfly pea**". Semicultivated ornamental plant.

Clusia rosea Jacq. Clusiaceae. "Matapalo", "Camé", "Renaquilla", "**Star fruit**". Bark decoction used for injuries and rheumatism. Leaves once used as message paper, due to paper scarcity (RVM).

Coccoloba barbeyana Lind. Polygonaceae. "Vino huayo". Fruit edible (VAG).

Coccoloba marginata Benth. Polygonaceae. "Vino huayo". Fruit edible (VAG).

Cochlospermum orinocense (HBK) Steud. Cochlospermaceae. "Sacha punga", "Shamburi". Wood used in handicraft. Bark provides fiber to make rope. Sometimes used as an ornamental. "Tikunas" use bark tea for fever (SAR). Used for bruises, and as a cicatrizant (RAR).

♀ *Cocos nucifera* L. Arecaceae. "Coco", "**Coconut**". Cultivated. Edible seeds. Leaves used for floral arrangements, and to decorate carriages during festivals. "Tikunas" use the leaf sheath to prevent miscarriage. Antiasthmatic, antigonorrheic, antiuteritic, diuretic, lactagogue (RAR). Mice fed on coconut milk had fewer mammary tumors than controls (SAR).

Codonanthe crassifolia (Focke) Morton. Gesneriaceae. "Shanen bana". Around Pucallpa, leaf decoction used internally or in washes for headache (VDF).

Codonanthe uleana Fritson. Gesneriaceae. "Picsho". The leaf infusion in water or aguardiente is used to treat swellings. "Tikunas" poultice the leaves on indolent sores and wounds (SAR). Venezuela's "Waika" value the root for wounds (SAR).

♀ *Coffea arabica* L. Rubiaceae. "Café", "**Coffee**", "Cafeto". Cultivated in small areas, mostly used locally. Brazilians mash leaves and beans to make a medicinal coffee to speed up labor (BDS). Used as antitussive in flu and lung ailments (RAR).

Coix lacryma-jobi L. var. *mayuen* (Romanet) Stapf. Poaceae. "Lágrima de Job", "**Job's tear**", "Lágrima de la virgen", "**Virgin's tear**", ""Trigo", "**Wheat**", "Mullo huayo". Cultivated. The first and second harvest is used to obtain all-purpose flour. After hardening, used for handicraft, e.g. in necklaces, curtains, bracelets, etc. In popular medicine, used for swellings, immunology, and as diuretic and antipyretic, taken orally, but studies indicate the activity may be nullified in the digestive tract (RVM). "Ketchwa" apply a cud of chewed leaves to toothache (SAR). (Fig. 72)

Coleus blumei Benth. Lamiaceae. "Entrada al baile", "Simorilla". Cultivated ornamental. Chopped leaves applied to inflammation (FEO). Closely related "Cimora oso" rubbed onto rheumatism.

Fig. 70. *Chrysophyllum caimito* (IIC)

Fig. 71. *Clarisia racemosa* (GAV)

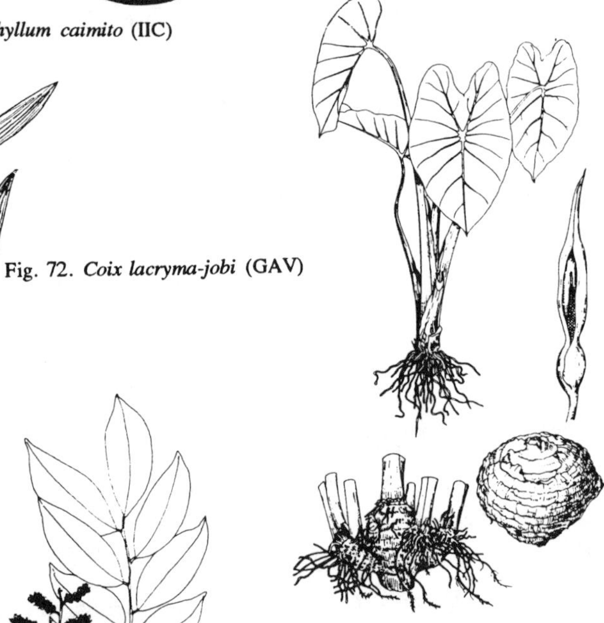

Fig. 72. *Coix lacryma-jobi* (GAV)

Fig. 73. *Colocasia esculenta* (IIC)

Fig. 74. *Copaifera reticulata* (GAV)

Colocasia esculenta (L.) Schott. Araceae. "Pituca", "Papa china". Cultivated. Roots eaten as a substitute for cassava and potato. The leaves are rinsed in boiling water and eaten as a vegetable. (Fig. 73)

Commelina erecta L. Commelinaceae. "Lancetilla blanca". Cultivated ornamental.

Copaifera paupera (Herzog) Dwyer. Fabaceae. "Copaiba". Wood used for lumber; parquets, construction, and to make decorative plaques. Oil used as liniment.

Copaifera reticulata Ducke. Fabaceae. "Copaiba", **"Copal"**. On Rio Solimoes, resin used as a cicatrizant, for gonorrhea, psoriasis, and sores (SAR); in Piura used for catarrh, syphilis, and urinary incontinence (FEO). Plotkin (1993) notes that the resin (copal) is used to coat tubules exposed by the dentist drill. Once employed in the US as disinfectant, diuretic, laxative, and stimulant, as well as in cosmetics and soaps (MJP). (Fig. 74)

Corchorus silicosus L. Tiliaceae. "Nobe rao". Leaves used as antiinflammatory around Pucallpa (VDF).

Cordia alliodora (R.&P.) Cham. Boraginaceae. "Añallo caspi", "Laurel", "Ajos quiro", **"Clammy cherry"**. The sawn wood is used for interior decorations.

Cordia bifurcata R. & P. Boraginaceae. "Mulla quillo". Around Pucallpa, used topically as antiecchymotic antiinflammatory (VDF).

Cordia nodosa Lam. Boraginaceae. "Añallo caspi". Edible fruit. Leaves poulticed on snakebite. "Wayãpi" use bark decoction for lung ailments and chest colds (GMJ). Vaupes natives use leaf paste to kill bot flies (SAR).

Cordyline terminalis (L.) Kunth. Liliaceae. "Barbusho", **"Good-luck plant"**, **"Ti palm"**. Cultivated ornamental.

Coriandrum sativum L. Apiaceae. "Culantro del país", **"Coriander"**. Cultivated herb. Chopped leaves inhaled for nosebleed, applied to the forehead for soroche (FEO).

Costus amazonicus (Loes.) McBr. ssp. *krukovii* Maas. Costaceae. "Iwajyu". The stalk provides drinking water (DAT).

♀ *Costus arabicus* L. Costaceae. "Cañagre", **"Bitter cane"**. Almost all *Costus* and *Dimerocostus* are used to reduce internal fever, cough, bronchitis, laryngitis, pharyngitis, and tonsilitis; also used in gargles for mouth sores. "Créoles" use decoction of stems or bulbs for leucorrhea. Stem infusion also used for blennorrhea. Stem juice mixed with honey for cough, colds, and whooping cough (GMJ).

Costus cylindricus Jacq. Costaceae. "Bocon taco", "Caña agria". Used for inflammations, e.g. gastritis and vaginitis, around Pucallpa VDF.

Costus erythrocoryne K.Schum. Costaceae. "Cañagre". Used like *Costus arabicus*. (Fig. 75)

Costus guanaiensis Rusby var. *guanaiensis*. Costaceae. "Caña agria" "Cañagre". Used to reduce internal fever, cough, bronchitis, laryngitis, pharyngitis, stomatitis and tonsilitis; "Cuna" use the leaf decoction for stomachache (RVM). Maxwell chewed the cane when she had a cough (NIC).

Costus lasius Loes. Costaceae. "Cañagre". Leaves considered hemostatic by the "Wayãpi", especially for arrow wounds (GMJ).

Costus longibracteolatus Maas. Costaceae. "Iwajyu". "Boras" use for cough and fever (DAT).

♀ *Costus scaber* R.& P. Costaceae. "Cañagre". Natives use for liver diseases (SOU). "Cuna" use the root decoction for stomachaches and snakebite. The "Waunana" drink the juice from the decoction of the leaves and flowers to get rid of intestinal worms. "Wayãpi" use the flower decoction in douches for vaginal infections. "Créoles" use it like *C. arabicus* (GMJ). (Fig. 76)

Couepia chrysocalyx (Poepp. & Endl.) Benth. ex Hook. Chrysobalanaceae. "Parinari", "Supay ocote". Fruit edible. Wood used for construction.

Couepia dolichopoda Prance. Chrysobalanaceae. "Parinari", "Hamaca huayo". Edible fruits.

Couepia obovata Ducke. Chrysobalanaceae. "Parinari". Wood for construction and charcoal. Ashes used in pottery making (SAR).

Couepia paraensis (Mart. & Zucc.) Benth. ssp. *glaucescens* (Spr. ex Hook.) Prance. Chrysobalanaceae. "Parinari". For dormers and posts.

Couepia parillo A.DC. Chrysobalanaceae. "Parinari". For beams and decks.

Couepia subcordata Benth. ex Hook. Chrysobalanaceae. "Parinari", "Supay ocote". Cultivated ornamental. Fruits edible.

Couma macrocarpa Barb. Apocynaceae. "Aso" "Capirona", "Fransoca", "Leche caspi", "Leche huayo", "**Milk tree**". Fruit edible. As "balata" and "leche caspi", thousands were "milked" to extract latex used in making gum. Latex potable, used for diarrhea caused by ameba; also used for skin irritations and asthma (RVM). Vaupes Indians occasionally treat the newborn umbilicus with latex. Amazonian Colombians take latex as purgative (SAR). "Witoto" sometimes use leaves as a coca substitute (SAR). Pucallpa natives use powdered bark as antiseptic, resolvent and vulnerary (VDF). There are possibilities for this tree as an ornamental (RVM). (Fig. 77)

Couratari macrosperma A.C.Smith. Lecythidaceae. "Papelillo", "Machimango cachimbo". Lumber for heavy construction, fiber used to make rope (RAR).

Couratari oligantha A.C.Smith. Lecythidaceae. "Machimango blanco", "Zorro caspi". For house construction; the fiber from the bark is used occasionally.

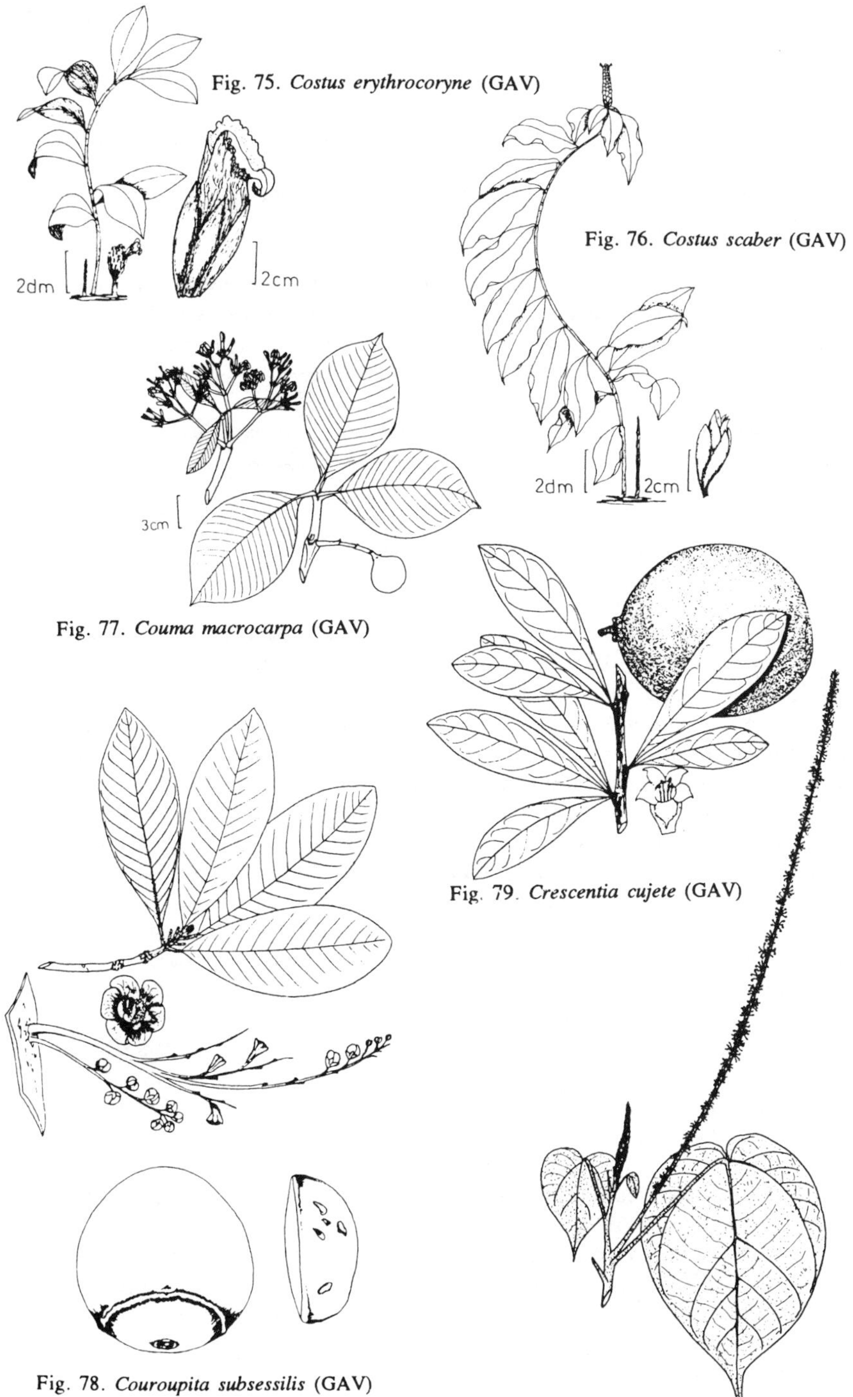

Fig. 75. *Costus erythrocoryne* (GAV)

Fig. 76. *Costus scaber* (GAV)

Fig. 77. *Couma macrocarpa* (GAV)

Fig. 79. *Crescentia cujete* (GAV)

Fig. 78. *Couroupita subsessilis* (GAV)

Fig. 80. *Croton lechleri* (GAV)

Couroupita guianensis Aubl. var. *surinamensis* (Mart.) Eyma. Lecythidaceae. "Ayahuma". For lumber, wood used for interior decorations, and rustic furniture. Fruit used to feed chickens. Fruit juice used to exorcise witchcraft. Natives along the Rio Tigre use leaf buds to relieve toothaches, by filling the cavities with the tip of the bud (RVM). "Piró use to treat serious dermatoses, bacterial, fungoid or viral (NIC). Considered contraceptive; used for acarosis, dysmenorrhea, gastralgia, rheumatism (RAR).

Couroupita subsessilis Pilger. Lecythidaceae. "Ayahuma". The clean wood is used for interior decorations. (Fig. 78)

Crataeva tapia L. Capparidaceae. "Tamara blanca", "Cahuara micuna", "Tapia". Ripe fruits edible, here used as fishbait (RVM). Pucallpa natives apply the bark topically for toothache and arthritis (VDF). "Witotos" use the leaf tea as a stomachic (SAR). Amazonian Brazilians apply the sap to rheumatic pain (SAR) and use the plant as a tonic.

Crepidospermum prancei Daily. Burseraceae. "Copal blanco". Arilb edible (VAG).

Crescentia amazonica Ducke. Bignoniaceae. "Huingo sacha", "**Wild calabash**". Ripe fruits used to make maracas and other handicrafts.

♀ *Crescentia cujete* L. Bignoniaceae. "**Calabash**", "Huingo", "Tutumo", "Pate". Cultivated ornamental used in making containers. Green fruits juice used for bronchial asthma. Shells of ripe fruits used as containers, and to carve Amazon scenes, sold as regional handicraft; also used in making maracas. Wood used to make ribs for the building of barges, and also to make drums. The branches are known as "huingo-rama", used to whip children when they misbehave. "Chami" eat the fruit, when green, they use it for diarrhea. "Créoles" use leaf decoction to stimulate bile secretions, and as a purgative. Juice of the young fruit used for diarrhea, and other intestinal problems. Decoction of the ripe fruit is used in the preparation of an abortive tea (GMJ). Leaves chewed for toothache (SAR). Plant also used for alopecia, douches, hernia, and sprains (RAR). (Fig. 79)

Crotalaria retusa L. Fabaceae. "Frejol del tunchi", "**Witches bean**". "Créoles use leaf and flower decoction to prepare a cold remedy. "Wayãpi" eat the fresh seeds as an analgesic when bitten by scorpions (GMJ).

Croton cajucara Benth. in Hook. Euphorbiaceae. Bark tea used in Amazonia for diabetes, diarrhea, and inflamed liver. Methanol extracts reduce growth of *Pectinophora gossypiella*. Antigrowth activity of the new nor-diterpene, cis-dehydrocrotonin, proved similar to that of trans-dehydrocrotonin against *P. gossypiella* and *Heiothis virescens*. With rat hepatocytes, the methanol extract showed hepatotoxicity rather than hepatoprotective activity. Thus it might be counterindicated in hepatitis. (PC30{8}:2545-2546. 1991)

♀ *Croton lechleri* Muell.-Arg. Euphorbiaceae. "Sangre de drago", "Sangre de grado", "**Dragon's blood**". The latex is used to heal wounds, and for vaginal baths before childbirth. It is also recommended for intestinal and stomach ulcers (RVM). It yields the hemostatic sap that accelerates wound healing (NIC). For leucorrhea, fractures, and piles (RAR). (Fig. 80)

Croton palanostigma Klotzsch. Euphorbiaceae. "Pashna huachana", "Señora Vara", "Sangre de drago", "Shambo quiro", "Topilla", "**False balsa**". Wood used for

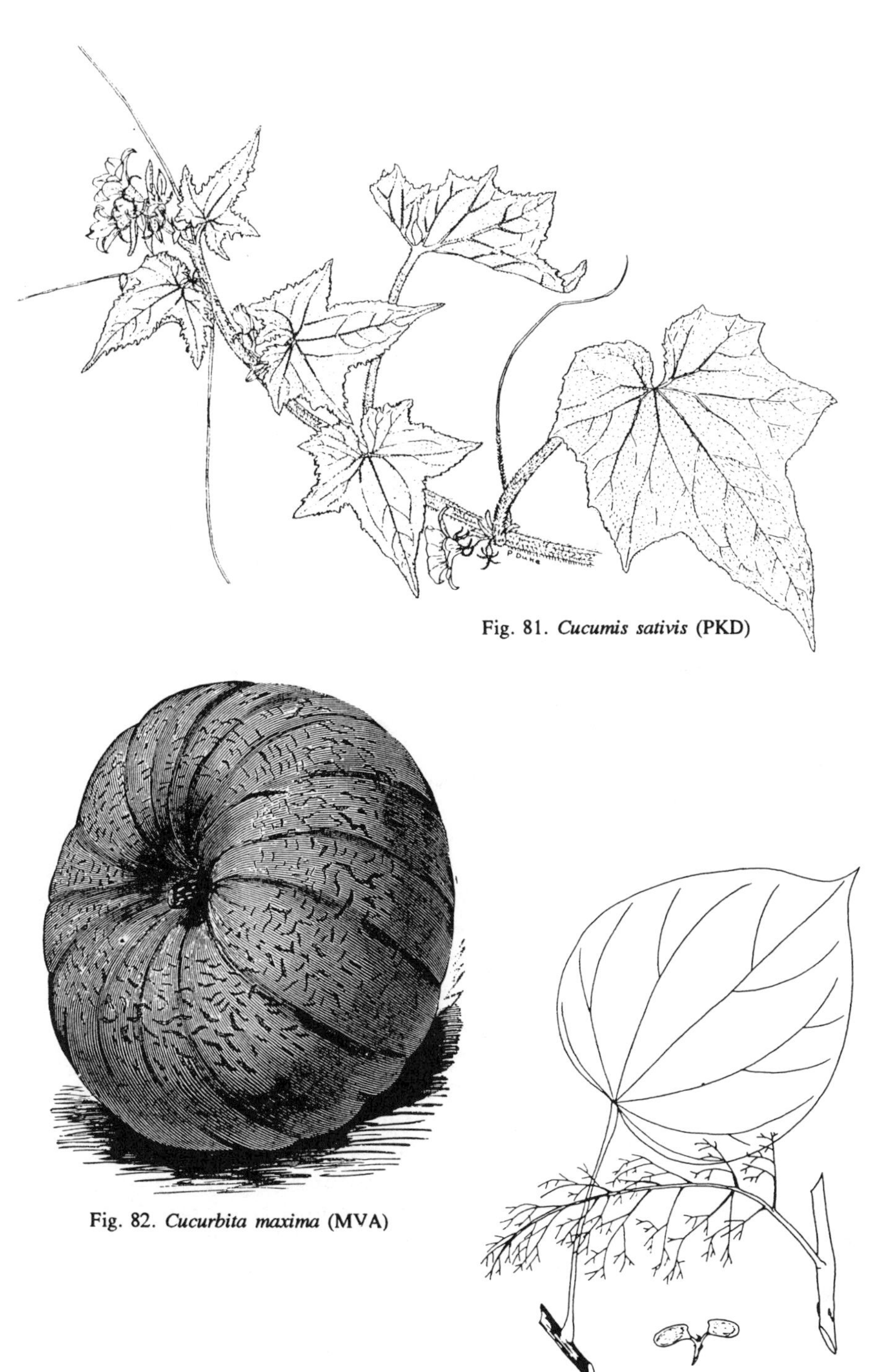

Fig. 81. *Cucumis sativis* (PKD)

Fig. 82. *Cucurbita maxima* (MVA)

Fig. 83. *Curarea toxicofera* (GAV)

house construction, beams, and decks (RVM). Country folk around Manaus apply the sap as an anodyne to boils and sores (SAR). Chopped leaves used as decongestant in snakebite (FEO). Latex contains taspine, an antiviral antiinflammatory (FNF).

Croton trinitatis Millsp. Euphorbiaceae. "Sacha pichana", "Sinchi pichana". "Boras" use to remove phlegm from their throats (DAT). "Tikunas" rub leaves onto bleeding gums (SAR). Leticia natives take a shot of leaf tea every hour for flu (SAR).

Cucumis anguria L. Cucurbitaceae. "Mashishe", "**Gherkin**". Cultivated. Edible fruits.

Cucumis melo L. Cucurbitaceae. "**Melon**", "**Cantaloupe**". Fruit edible.

Cucumis sativus L. Cucurbitaceae. "Pepino", "**Cucumber**". Cultivated for edible fruits. Mainly used as an antiinflammatory, applying slices of fruit on swollen areas; also around the eyes. (Fig. 81)

Cucurbita maxima Duch. Cucurbitaceae. "Zapallo", "Calabazo", "**Pumpkin**". Cultivated. Edible fruit. Elsewhere the seeds are recommended for prostate troubles (JAD). (Fig. 82)

Cucurbita pepo L. Cucurbitaceae. "Zapallo", "zapayo", "**Squash**". Cultivated for edible fruit. Brazilians believe the latex from the fruit husk will remove scars (BDS).

Curarea tecunarum Barneby & Krukoff. Menispermaceae. "Sacha ampihuasca", "**Wild curaré**". Sometimes used in preparation of "curaré". Used as a male contraceptive. Tests in rats show a decrease in testosterone levels (RVM). Ecuadorean "Waoranis" use it for dermatoses and fungal infections.

Curarea toxicofera (Wedell) Barneby & Krukoff. Menispermaceae. "Sacha ampihuasca". Used in making "curaré"; contains curine, curarine, isochondodendrine, and toxicopherine (AYA). "Waunana" use the leaves for casting spells on people (RVM). "Tikunas" treat itch and wounds with the bark infusion (SAR). (Fig. 83)

Curcuma longa L. Zingiberaceae. "Guisador", "Azafran", "Palillo", "**Turmeric**". Cultivated. Rhizome frequently used as spice. Used for hepatitis. Rhizomes are crushed fresh and mixed with water. This juice is taken one spoon for children and 1 to 2 for adults, once a day for 10 to 15 consecutive days for hepatitis. Some people bathe in this extract. "Créoles" use it to treat injuries. Crushed rhizome, mixed with the leaves of *Siparuna guianensis* and of *Justicia pectoralis,* salt and rum, is poulticed on bruises on their backs. (Decoction of the three plants, taken 3 times a day, adding to this 3 drops of arnica tincture, and some sugar) (GMJ). The root contains at least 3 antiinflammatory compounds, cucurmin, feruloyl, 4-hydroxy-cinnamoyl methane, and bis-14-hydroxy-cinnamoyl methane, dose dependent up to 30 mg/kg (Indian J. Mod. Res. 75:574. 1982). (Fig. 84)

Cyathea sp. Cyatheaceae. "Shapunga", "Helecho arborescente". Trunk used as a substrate for epiphytic orchids. Exudates are used for healing small cuts.

Cybistax quinquefolia (Vall.) MacBr. Bignoniaceae. "Achichua achihua". Diuretic branch decoction used for syphilis in Piura (FEO).

Cycas circinalis L. Cycadaceae. "Palma brashica", "Pelo ponte". Cultivated ornamental.

Cyclanthera pedata (L.) Schrad. Cucurbitaceae. "Caigua". Cultivated. Fruit edible. It has various medicinal usages. The tea of the seeds is well known for controlling high blood pressure (RVM). De Feo suggests that the decoction of the epicarps is also antidiabetic (FEO). (Fig. 85)

Cyclanthus bipartitus Poit. Cyclanthaceae. "Calzón panga". "Boras" use burned leaves mixed with other plants to produce salt. They also use the leaves to wrap fish (DAT). The "Cuna" use the cooked roots to treat severe stomachaches (RVM). Rio Pastaza natives grate the inner stem for snakebite (SAR). Peruvian "Boras" and Ecuadorian "Ketchwas" use the juice for ant stings (SAR). Vermifuge (SAR).

Cyclopeltis semicordata (Sw.) J.Sm. Aspleniaceae. "Shapumba". "Cuna" use cooked fronds and buds for liver disease and stomachache (RVM).

♀ *Cymbopogon citratus* (DC.) Stapf. Poaceae. "Yerba Luisa", "María Luisa", "**Lemongrass**". Cultivated. Leaves yield "citronela" essence used in making carbonated drinks. Used as a digestive for acid indigestion and upset stomach. Natives from Risaralda use the leaves and roots for intestinal irritations, and the juice extracted from the roots to stabilize menstrual irregularities. Leaves are used by "Créoles", "Wayãpi", and "Palikur" to reduce fever (GMJ). "Sionas" use the tea for stomachache, "Tikunas" for fever, flu and headache (SAR). In Piura, the decoction considered contraceptive (FEO).

Cynometra spruceana Benth. Fabaceae. "Azúcar-huaillo", "pate-vaca". Wood used for house construction and dormers. (Fig. 86)

Cyperus articulatus L. Cyperaceae. "Piripiri de víbora". Cultivated. For snakebite, they chew the pulp, swallow the juice and poultice the cud onto the bite after it has bled. It is also considered abortive. Native Americans poke crushed stems in their nose to alleviate snoring (GMJ). "Secoyas" mixed ground rhizome with water for fever, flu and fright (SAR). In Piura, the chopped shoots are considered hemostatic and vulnerary (FEO).

Cyperus diffusus Vahl. Cyperaceae. "Piripiri". The inflorescence stalk is sometimes used as a paintbrush. The whole plant is used by the "Wayãpi". The roots mixed with other plants are used in baths to reduce fever (GMJ). Tapojos natives used the root for headache (BDS).

♀ *Cyperus luzulae* L. Cyperaceae. "Piripiri". "Chami" mash the plant in cold water, and take for diarrhea and stomachaches (RVM). "Tikunas" used crushed fruits to induce labor (SAR).

♀ *Cyperus prolixus* H.B.K. Cyperaceae. "Piripiri". Cultivated by the Achuar of Loreto. Males tell you that the rhizomes are scraped and, sometimes mixed with water, used as an oxytocic to speed up parturition. Females say that the topi (actually fungal celrotia) were used.

Cyperus sp. Cyperaceae. "Tobi uaste". Rhizome poulticed onto rheumatic pain (VDF).

Cyphomandra hartwegii Dunal. Solanaceae. "Asna panga", "Gallinazo panga", **"Tree tomato"**. Fruits edible. "Wayãpi" use the decoction for baby baths to protect the infant against progressive weakness, caused by baby having direct contact with father right after birth (the father had killed a wild animal, violating a taboo) (GMJ).

Cyphomandra obliqua Dunal. Solanaceae. "Cupa sacha", "Poshno rao". Around Pucallpa, used as an ecchymotic resolvent. Also used in febrifugal baths (VDF).

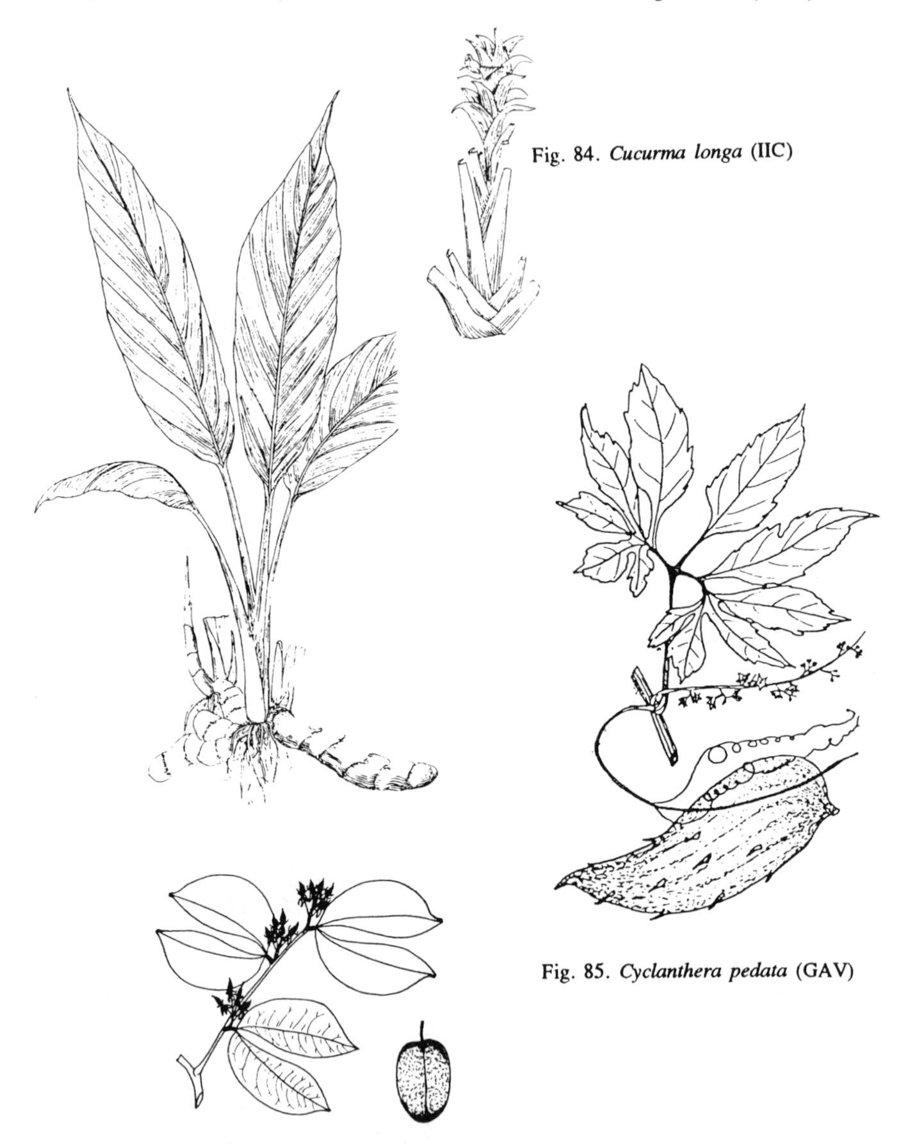

Fig. 84. *Cucurma longa* (IIC)

Fig. 85. *Cyclanthera pedata* (GAV)

Fig. 86. *Cynometra spruceana* (GAV)

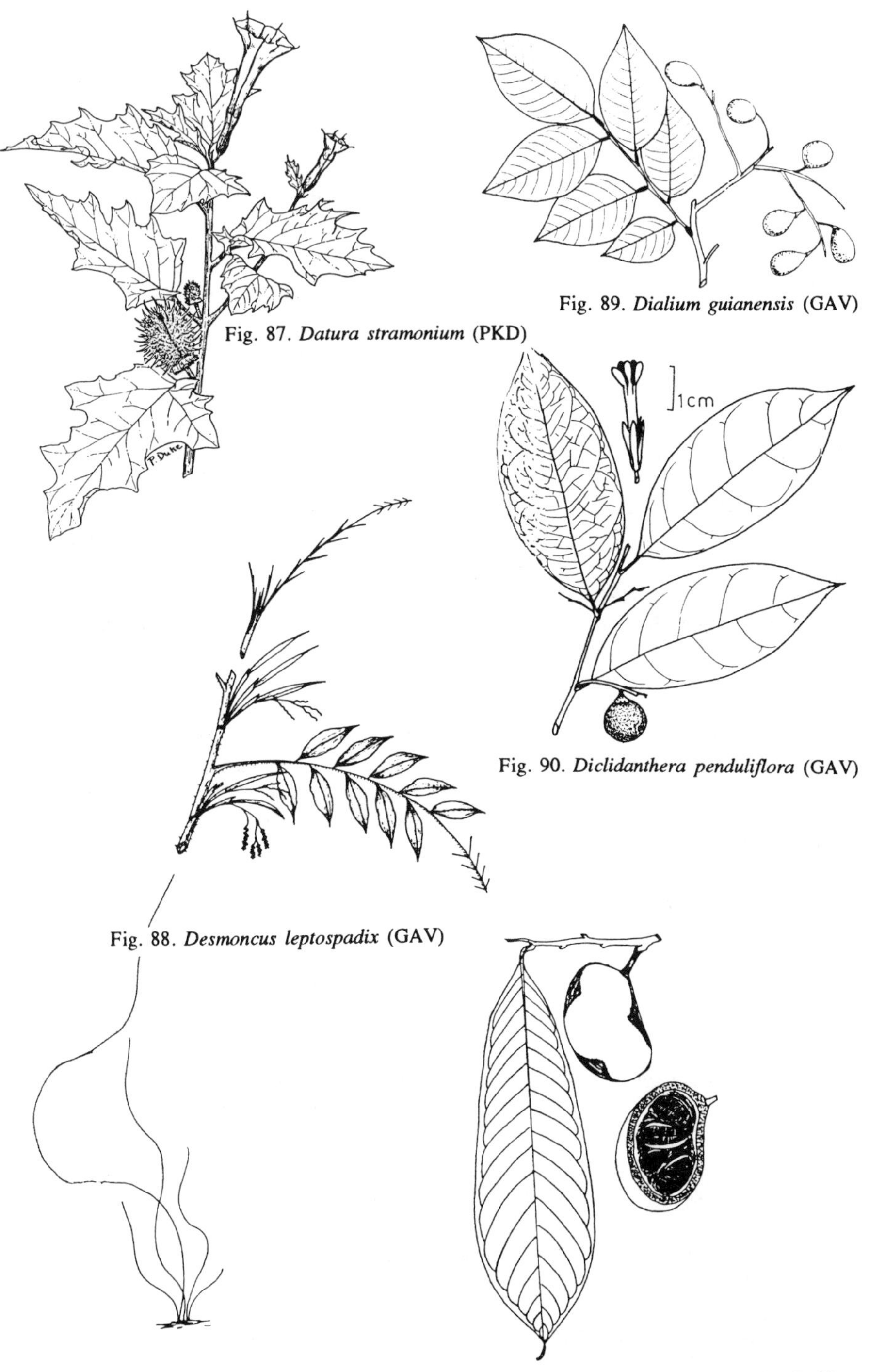

Fig. 87. *Datura stramonium* (PKD)

Fig. 88. *Desmoncus leptospadix* (GAV)

Fig. 89. *Dialium guianensis* (GAV)

Fig. 90. *Diclidanthera penduliflora* (GAV)

Fig. 91. *Diclinanona tessmannii* (GAV)

- D -

Dacryodes peruviana (Loes.) Lam. Burseraceae. "Copal". The latex is mixed with other "copals" to caulk boats.

Dalbergia monetaria L.f. Fabaceae. "Cushqui-huasca". The "Palikur" use macerated bark to treat diarrhea (GMJ). Tonic, used in lung ailments (RAR).

Datura arborea L. Solanaceae. "Toé". Nicole Maxwell mentions permanent brain damage in one gringo habitue of this "giant jimsonweed" (NIC).

♀ *Datura stramonium* L. Solanaceae. "Chamico", "**Jimsonweed**". Chopped leaves are applied to dermatitis, the decoction used as an antiseptic in vaginitis (FEO). (Fig. 87)

Daucus carota L. Apiaceae. "Zanahoria", "**Carrot**". Cultivated. More temperate than tropical, the carrot is a major source of the cancer-preventive beta-carotene (JAD).

Davilla kunthii St. Hil. Dilleniaceae. "Paujil chaqui". Stems contain potable water. They cut a one-meter section of the stem and stand it vertically, the water then flowing.

Dendropanax tessmannii (Harms) Harms. Araliaceae. "Cheriz". The natives from the Putumayo chew the fresh leaves in order to make their teeth strong, and protect them from cavities (RVM).

♀ *Desmodium adscendens* (Sw.) DC. Fabaceae. "Amor seco", "**Beggar-lice**", "Margarita". The plant infusion is given to people who suffer from nervousness. It is also is used in baths to treat vaginal infections. Because they believe this plant has magic powers, it is given to the lover who has lost interest in his mate, to make him/her come back. It is also used as a contraceptive (RVM). Rio Pastaza natives wash the breast of dry mothers with the leaf tea (SAR).

Desmodium axillare (Sw) Kuntze. Fabaceae. "Amor seco", "Pega pega". Similar to *D. adscendens*.

Desmodium cajanifolium (HBK) Kuntze. Fabaceae. "Amor seco". Used in cases of shock (RVM).

Desmoncus leptospadix Mart. Arecaceae. "Vara cacha", "**Rattan palm**". The split stems are used in making baskets, mats, and other handicrafts. Used like rattan, around Iquitos (RVM). (Fig. 88)

Desmoncus cf. *macroacanthos* Mart. Arecaceae. "Vara casha". Used in basketry.

Desmoncus prunifer Poepp. Arecaceae. "Vara casha". Used in basketry. Fruit edible (RAR).

Desmoncus vacivus Baily. Arecaceae. "Vara casha". Similar to *D. leptospadix*.

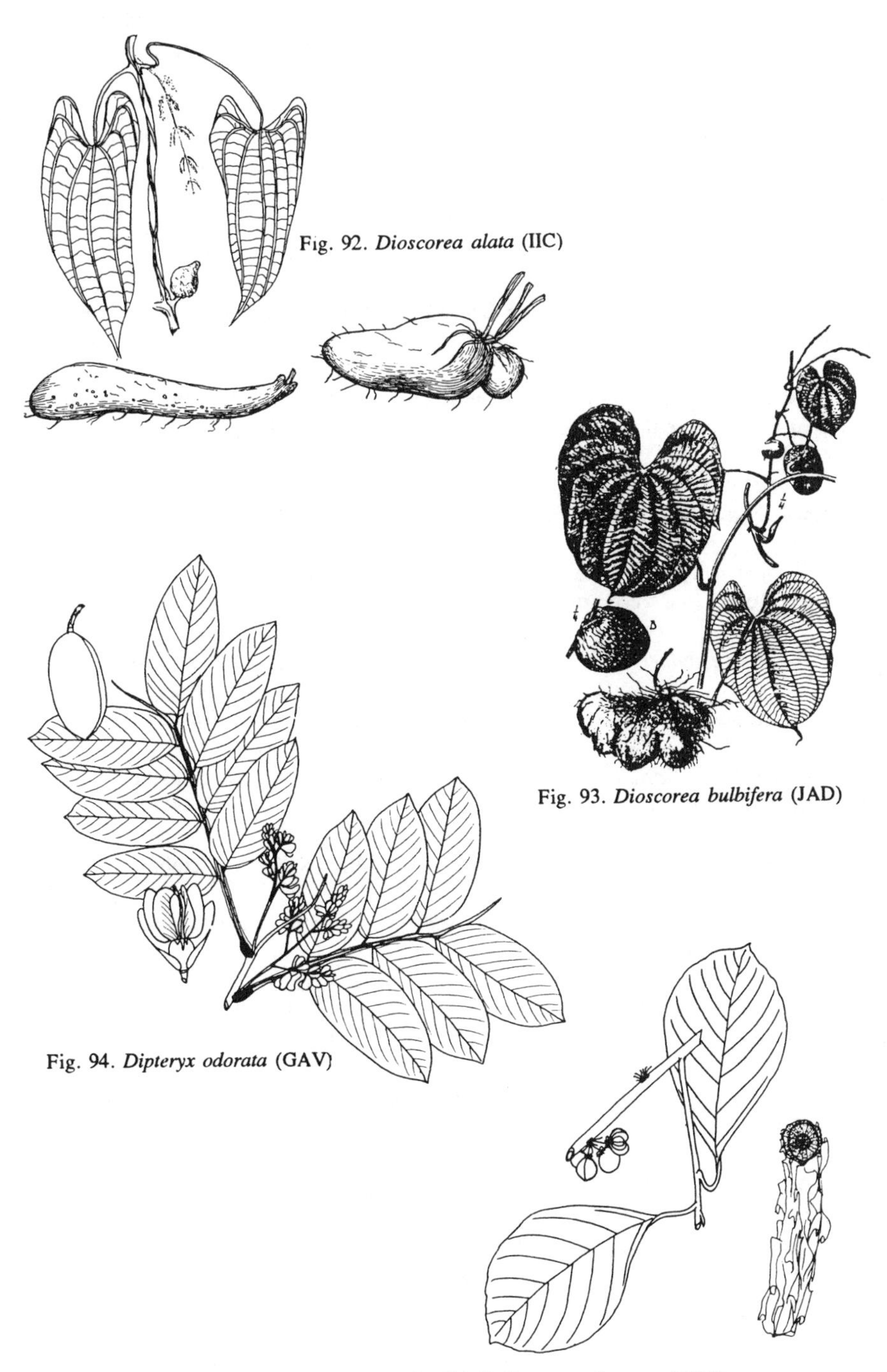

Fig. 92. *Dioscorea alata* (IIC)

Fig. 93. *Dioscorea bulbifera* (JAD)

Fig. 94. *Dipteryx odorata* (GAV)

Fig. 95. *Doliocarpus dentatus* (GAV)

Dialium guianensis (Aubl.) Sandw. Fabaceae. "Palo sangre". For lumber; in the manufacturing of decorative plaques, dormers, posts. The powder (aril) around the seeds is edible (RVM). (Fig. 89)

Dichorisandra hexandra (Aubl). Steudel. Commelinaceae. "Pishco huichi". Juice used as a laxative. "Wayãpi" use the decoction in good luck baths for hunting *Crax alector* (GMJ).

Diclidanthera penduliflora Polygalaceae. "Coto huayo". Edible fruit. (Fig. 90)

Diclinanona calycina (Diels) Fries. Annonaceae. "Espintana". Wood used for beams and decks.

Diclinanona tessmannii Diels. Annonaceae. "Tortuga blanca", **"White tortoise"**. Wood used for construction, beams, decks, columns. Fruit edible. (Fig. 91)

Dictyoloma peruvianum Planch. Rutaceae. "Barbasco negro", "Huamanzamana". The crushed leaves are used to catch fish in still waters (RVM).

Dieffenbachia spp. Araceae. "Patiquina", "Planta china". Cultivated as ornamentals. Juice from fresh crushed leaves used to extract parasites.

Digitaria insularis (L.) Mez. ex Ekman. Poaceae. "Torurco". As forage.

Dilkea acuminata Mart. Passifloraceae. "Granadilla caspi". Fruit edible.

Digitaria sanguinalis (L.) Scop. Poaceae. "Gramilla", "Pata de gallina", "Pasto de cuaresma". Common forage weed (RAF).

Dilkea wallisii Mart. Passifloraceae. "Granadilla caspi". Fruit edible.

Dimerocostus strobilaceus O.Kuntze. ssp. *gutierrezii* (O.Kuntze) Maas. Costaceae. "Caña agria". See *Costus arabicus*.

Dioclea ucayalina Harms. Fabaceae "Tikunas" rub leaves on forehead for headache (SAR).

Diodia sp. Rubiaceae. "Ai pana" Leaves used in head baths for gastritis around Pucallpa (VDF).

Dioscorea alata L. Dioscoreaceae. "Chami papa", "Ñame", **"Yam"**. Cultivated. Edible tubers of the genus *Dioscorea* may contain diosgenins, starter materials for steroids. Wild species may contain larger quantities of diosgenin. The nutritious species usually have less (RVM). Elsewhere a folk remedy for fever, gonorrhea, leprosy, piles and tumors (DAW). (Fig. 92)

Dioscorea bulbifera L. Dioscoreaceae. "Ñati papa", "huayra papa", **"Air potato"**. Cultivated. The tubers are edible. The crushed raw pulp is poulticed onto boils (RVM). Tubers considered alexeteric, antidotal, antiinflammatory, diuretic (RVM), hemostatic, even POISONOUS, and used for cancer, dysentery, fever, goiter, hernia, piles, sores, syphilis and tumors (DAW). (Fig. 93)

Dioscorea decorticans Presl. Dioscoreaceae. "Macaquiño". Cultivated ornamental, occasionally escaping.

Dioscorea trifida L. Dioscoreaceae. "Sacha papa", "Sacha papa morada", **"Wild potato"**. Cultivated tubers are edible. Raw tubers are poulticed on dermal inflammations (RVM).

Diospyros sp. Ebenaceae. "Caimitillo". Fruit edible fresh. "Cuna" drink the root decoction to treat stomachaches (RVM).

Diplotropis martiusii Benth. Fabaceae. "Chonta quiro". Wood for posts, forked poles, keel plates for boats, and dormers (RVM). "Kubeos" use powdered leaves with yuca flour for hematochezia (SAR). Leaf ash mixed with coca leaves (SAR).

Diplotropis purpurea (Rich.) Amshf. Fabaceae. "Chonta quiro". Wood for posts, forked poles, dormers, jam posts for bridges, keel plates for boats, and parquets (RVM). Piscicide (SAR).

Dipteryx micrantha (Harms) Ducke. Fabaceae. "Charapilla", "Shihuahuaco". Wood for posts, forked poles, jam posts for bridges, dormers, decorative plaques. Well-formed trunks used by "Boras" for making drums. The seeds are edible when cooked. Fruits used in waist ornaments that rattle as you dance (RVM).

Dipteryx odorata Aubl. Fabaceae. "Charapilla del murciélago", "Shihuahuaco". The wood is used for bridges, dormers, posts, etc. (RVM). Seeds soaked in rum are used by the "Créoles" for snakebite, shampoos, contusions and rheumatism. The "Wayãpi" use the bark decoction as antipyretic baths, and the "Palikur" use it as fortifying baths for infants and small children (GMJ). Brazilians make a cough pill by balling up the crushed seed (BDS). Elsewhere used as anticoagulant, antidyspeptic, antitussive, cardiotonic, diaphoretic, febrifuge, fumigant, narcotic, stimulant and stomachic DAW. The coumarin explains its anticoagulant activity (JAD). (Fig. 94)

Doliocarpus dentatus St.Hil. Dilleniaceae. "Puca huasca", **"Watervine"**. Drinkable water. "Kubeos" believe the water helps the after effects of malaria (SAR). (Fig. 95)

Doliocarpus macrocarpus Mart. ex Eichl. Dilleniaceae. "Puca huasca", **"Watervine"**. Drinkable water.

Doliocarpus major J.F.Gmel. ssp. *major*. Dilleniaceae. "Puca huasca", **"Watervine"**. Drinkable water.

Dracaena fragrans (L.) Ker-Gawl. Liliaceae. **"Corn plant"**, "Flor de cementerio", **"Cemetery flower"**. Cultivated ornamental, especially in cemeteries.

Dracontium loretense Krause. Araceae. "Hierba del jergon", "Jergón sacha", **"Fer-de-lance"**. Tuber believed to help snakebites perhaps on account of the snakeskin like mottling of the petiole. Some people whip their feet and legs with the branches to repel snakes. The corms are used to control and steady the hands. The roots are reported to be edible (DAT).

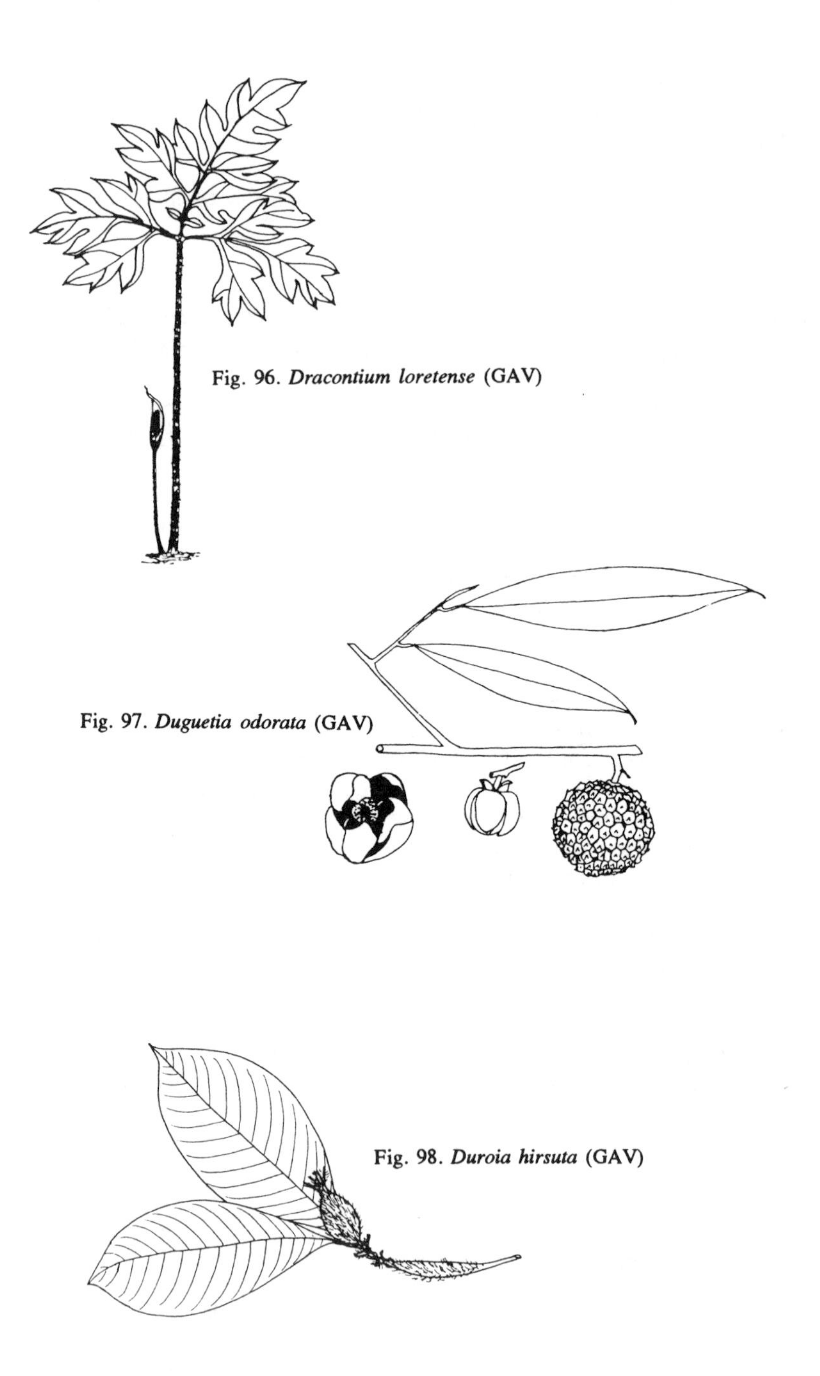

Fig. 96. *Dracontium loretense* (GAV)

Fig. 97. *Duguetia odorata* (GAV)

Fig. 98. *Duroia hirsuta* (GAV)

Drymonia pendula (Poepp) Wiehl. Gesneriaceae. "Delia". Ornamental.

Duckesia cf. *verrucosa* (Ducke) Cuatr. Humiriaceae. "Loro shungo", "manchari caspi". Wood for forked poles, jam posts for bridges, and dormers.

Duguetia hadrantha (Diels) Fries. Annonaceae. "Tortuga caspi". Wood for beams, decks, and columns.

Duguetia latifolia R.E.Fries. Annonaceae. "Tortuga caspi". Wood for decks and columns (RVM).

Duguetia macrophylla R.E.Fries. Annonaceae. "Tortuga caspi", "Júúmyba". Fruit edible (VAG). Bark provides a balm used for sore muscles, pains and aches (RVM).

Duguetia odorata Macbr. Annonaceae. "Tortuga caspi". Wood for beams, decks, wood strips and sheets. (Fig. 97)

Duguetia quitarensis Benth. Annonaceae. "Tortuga caspi". Wood for beams and decks (RVM).

Duguetia spixina Mart. Annonaceae. "Tortuga caspi". Wood for beams and decks (RVM). "Tikunas" use the tea of leaves and bark to wash leg ulcers (SAR).

Duguetia stenantha R.E.Fries. Annonaceae. "Tortuga caspi". Wood for beams and decks. Clean wood is used for interior decorations. Fruit edible (RVM).

Duguetia tessmannii R.E. Fries. Annonaceae. "Tortuga caspi". Wood for beams and decks (RVM).

Duroia hirsuta (Poepp. & Engl.) Schum. Rubiaceae. "Huitillo del supay". These shrubs, associated with ants, grow in small homogeneous stands called "Supay chacra" (Devil's fields). Other plant species with ant symbioses: *Cecropia* spp., *Cordia nodosa, Toccoca* spp., and *Triplaris* spp. The soil around *Duroia* is usually free of weeds, possibly because of the ants. Gentry and Blaney (pers. comm.) think it may be due to secretions or micro-organisms associated with the ants that prevent the growing of weeds and other plants. The forked stakes are occasionally used in construction. Rural people, superstitious about the "Supay chacra", avoid walking nearby. Some rural Colombians chew the fruits to prevent dental caries (RVM). "Waoranis" rub the ant pheromones inside their cheeks for oral aphthae (SAR). Putumayo natives bind a bark strip on the arm, both staining and scarring the area (SAR). (Fig. 98)

Duroia paraensis Ducke. Rubiaceae. "Pampa remo caspi". Wood for beams and decks (RVM).

Duroia saccifera (Mart.) Schum. Rubiaceae. "Hormiga caspi". Fruit edible cooked (VAG).

- E -

Ecclinusa lanceolata (M.&E.) Pierre. Sapotaceae. "Balata". Fruit edible. Wood for rural construction; latex for gum industry (RVM).

Ecclinusa guianensis Eyma. Sapotaceae. "Balatilla". Fruit edible (VAG).

Ecclinusa ramiflora Mart. Sapotaceae. "Balatilla". Fruit edible (VAG).

Echinochloa polystachya (HBK) Hitchc. Poaceae. "Gramalote capo". Forage for water buffalos.

Echinodorus tunicatus Small. Alismataceae. "Amanso". Cultivated in ponds as an ornamental.

Eclipta alba (L.) Hassk. Asteraceae. "Huanguilla", "Naparo cimarron", "Shobi isa sheta", "Naparo cimarron". Around Pucallpa, leaf maceration used for headache (VDF). In Brazil the plant is used as an antiasthmatic and as a depurative. "Créoles" rub the leaf decoction on children for skin blemishes. It is also used for albuminuria (GMJ). Folk remedy elsewhere for catarrh, copremia, cough, dyspepsia, elephantiasis, enterorrhagia, headache, hemorrhage, hepatitis, jaundice, lumbago, marasmus, pertussis, splenitis, toothache, and vertigo. Also considered estrogenic and insecticidal (DAW). Being seriously studied as a remedy for snakebite (JAD). The active ingredient wedelolactone is antiinflammatory and inhibits hemorrhage and the liberation of creatinine kinase induced by snake venom (Mem. Inst. Oswaldo Cruz 86 {Suppl.II}:203-5).

Eichhornia azurea Kunth. Pontederiaceae. "Putu-putu", **"Water hyacinth"**. Semicultivated. Cultivated in ponds as an ornamental.

Eichhornia crassipes Solms. Pontederiaceae. "Putu-putu", **"Water hyacinth"**. Semicultivated. Cultivated in ponds as an ornamental.

Elaeis guineensis Jacq. Arecaceae. "Palma aceitera", "palma africana", **"African oilpalm"**. Cultivated in industrial plantations. Source of palmoil for cooking and its byproducts; also used as an ornamental. Excellent source of beta-carotene and tocotrienol (Vitamin A and Vitamin E).

Elaeis oleifera (HBK.) Cortes. Arecaceae. "Puma yarina", "peloponte", **"American oilpalm"**. Fruit edible when cooked. Leaves used to roof houses. Managed properly, this could be better than African oilpalm for oil extraction (RVM).

Elephantopus scaber L. Asteraceae. "Lengua de vaca", **"Cow's tongue"**. Considered astringent, emollient, febrifuge, and sudorific; used for elephantiasis, itch, sores (BDS, RAR).

Eleusine indica (L.) Gaertn. Poaceae. "Grama", "Pasto estrada", "Pata de gallina", **"Goose grass"**. As forage. Grains edible. Before it blooms, it is used in infusions to treat colds and flu. The root infusion is used for diarrhea; leaf decoction for dysentery, swellings and oliguria. In Zaire, used for chickenpox and colds. Leaf infusion rubbed on skin for measles (RVM). Trinidad natives use the leaf decoction for cystitis and pneumonia. Seed used for dysentery, diarrhea, and contusions (RVM). "Cuna" rub the decoction onto rheumatic areas. "Créoles" use it as a refreshing herbal tea and antidiarrheic (GMJ). (Fig. 99)

♀ *Eleutherine bulbosa* (Miller) Urb. Iridaceae. "Jasin huaste", "Pachahuasten", "Yahuar piri piri". Cultivated. Bulb commonly used for diarrhea (RVM) and worms (SAR). Around Pucallpa, used as ecchymotic (VDF). Dominicans use it for dysmenorrhea and menopause (DAW). Around Iquitos the bulb decoction is used for stomachaches and diarrhea caused by bacteria or *Entamoeba hystolitica* (RVM). "Créoles" use the bulb in poultices to treat epilepsy and twitches. The crushed bulb mixed with a "cockroach" *Periplaneta americana,* roasted with a little bit of oil, and applied hot, is believed to help infected wounds caused by rusty nails. The bulb, macerated with red wine is used as an abortive (GMJ).

Eleutherine plicata Herb. Iridaceae. "Picuru inchi". Peruvians regard the seed as antidiarrhetic, antidysenteric, hemostat, vulnerary (RAR); bulb grated into tea for stomachache or dysentery (BDS).

Endlicheria anomala Nees ex Meissn. Lauraceae. "Yacu Moena". Wood for beams, decks (RVM).

Endlicheria formosa A.C. Smith. Lauraceae. "Moena", "Pampa moena". As saw wood.

Endlicheria sericea Nees. Lauraceae. "Moena". As saw wood.

Entada polyphylla Benth. Fabaceae "Pashaco". Bark yields a yellow resin used for dyeing leather black (MAC).

Enterolobium barnebianum Mesquita & Da Silva. Fabaceae. "Pashaco oreja de negro", "Plantilla pashaco". Wood for lumber; parquets, and construction in general. (Fig. 100)

Epiphyllum phyllanthus (L.) Haw. var. *phyllanthus*. Cactaceae. "Lagarto shupa". Semicultivated as an ornamental. Considered cardiactive (DAW).

Episcia hirsuta (Benth.) Hanst. Gesneriaceae. "Llama plata". Cultivated ornamental.

Episcia reptans Mart. Gesneriaceae. "Llama plata". As an ornamental.

Eragrostis pilosa (L.) Beauv. Poaceae. "Grama". As forage.

Erisma bicolor Ducke var. *macrophyllum* (Ducke) Stafleu. Vochysiaceae. "Quillo sisa", "Mauba". As saw wood for general construction.

Erisma calcaratum (Link) Warm. Vochysiaceae. "Cacahuillo", "Quillo sisa". The wood is used for keel plates on canoes. Macerated seed oil used as liniment for aches and pains. Oil used in making soaps (DAW). Brazilians use for tumors (DAW).

Eryngium foetidum L. Apiaceae. "Siuca culantro", "Sacha culantro", "Suico", **"Wild coriander"**. Cultivated. Used mainly as spice. Leaf infusion used for stomachaches.

Around Pucallpa, culantro with meat broth, is taken for bronchitis and fever (VDF). "Chamis" braise the dried fruits and have the children inhale smoke to treat diarrhea. Green fruits are crushed and mixed with food to treat insomnia (RVM). "Créoles" drink the decoction for colds and flu; they rub crushed leaves over the body to reduce high temperature. Leaf decoction used by "Créoles" as an antipyretic (GMJ). Elsewhere used for arthritis, colds, colic, cough, diabetes, diarrhea, fever, fits, flu, herbicide, hypertension, malaria, nausea, pneumonia, rheumatism, and tumors (DAW, RAR). Antimalarial, antispasmodic, carminative, and pectoral activities reported (DAW).

Erythrina amazonica Krukoff. Fabaceae. "Amasisa", "Huayruro amasisa", "**Amazonian coraltree**". Sometimes planted as an ornamental tree. Seeds used in handicrafts, e.g., necklaces, curtains, purses, etc. (RVM).

Erythrina fusca Lour. Fabaceae. "Amasisa", "Gallito", "**Swamp immortelle**". Semicultivated. Soil conservation species, adding nitrogen to the soil, used as ornamental and living fence. Bark decoction used to wash infected wounds to treat fungal dermatoses. Effective in a skin infection called "arco". Créoles" use the root decoction as a sudorific to reduce fever caused by colds and malaria. Flowers in decoction regarded as antitussive. "Palikur" use bark of trunk and roots mixed with the bark of *Parkia pendula* to purify waters. Trunk bark put in hot water and poulticed onto migraine headaches (GMJ). Hartwell mentions its use for cancer (DAW). (Fig. 101)

Erythrina glauca Willd. Fabaceae. "Amasisa", "Assacu" "Assasu-rana". Tikunas bathe aches and wounds with the reddish liquid. A half-cup of decoction is suggested for malaria (SAR). Brazilians, considering the plant narcotic and purgative, use it for hepatosis and rheumatism (SAR).

Erythrina peruviana Krukoff. Fabaceae. "Huayruro", "amasisa". Seeds used in handicrafts, necklaces, purses, etc. (RVM).

Erythroxylum coca var. *coca*. Erythroxylaceae. "Coca". Leaves chewed for "aire"; ecocted for parturition and respiratory ails, poulticed for headache, rheumatism (FEO). Coca leaf is described by Plotkin as "one of the world's most effective medicinal plants, particularly valuable for the treatment of stomachache and altitude sickness" (MJP). In a study of 15 nutrients, coca leaves were compared with the average of 50 Latin American foods. Per 100 g, coca was higher in: calories 305 in coca, vs. 279 units of the average on the 50 foods; proteins 18.9 g vs. 11.4 g; carbohydrates 46.2 g vs 37.1 g; fiber 14.4 g vs 3.2 g; calcium 1,540 mg vs. 99 mg; phosphate 911 mg vs 279 mg; iron 45.8 mg vs 3.6 mg; vitamin A 11,000 IU vs 135 IU, and riboflavin 1.91 mg vs 0.16 mg (RVM). Phytochemicals occuring in the four narcotic varieties of coca are detailed in Duke (1992b). Cocaine yields (in dried leaves) 0.63% in *Erythroxylum coca* var. *coca*; some other authors report 0.96% cocaine from the coca coming from Chinchao, (Huanuco), 0.25% for *E. coca* var. *ipadu*; 0.77% for *E. novogranatense* var. *novogranatense*; 0.72% for *E. novogranatense* var. *truxillense*; a sample collected by Plowman (#5600) in Trujillo contained 1.02% cocaine (RVM).

Erythroxylum coca var. *ipadu* Plowman. Erythroxylaceae. "Ipadú". Cultivated, especially by the Amazonian ethnic groups of Peru, Brazil and Colombia. Cultivated by the "Boras" along the Rio Yaguasyacu; the "chacchado" or "chaw" is enjoyed during parties, work or spare time. To prepare leaves for chewing, they roast them slowly in a

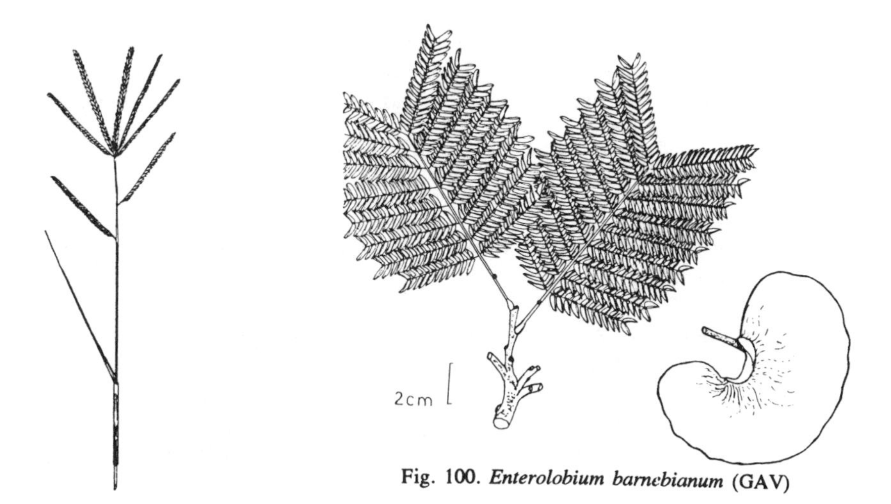

Fig. 100. *Enterolobium barnebianum* (GAV)

Fig. 99. *Eleusine indica* (GAV)

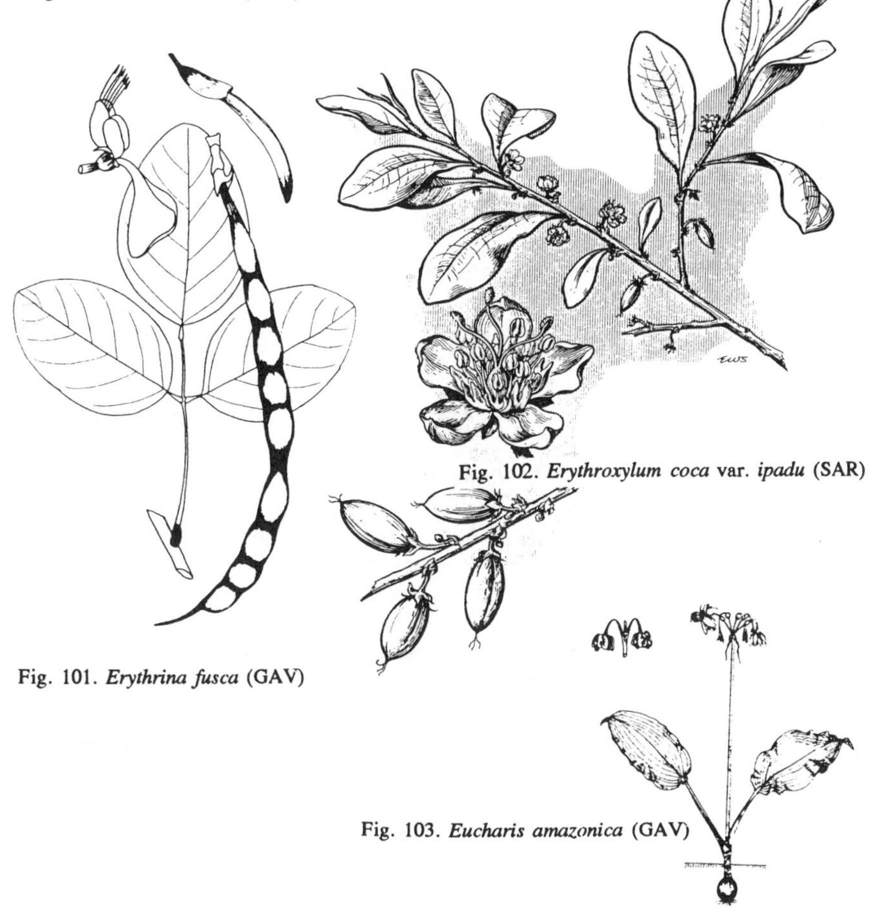

Fig. 102. *Erythroxylum coca* var. *ipadu* (SAR)

Fig. 101. *Erythrina fusca* (GAV)

Fig. 103. *Eucharis amazonica* (GAV)

clay pot; they fill their mouths with these leaves, occasionally adding ashes from *Cecropia* leaves and other plants to give a strong and better flavor. Chewing gives the sensation ofincreased energy and strength, leaving behind fatigue and hunger; also leads to euphoria and good disposition. Leaf infusion taken for gastrointestinal problems like diarrhea and indigestion. Coca is common in the religious and social life of Amazonian Peru (RVM). It is used in diarrhea and to help the mother get rid of unwanted blood after childbirth (RVM). (Fig. 102)

Erythroxylum spp. Erythroxylaceae. "Sacha hayo", **"Wild coca"**. Used as a coca substitute by Indians.

Eschweilera spp. A.DC. Lecythidaceae. "Machimango blanco", "Machimango negro". Wood for contruction poles, beams, decks; bark fiber for temporary rope (RVM).

Eschweilera gigantea (R. Knuth) Macbr. Lecythidaceae. "Machimango blanco hoja grande". Wooden part of fruit used for ashtrays, or to store small things.

Eucharis castelnaeana (Baill.) MacBr. Amaryllidaceae. "Delia", "Lilea", **"Amazon Lily"**. Cultivated ornamental (RVM). Bulb or whole plant boiled to make an emetic tea (SAR). (Fig. 103)

Eucharis grandiflora Planch. & Linden. Amaryllidaceae. "Barba de chivo". Cultivated ornamental.

Eugenia inundata DC. Myrtaceae. "Juanache", "Mishquina". Fruit edible fresh or in beverages.

Eugenia patrisi M.Vhl. Myrtaceae. "Sacha guayaba". Fruit edible (VAG).

Eugenia stipitata McVaugh. Myrtaceae. "Araza", "Guayaba brasilera". Cultivated fruits are used in refreshments. (Fig. 104)

Eugenia uniflora L. Myrtaceae. "Ceresa", "Pitanga". Fruit edible (VAG).

Euphorbia cotinifolia L. Euphorbiaceae. "Yuquilla", **"Red spurge"**. Cultivated. Used as an ornamental plant because of the colorful foliage. Used for curaré. Emetic, purgative (RAR). The latex is caustic, ichthyotoxic, and insecticide. "Créoles" use the leaf maceration to kill leaf cutter ants (*Solenopsis* spp. and *Atta* spp.) (GMJ).
Folklorically, used for condylomata, sores and wounds (DAW).

Euphorbia hirta L. Euphorbiaceae. "Golondrina", "Yerba colorada". Latex used in massage for arthritis and rheumatism (VDF).

Euphorbia huanchahana (Kl.&Gke.) Boiss. Euphorbiaceae. "Huanchangana". Root decoction, a drastic purge and urinary antiseptic, is used to reform alcoholics and treat jaundice (FEO, RAR).

Euphorbia pulcherrima Willd. Euphorbiaceae. "Poinseta", **"Poinsettia"**. Cultivated ornamental, folklorically used as bactericide, depilatory, emetic, lactagogue, and for dermatoses, erysipelas, toothache and warts (DAW).

Euphorbia tirucalli L. Euphorbiaceae. "Planta navideña", **"Milkbush"**. Cultivated ornamental. Elswehere, recommended as a source of energy hydrocarbons (JAD). Elsewhere used for abscesses, asthma, cancer, colic, cough, earache, gastralgia, neuralgia, rheumatism, toothache and warts (DAW).

Euterpe oleracea Mart. Arecaceae. "Asahí", "Palmito". Cultivated. Used mainly as an ornamental. An ideal palm that could be highly profitable in the development of the palm heart industry. A single seed can grow a plant providing 25 shoots growing individually (RVM). In Brazil, where known as "Acai", the root tea is drunk as a blood medicine for jaundice (BDS); seeds roasted and made into coffee-like beverage drunk for fever. (Fig. 105)

Euterpe precatoria Mart. Arecaceae. "Chonta", "Huasahí", "Palmito", **"Heart palm"**. The buds are edible fresh, in salads or in preserves; fruit used for refreshments. Trunks, cut into strips, are used in interior decoration. Fiber for making fabrics. The whole plant is used in ceremonial wreaths during some Amazonian fiestas. The palm-heart industry, concentrated in one factory in Iquitos, could disappear shortly, due to poor management. Used for dysmenorrhea (RAR). (Fig. 106)

Euterpe sp. Arecaceae. "Huasahí del varillal". The buds are edible.

Duke coined the word suriculture in 1992 for the cultivation of what Peruvians call "suri", grubs or larvae of the palm beetle (*Rhynchophorus palmarum*). He suggested that using the 95%+ of palm-heart palm (that is wasted) to produce edible protein (entomophagous delicacy) could give a "green" seal of approval to the palm heart industry. Some entrepreneurs say that the palm-heart industry is sustainable for centuries. Many botanists disagree. Clearly, more than 95% of the palm is wasted when a palm heart is harvested, renewably in certain species of *Bactris*, destructively in the *Euterpe* we have enjoyed around Iquitos.

Many interesting questions need answers. Could we sustainably harvest both palm hearts and protein? Could the waste palm be piled up on the soil, or in pits, and used for larval production? Can the larvae be cultivated in unturned piles or pits? What are the temperature tolerances of the larvae? Would beetles lay eggs on decaying palm debris. Would they lay in the waste of any palm? (How about the famous hat palm, not a true palm? Ivory Palm? Thatch Palm? Stilt Palm?). Does urinating on the palm really attract the *Rhynchophorus*. How long will the larvae stay alive for shipment in palm mulch? How long will smoked suri remain wholesome unrefrigerated under ambient tropical conditions? Would suri be legal imports into the US? Would an environmental impact statement be necessary and worthwhile? Do the larvae contain squalene or any of the "healing" zoochemicals alleged for shark cartilage?

The Yanomamo Indians fell trees deliberately to provide fodder for the larvae. When they cut the tree, they eat the palm heart. One large palm can yield up to 50 pounds palm heart. A palm trunk can yield three or four pounds of grubs. There are descriptions of excellent palm "butter" made by melting and clarifying the fat of the larvae. Smoked larva in the Food Composition Table for Use in Africa (#1099) is reported to contain 20.4% water, 62.3% protein, 4.6% fat, 6.5% carbohydrate, 2.2% fiber; 6.2% ash and, per 100 g, 513 mg Ca, 471 mg P., 6.9 mg Fe, 0 mg beta-carotene, 0.1 mg thiamin, 0.12 mg riboflavin, 4.2 mg niacin and 0 mg ascorbic acid. Dried larva (#1097) is reported at, per 100 g edible portion, 430 calories, 7.8-10.1% water, 51.1-54.9% protein, 13.8-17.5%

fat, 16.9% carbohydrate, 1.4-7.4% fiber, 3.8-8.4% ash, 124-270 mg Ca, 142-854 mg P, 2-3 mg Fe, 50 μg beta-carotene (?), 0.09-0.35 mg thiamin, 0.18-2.87 mg riboflavin; 3.8-11.2 mg niacin and 0 mg ascorbic acid.

Our summary of the literature suggests that 100 g of larvae could supply twice the RDA for thiamin, 1.5 times the RDA for zinc, 1.3 times the RDA for riboflavin, about 70% of copper and iron requirements, 40% of niacin, 30% of phosphorus, ca 20% of protein and calcium, but less than 10% of daily requirements of magnesium. Insect fatty acids, in general, are highly unsaturated.

We need more precise analytical data, not only on palm beetle larvae, but also on palm hearts, and other palm products. We could surely devise a nutritionally complete package based solely on renewable palm products, a TV-dinner or Palm Sunday Brunch, if we include the "suri". We think that the MUFA's, tocotrienols and beta-carotene make palm oils more attractive as health-food items than the North American press would have us believe. Oil palm is the best reported source of tocotrienols which some scientists regard as better than tocopherol in Vitamin E antioxidant activity. Duke has proposed an ACEER Amazonian Antioxidant salad dressing embracing wholesome Amazonian palm oils (best sources of tocotrienols and MUFA's and good source of beta-carotene), brazilnut (best source of selenium) camu-camu (best source of vitamin C), chile (best source of capsaicin), and puree of beans (good source of genistein). Try renewable palm hearts, drenched in antioxidant salad dressing, with a few smoked suri, hopefully contributing to your longevity and that of the rain forest!

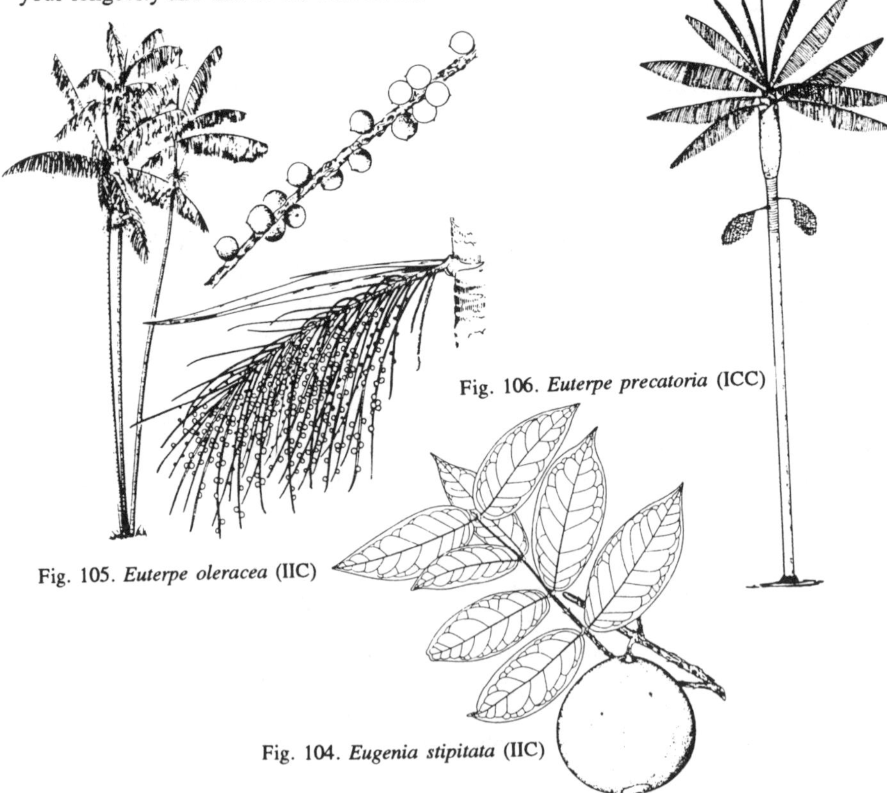

Fig. 106. *Euterpe precatoria* (ICC)

Fig. 105. *Euterpe oleracea* (IIC)

Fig. 104. *Eugenia stipitata* (IIC)

- F -

Faramea anisocalyx Poepp. & Endl. Rubiaceae. "Sanango". Leticia natives use the bark decoction as an emetic in food poisoning (SAR).

Faramea capillipes Stand. Rubiaceae. "Quinquincaca", "Yacu sanango". The root decoction is used for antipyretic baths. The alcoholic maceration of the flowers is used to treat colds; maceration on the forehead for migraine headaches (RVM).

Faramea maynensis Spruce. Rubiaceae. "Caballo sanango". Similar to previous species. Fruit yield a blue tint used for handicrafts (RVM).

♀ *Faramea occidentalis* (L.) A.Rich. Rubiaceae. "Caballo sanango". "Cuna" use the root decoction to stop menstrual hemorrhages, by taking two cups a day; one in the morning, and one at night (RVM).

Fevillea albiflora Cogn. Cucurbitaceae. "Jiman rao". Dried powdered leaves and/or resin used as a resolvent antiecchymotic around Pucallpa (VDF).

Fevillea cordifolia L. Cucurbitaceae. "Cabalonga", "Habilla". The ripe dried seeds (from this and other species of *Fevillea)* are used in making necklaces and "shacapas" which they use around the waist and knees during native dances. The infusion of mashed seed is a strong cattle purge (SOU). Venezuelans use the plant as antidote to *Hippomane*, for constipation, leprosy and snakebite (DAW). Haitians use for dermatoses, dropsy, erysipelas, hepatitis, jaundice, and rheumatism (DAW). "Campa" use the seeds as substitute for candles. This plant has more useful oil per fruit than any other Dicotyledon (RVM). The seed oil is applied to the face before *Bixa* (SAR). "Kofán" make necklaces from the seed, polishing them with the seed oil (SAR). (Fig. 107)

Ficus americana Aubl. complex. form. *guianensis*. Moraceae. "Renaco", "Matapalo", "**Strangler fig**". "Palikur" poultice latex on stomachaches (GMJ). Fruit edible (VAG).

♀ *Ficus carica* L. Moraceae. "Higuera", "Higo", "**Fig**". Cultivated; fruit edible. Leaves are used to improve the aroma and taste of corn "chicha", by cooking the leaves with the meal. Fruit decoction used as cough suppressant; latex used on warts. Women take two cups a day of this decoction a few days before they go into labor to make delivery easier (RVM). Latex is used in setting bones, the dried inflorescence for dandruff, diarrhea, and hernia (FEO). Considered mnemonic, purgative, stimulant, toxic (RAR).

Ficus citrifolia P. Mill. Moraceae. "Renaco". Panama "Cuna" apply latex to infected wounds, elsewhere used as a vermifuge (JAD).

Ficus elastica Roxb. Moraceae. "Cauchillo", "**Indian rubber tree**". Cultivated ornamental, potential rubber source.

Ficus insipida Willd. var. *insipida* Moraceae. "Ojé", "Doctor ojé". Locals take latex as vermifuge, drinking one cup fresh mixed with orange juice, or with sugar cane juice. Those who take this purge must avoid greasy and salty foods for a week; they can not receive direct sun, and must avoid being seen by strangers to the family. Those not following this diet become "overo" (with white skin pigmentation) (RVM). Pucallpa residents rub the latex onto rheumatic inflammations (VDF). "Cuna" mix some latex with a liter of water, and drink some of this mixture every other day to get rid of intestinal parasites. In Piura, the leaf decoction is used for anemia and tertian fever. Contains phyllosanthine, beta-amyrin or lupeol; lavandulol, phyllanthol, and eloxanthine (AYA). (Fig. 108)

Ficus maxima Miller. Moraceae. "Sacha ojé", "Sacha ojé del cauchero". Bark pounded to make cloth. "Campas" use the tree as a blind when hunting game animals (RVM). "Wayãpi" use the latex as an antirrheumatic (GMJ).

Ficus paraensis Miq. Moraceae. "Renaco". "Wayãpi" for diarrhea; "Palikur", for infected wounds (GMJ). "Makunas" and "Puinaves" use the latex for worms (SAR).

Ficus radula Willd. Moraceae. "Yanchama caspi". "Witoto" mix the latex with mud for stomachache (SAR). Mixed with yuca flour, it is packed into painful caries and wounds (SAR). Used to make bark cloth (RAR).

Ficus trigona L. Moraceae. "Renaco". Used by the "Wayãpi" as an antidiarrheic (GMJ).

Ficus yoponensis Desv. Moraceae. "Ojé de hoja menuda", "Ojé macho". The bark, removed carefully after pounding, is dried and used as a canvas to paint amazonian scenes, sold as crafts. Latex ingested, as *F. insipida*, for diarrhea and worms (SAR).

Fittonia verschaffeltii E.Cogn. Acanthaceae. "Motelillo". Cultivated ornamental (RVM). Ecuadorian "Ketchwa" use the leaf infusion for toothache (SAR).

Floscopa robusta (Schub.) Clarke, var. *sprucei* Clarke. Commelinaceae. "Piñaraño". Ornamental.

Fourcroya andina Trec. Amaryllidaceae. "Penca". Cultivated ornamental.

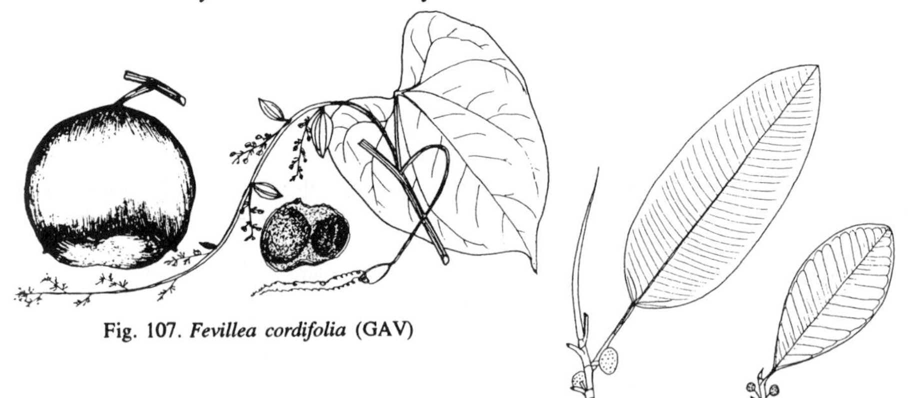

Fig. 107. *Fevillea cordifolia* (GAV)

Fig. 108. *Ficus insipida* (GAV)

- G -

Gallesia integrifolia (Spreng.) Harms. Phytolaccaceae. "Palo de ajo", "Palo de cebolla". Timber (RAR). Bark infusion used for external baths to protect against bad luck and witchcraft (RVM).

Garcinia acuminata (R.&P.) Planch.& Triana. Clusiaceae. "Charichuelo". Wood used for house construction, beams, decks, and columns.

Garcinia benthamiana Planch. & Triana. Clusiaceae. "Charichuelo". Fruit edible. Wood used for house construction.

Garcinia macrophylla Mart. Clusiaceae. "Charichuelo hoja grande". Cultivated as fruit tree; wood used for house construction.

Garcinia madruno (Kunth.) Hamel. Clusiaceae. "Charichuelo". Fruit edible. In Panama, used for cholera, ulcers, and yellow fever (JAD).

Gasteranthus corallinus (Frist) Will. Gesneriaceae. "Puca cabaciña". As an ornamental plant.

♀ *Genipa americana* L. Rubiaceae. "Huito", "Huitol", "Jagua", **"Genipap"**. Fresh fruit eaten for bronchitis; also used to make spiritous drinks. Cooking with brown sugar and aguardiente makes a nice dessert. Green fruit used to dye clothes, also used to paint and decorate their faces. Wood used in carpentry. Some people affirm that the fruit decoction is abortifacient. Don Antonio Montero claims that the strained fruit juice is good for cancer of the uterus. "Achuales" from Pastaza use the green pericarp to extract decayed teeth. "Achuales" and peasants near Iquitos cook the fruit and seeds; this decoction is use on baths for female genital inflammations. It also reduces swelling of the respiratory mucous membranes. "Kayapo" eat the fruit and use it to decorate their bodies. "Créoles" prepare a cathartic and antidiarrheic decoction; the same decoction is used in poultice to treat ulcers (GMJ). Haitians use for anemia, aphrodisia, blenorrhagia, diarrhea, gonorrhea, hepatoses, and tumors (DAW). Brazilians express the fruit juice, let stand overnight, and drink a small cup each day for 2 or 3 days for jaundice (BDC). Contains: genipin, mannitol, tannin, methyl-ethers, caterine, hydantoin, and tannic acid (RVM). (Fig. 109)

Genipa spruceana Steyer. Rubiaceae. "Yacuruna huito". Used by the "Créoles" like *G. americana,* but inferior (GMJ).

Gentianella alborogea (Gilg.) Fabr. Gentianaceae. "Chavin", "Harcapura", "Hercampuri". Shoot decoction considered cholagogue, diuretic; used for jaundice and weight loss (FEO).

Geogenanthus rhizanthus (Ule) Bruckn. Commelinaceae. "Mishquipanguilla". Occasionally planted as an ornamental (RVM). "Kofán" rub the hot water infusion of *G. ciliatus* on swollen knees (SAR). "Secoya" use cold water infusion for parasites in children. Rio Pastaza Indians pat the leaf on the buttocks of those perturbed by excessive flatulence (SAR).

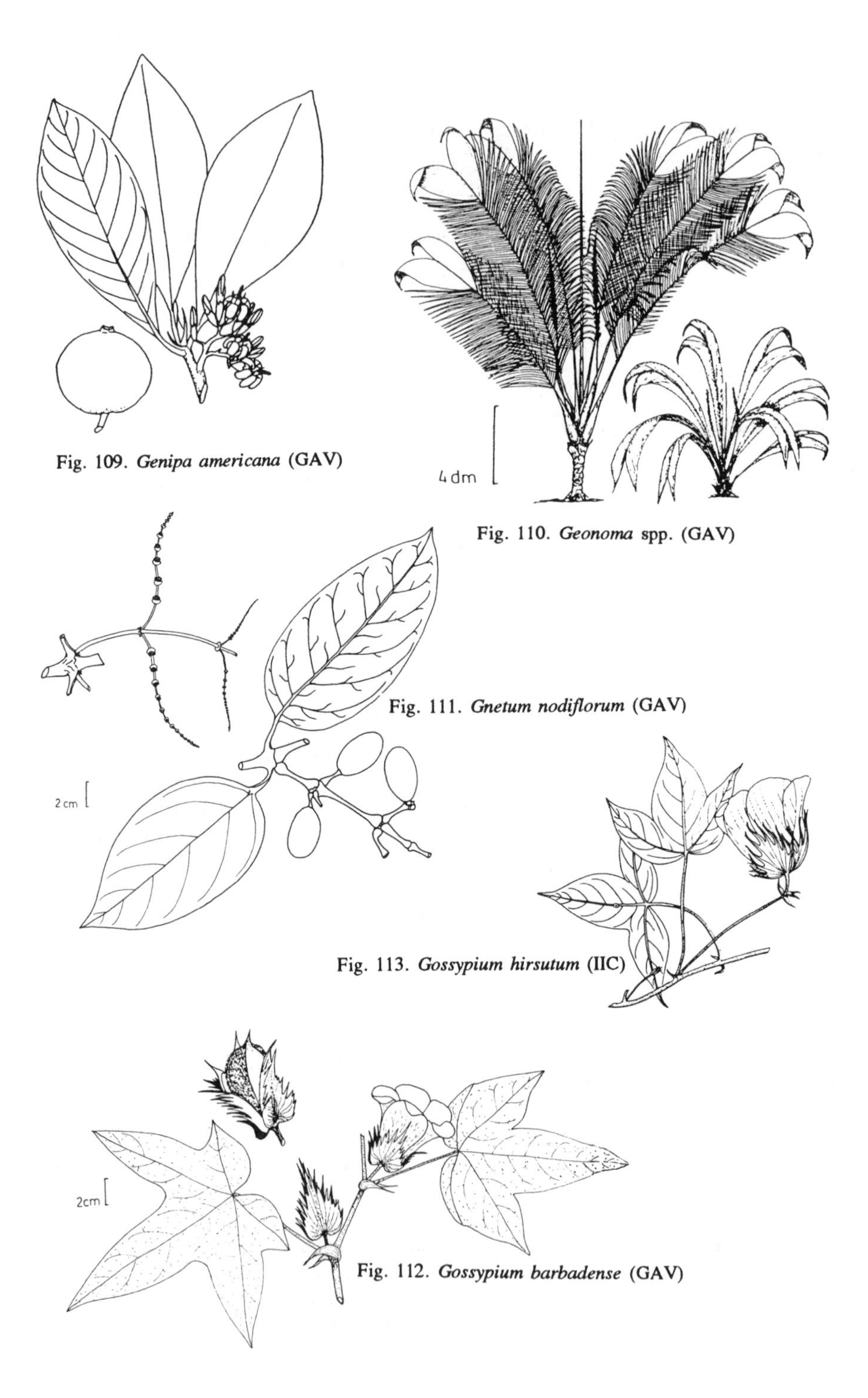

Fig. 109. *Genipa americana* (GAV)

Fig. 110. *Geonoma* spp. (GAV)

Fig. 111. *Gnetum nodiflorum* (GAV)

Fig. 113. *Gossypium hirsutum* (IIC)

Fig. 112. *Gossypium barbadense* (GAV)

Geonoma juruana Dammer. Arecaceae. "Palmicha". Leaves used for roofing; used by "Boras" to extract vegetal salts (DAT).

Geonoma spp. Mart. Arecaceae. "Palmicha". Semicultivated ornamental. (Fig. 110)

Geophila repens (L.) I.M.Johnston. Rubiaceae. "Poroto huangunillo". "Palikur" use the fruit to treat dermatosis; very effective as an antimycosic (RVM). "Ketchwa" also use it for fungal infection (SAR). In Fiji, it is even used for leprosy (DAW).

Geophila sp. Rubiaceae. "Maima pash". Powdered leaves applied locally as antiinflammatory (VDF).

Gloeospermum sphaerocarpum Tr. & Pl. Violaceae. "Tamarillo". Edible fruit. Leaf maceration used by "Waunana" as a ceremonial hallucinogen (RVM).

Glycine max (L.) Merr. Fabaceae. "Soya", **"Soybean"**. Cultivated. Contemplated as an alternative crop for coca, at least in Bolivia, the soybean has gotten good press recently as a cancer preventive. Among the compounds found in soy, isoflavones, like genistein and daidzein; Bowman-Birk inhibitors, phytic-acid, saponins and phytosterols have been suggested as potentially contributing to this cancer-preventive activity (JAD).

Gmelina arborea Roxb ex J.E.Sm. Verbenaceae. "Melina". Cultivated. Asian native, introduced and cultivated in Latin America as a paper pulp and ornamental. There are three varieties: *C. arborea* var. *canescens*; var. *dentata*, and var. *glaucescens* (RVM).

Gnetum leyboldi Tul. Gnetaceae. "Bala huayo", "Hambre huayo", "Paujíl ruro". Seeds edible roasted.

Gnetum nodiflorum Brong. Gnetaceae. "Bala huayo", "Hambre huayo", "Paujíl ruro". Seeds edible roasted (RVM). "Puinave" use the gummy bark decoction hot to reduce swellings caused by damaged muscle or tendon (SAR). Surinam "Tirio" use decoction to wash headache (SAR). "Wayana" use water from the vine for debility and inappetence (SAR). (Fig. 111)

Gomphrena globosa L. Amaranthaceae. "Siempre viva", "Manto de Cristo", **"Globe amaranth"**. Cultivated ornamental (RVM). elsewhere, considered depurative and used for cough, diabetes, heart problems, hypertension, nosebleed and oliguria (DAW). Febrifugal, herbicide (RAR).

Gordonia planchonii H. Keng. Theaceae. "Aripay". Timber used for general construction.

♀ *Gossypium arboreum* L. Malvaceae. "Algodón", **"Cotton"**, "Algodonero", **"Tree cotton"**. Cultivated in small amounts. Seeds lactagogue (SAR). Root bark for amenorrhea and dysmenorrhea in Amazonian Brazil (SAR). Uses of cotton are many, here and in industrialized society.

♀ *Gossypium barbadense* L. Malvaceae. "Algodón", **"Cotton"**, "Algodonero". Cultivated in small amounts. Ashes from dried buds used for diaper rash, and infected wounds. Leaf decoction used as oxytocic. Flower decoction used for hepatitis (RVM).

Flower buds are used by the "Wayãpi" for earache. Leaves used for parasites, to eliminate filaria (GMJ). (Fig. 112)

Gossypium herbaceum L. Malvaceae. "Algodón", **"Cotton"**, "Algodonero". Cultivated in small amounts. Amazonian Brazilians consider the root diuretic (SAR).

♀ *Gossypium hirsutum* L. Malvaceae. "Algodon blanco", "Cotton". In Brazil, bathing with the leaf tea or drinking the flower tea is considered useful for uterine problems (BDS). Brazilians also place mashed cottonseed into aching caries (BDS). (Fig. 113)

Goupia glabra Aubl. Celastraceae. "Muena rifarillo". Very good-quality wood has multiple uses, including rural construction, beams, and decks. Juices from macerated leaves used for eye disorders (RVM). "Créoles" use bark decoction as an oral analgesic (GMJ). "Andokes" use leaf decoction for cataracts and to dye the skin and hair (SAR).

♀ *Grias newberthii* Macbr. Lecythidaceae. "Sacha mangua", **"Anchovy pear"**. Ornamental (RVM); fruits roasted and eaten (SAR). "Siona" grate the fruit into water as a purgative (SAR). Rio Chico natives of Ecuador use the cambium as an emetic in delivery, inappetence and malaria (SAR). Seed used in enema for dysentery (SAR). (Fig. 114)

Grias peruviana Miers. Lecythidaceae. "Sacha mangua", "Mancoa". Edible fruit.

Guadua weberbaueri Pilger. Poaceae. "Marona". Cultivated ornamental. Trunks used for house construction; with the internodes they make musical instruments and containers (RVM).

Guarea cinnamonea A. Juss. Meliaceae. "Requia". The saw wood is used for general construction.

Guarea gomma Pulle. Meliaceae. "Requia". Timber used for decks and columns (RVM). Bark decoction used by "Palikur" as an emetic, and for liver diseases (GMJ).

Guarea grandifolia D.C. Meliaceae. "Bola requia". Wood used for columns. "Wayãpi" believe that adding a drop of toxic sap from the trunk increases the alcoholic level of their cassava chicha (GMJ).

♀ *Guarea guidonia* (L.) Sleumer. Meliaceae. "Requia", "Latapi". Wood used for beams, etc. Early authors believed this plant very TOXIC and dangerous. Women of San Martin use the decoction in enemas to increase fertility (RVM). Hispaniolans use the plant for enterorrhagia (DAW).

Guarea kunthiana A. Juss. Meliaceae. "Requia", "Paujil ruro". The wood is used for beams and decks and sometimes sawwood.

Guarea pubescens (Rich) A.Juss. ssp. *pubescens*. Meliaceae. "Requia". Wood used for columns and decks. Used by the "Wayãpi" to POISON their arrows (GMJ).

Guarea purusana C.DC. Meliaceae. "Latapi", "Requia". The sawwood is good for furniture.

Guatteria citriodora Ducke. Annonaceae. "Espintana". Wood used for construction; beams, decks, and columns. It is occasionally cleaned for interior decorations.

Guatteria decurrens R.E.Fries. Annonaceae. "Carahuasca". Sawwood used for construction, beams, decks, and columns.

Guatteria dielsiana R.E.Fries. Annonaceae. "Carahuasca". The wood is used for decks and columns.

Guatteria elata R.E.Fries. Annonaceae. "Carahuasca". The wood is used for beams, for decks and columns. Bark decoction taken once a day for epilepsy and gout (?gota coral) (ACEER).

Guatteria exellens R.E. Fries. Annonaceae. "Carahuasca". For columns.

Guatteria hyposericea Diels. Annonaceae. "Carahuasca". The sawwood is used for construction.

Guatteria megalophylla Diels. Annonaceae. "Carahuasca", "Boa". The sawwood is used for construction (RVM). "Kofáns", believing the species TOXIC, once used it in curaré (SAR).

♀ *Guatteria modesta* Diels. Annonaceae. "Carahuasca" Peruvians use in a contraceptive, boiling bark for a minute, then cooling, drinking one cup in the AM, one in the PM, during menstruation (SAR).

Guatteria multivenia Diels. Annonaceae. "Carahuasca-millua". The wood is used for house construction and decks.

Guatteria pteropus Benth. Annonaceae. "Zorro caspi". Wood for house construction, beams, and decks; not subject to attack by xylophagous insects (RVM).

Guatteria tomentosa Rusby. Annonaceae. "Espintana". Wood used for house construction, beams, and decks; occasionally hewn to use in interior decorations.

Guatteria sp. Annonaceae. "Anonilla", "Carahuasca". Sawwood used for construction, beams, and decks.

Guatteria sp. Annonaceae. "Carahuasca". Sawwood used for construction, beams, and decks. Fruit provides a garnet tint used in handicrafts.

Guazuma crinita Mart. Sterculiaceae. "Bolaina blanca", "Bolaina". Sawwood used for house construction and interior decorations; also for lollypop sticks, paper pulp. Bark used to make different kind of ropes.

Guazuma ulmifolia Lam. Sterculiaceae. "Bolaina", "Atadijo", "**West Indian elm**". Wood and bark for construction and ropes. Ripe fruits have a strong honey scent. Some people even chew the fruit to extract the sweet juice, spitting out the remainder. The macerated fruit mixed with aguardiente is used to scent the "siricaipe" or "mapacho". In Jamaica the bark is used to feed silkworms. Leaf decoction used for baldness, the bark decoction for dysentery (SOU). Elsewhere regarded as astringent, depurative, diaphoretic, emollient, pectoral, refrigerant, stomachic, styptic, and sudorific; used for alopecia, asthma, bronchitis, dermatosis, diarrhea, dysentery, elephantiasis, fever, hepatitis, leprosy, malaria, nephritis, pulmonosis, and syphilis (DAW, RAR).

♀ *Gurania spinulosa* (Poepp. & Endl.) Cogn. Cucurbitaceae. "Zapallito". "Palikur" cut the stems in small pieces and use the decoction as a remedy to "clean the bile" (GMJ). Ecuadorians use the root tea for dysmenorrhea and apply leaf decoction to cuts and wounds (SAR). Trinidadians take for constipation (DAW).

Gustavia angusta L. Lecythidaceae. "Sacha chopé". Semicultivated ornamental; fruit edible. Wood useful. Plant considered emetic, piscicidal, purgative; for the liver.

Gustavia hexapetala (Aubl.) Sm. Lecythidaceae. "Chopé cimarron". Wood used for house construction, beams, decks, and columns.

Gustavia longifolia Poepp. & Endl. ex Berg. Lecythidaceae. "Chopé", "Sacha chopé". Fruit is edible (RVM). "Sionas" once used the bark in curaré, considering the seed purgative (SAR).

Gustavia mexicana Knth. Lecythidaceae. "Chopé". Cultivated ornamental. Fruit edible.

Gynerium sagittatum (Aubl.) Beauv. Poaceae. "Caña brava", "Caña isana", "**Giant cane**". Used in house construction; fresh stems, used as fences for gardens, often become living fences. When the floriferous buds ripen, they are used to make arrows and darts for fishing. They are also used for handicrafts. Natives use them as pendents. Tender leaf decoction used for asthma, before or at onset of an attack. Rhizome decoction diuretic (SOU). Elsewhere used for alopecia and corns (DAW). (Fig. 115)

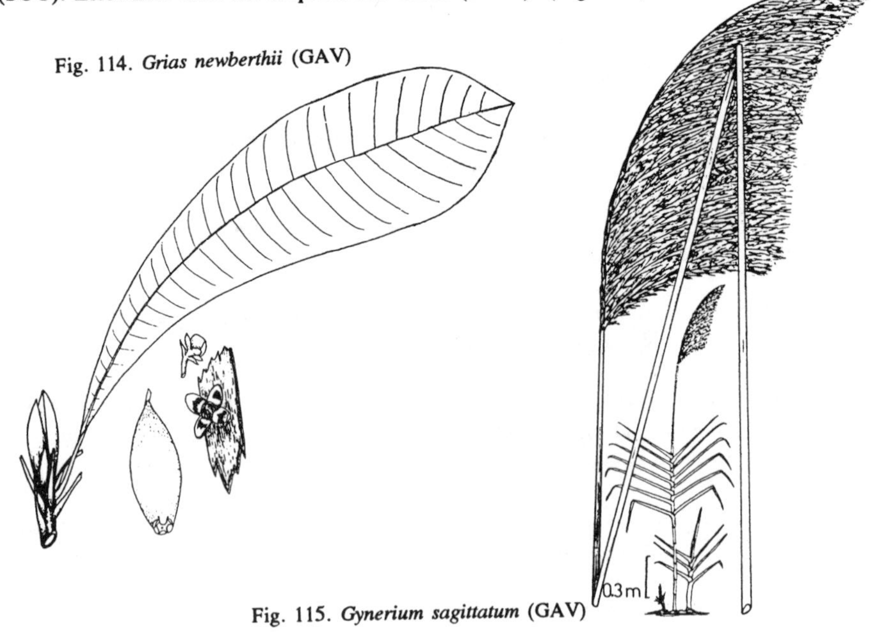

Fig. 114. *Grias newberthii* (GAV)

Fig. 115. *Gynerium sagittatum* (GAV)

- H -

Hamelia patens Jacq. Rubiaceae. "Benzen yuca", "Usia-ey", "Yoshin coshqui rao", "Yuto banco". Used as an antiflammatory and antipyretic in Peru (VDF). Leaf decoction used for skin diseases and rheumatism. Fruit edible (RAR). The fruit syrup is used for scurvy and dysentery; also used in making a fermented drink. The plant is believed POISONOUS (SOU). "Chocó" drink the leaf infusion for fever and diarrhea with blood. The roots are used as purge (DAW). The "Waunana" drink the juice from macerated leaves and flowers for cholera. "Ingano" value leaf tea as vermifuge (SAR). Elsewhere used for cancer, constipation, dermatoses, diarrhea, dysentery, erysipelas, fever, headache, jaundice, malaria and sores (DAW). Also used for syphilis. Used in Venezuela to avoid sunstroke (RVM). Around Piura, the depurative leaves are used for dysentery, rheumatism, and scurvy (FEO). Warm leaves are used as an analgesic in pharyngitis. Contains rosmarinic acid, narirutin, and 5,7,2',5'-tetrahydroxyflavanone-7-rutinoside (PC 29(7):2358, 1990). Chaudhuri and Thakur reported 500 ppm ephedrine (PM 57:199. 1991).

Haploclathra cordata Vasquez. Clusiaceae. "Boa caspi", "Balatillo". Good quality wood used for house construction.

Haploclathra paniculata Mart. Clusiaceae. "Boa caspi", "Palisangrillo". The wood is used for house construction; beams, decks and columns, and occasionally for posts.

Hedychium coronarium Koenig. Zingiberaceae. "Lirio", **"Butterfly lily"**, **"Garland flower"**, **"Ginger lily"**. Cultivated ornamental. Roots macerated in boiling water are used by the "Chami" as a soothing bath to relieve pains and aches (RVM). "Kubeo" take root decoction for arm and chest pains (SAR). "Tukano" take leaf tea for abdominal pain (SAR). Elsewhere, regarded as carminative, excitant, gargle, stomachic and tonic; used for halitosis, rheumatism, rhinitis, swellings, and tumors (DAW).

Hedyosmum huascari Macbr. Chloranthaceae. "Supinini". Resin used in rheumatism massage (FEO).

Heisteria iquitensis Sleumer. Olaceaceae. "Yutubanco". Wood used for house construction.

Heisteria pallida Engl. Olacaceae. "Chuchohasi", "Chuchuhuasha". Timber species. Stem infusion vulnerary, applied for inguinal hernia. Ingested as antirheumatic, aphrodisiac, tonic (FEO). For fever, hepatitis (RAR).

♀ *Helianthus annuus* L. Asteraceae. "Girasol", **"Sunflower"**. Cultivated ornamental (RVM); elsewhere used to extract oil, and for aftosa, anodyne, aphrodisia, bronchitis, cancer, carbuncles, catarrh, colds, cough, diarrhea, dysentery, dysuria, epistaxis, fever, flu, fractures, inflammation, laryngitis, malaria, menorrhagia, pleuritis, rheumatism, snakebite, splenitis, whitlows and wounds (DAW). (Fig. 116)

Heliconia chartacea Lane ex Bar. Musaceae. "Situlli". Leaves are generally used for wrapping, occasionally for roofing. All Heliconias could be used as ornamentals (RVM).

Heliconia episcopalis Velloz. Musaceae. "Situlli". As preivous species.

Heliconia hirsuta L.f. Musaceae. "Millua situlli". Leaves for wrapping and roofing (RVM). "Tatuyos" eat and make a fermented beverage with the root (SAR).

Heliconia rostrata R.&P. Musaceae. "Situlli". As the previous species.

Heliconia spp. Loes. Musaceae. "Sutilli". Ornamental; leaves for roofing and wrapping.

Helicostylis elegans (MacBr.) C.C.Berg. Moraceae. "Misho chaqui". Seed edible cooked (VAG).

Helicostylis scabra (Macbr.) C.C.Berg. Moraceae. "Misho chaqui". Fruit edible (RVM). "Makuna" use latex for intestinal parasites, though believing the plant POISON (SAR). "Puinaves" paint the latex onto fungal infections (SAR).

Helicostylis tomentosa (Poepp. & Endl.) Rusby. Moraceae. "Misho chaqui". Seed edible roasted or cooked. Inner bark used as an hallucinogen; in experimental rats, the effects are similar to symptoms produced by *Cannabis* (SAR).

Heliotropium curassavicum L. Boraginaceae. "Alacran", "**Seaside heliotrope**". Leaf decoction applied externally as antiseptic vulnerary, internally for rheumatism. Powdered branches used for hemorrhoids (FEO). For eczema, gonorrhea, kidneystones (RAR).

♀ *Heliotropium indicum* L. Boraginaceae. "Alacransillo", "Ihuin rao", "Ucullucui sacha". Around Pucallapa, used for scorpion stings and rheumatism (VDF). Elsewhere regarded as abortifacient (and ironically antiabortive), anodyne, astringent, diuretic, emmenagogue, emollient, pectoral, stomachic, and vulnerary; used for aftosa, asthma, boils, bugbites, calculus, cough, dermatoses, eczema, erysipelas, fever, furuncle, hyperuricemia, inflammation, itch, kidney stones, laziness, leprosy, myalgia, nephritis, ophthalmia, pharyngitis, rheumatism, scabies, sores, tumors, and warts (DAW). Folk remedy for cancer that contains an antitumor compound, indicine-N-oxide (JAD).

Helosis cayennensis (Sw.) Spruce. Balanophoraceae. "Mai toco". Juices used as an antiecchymotic and antiinflammatory around Pucallpa (VDF).

Helosis cf *guyannensis* L.C.Rich. Balanophoraceae. "Aguajillo". Common around Explorama, believed to be a good hemostat (JAD, SAR). Amazonians value the decoction for diarrhea and dysentery (SAR).

Hemarthria altissima (Poir.) Stapf & Hubb. Poaceae. "Grama playa". Forage (RVM).

Herrania nitida (Poepp.) R.E.Schultes. Sterculiaceae. "Cacahuillo". Fruits edible (RVM). "Karijona" use roasted seed for stomachache (SAR).

Heteropsis jenmannii Oliv. Araceae. "Tamshi", "Tamshi delgado". The strong and flexible roots are used for different kinds of bindings, but mainly to tie together the wooden structures in the local houses. Also used in making baskets, purses, and other handicrafts (RVM). (Fig. 117)

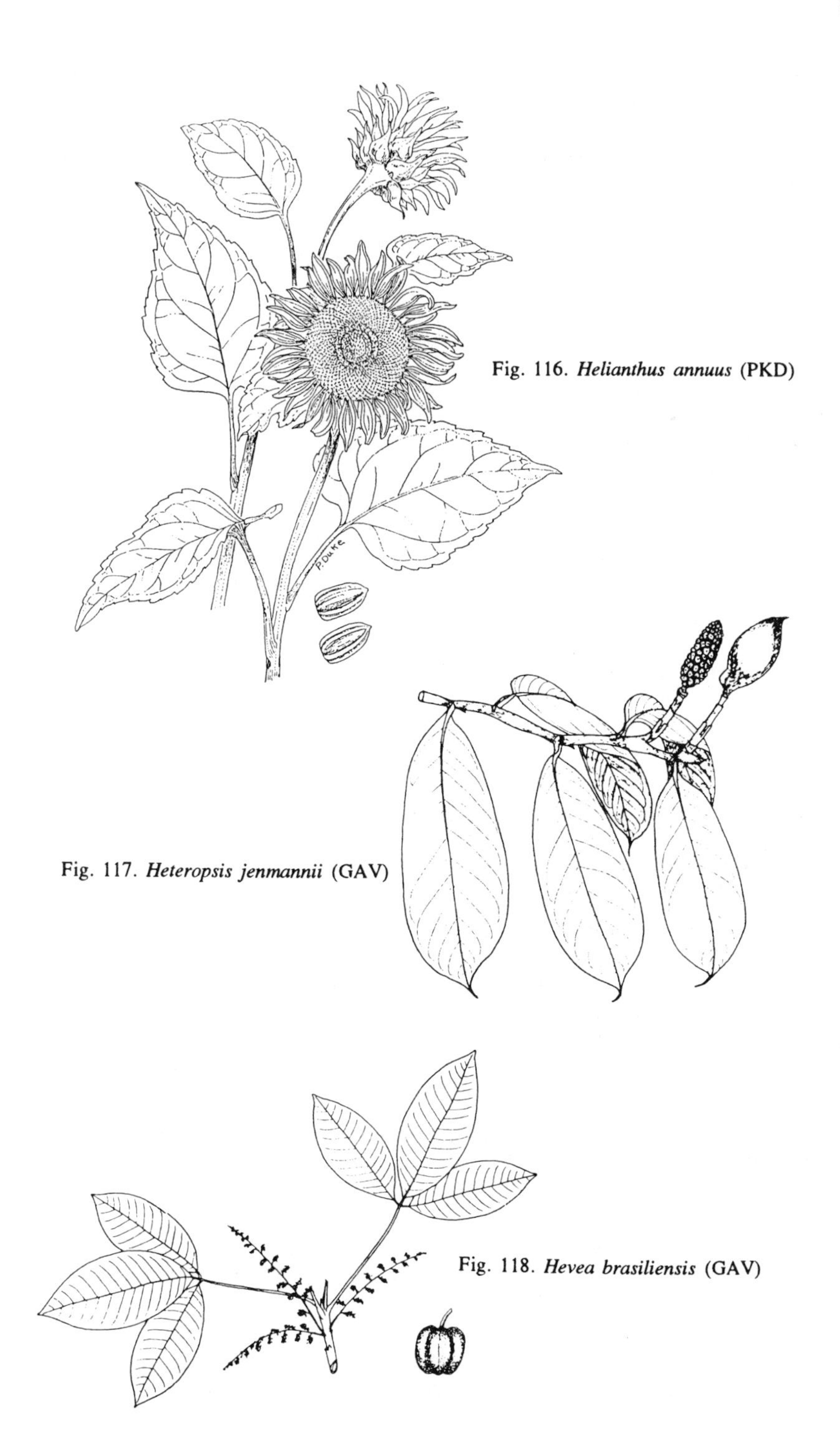

Fig. 116. *Helianthus annuus* (PKD)

Fig. 117. *Heteropsis jenmannii* (GAV)

Fig. 118. *Hevea brasiliensis* (GAV)

Heteropsis oblongifolia Kunth. Araceae. "Tamshi". The roots are used for handicrafts, and to bind the columns of rustic houses (RVM).

Heteropsis spruceana Schott. var. *spruceana*. Araceae. "Tamshi". As the previous species. A similar species is used in Brazil's Tapajos region in teas for asthma.

Heteropsis sp. Araceae. "Tamshi", "Tamshi canastero". Mainly used in basketry.

Hevea brasiliensis (Willd. ex A. Juss.) Muell.-Arg. Euphorbiaceae. "Shiringa", "Jebe débil fino", "Sernambi", "**Rubber**". Latex for rubber production. Cooked seeds are edible. The rubber provided by *Hevea* species cannot be matched by synthetic rubber for some applications; high quality tires require 40% natural rubber on radial tires, to 95% in space vehicles (RVM). According to Plotkin, Amazonian Indians not only ate the seeds but the "dipped their feet in the latex and dried them over the fire, thus creating the first custom-made sneakers" (MJP). (Fig. 118)

Hevea guianensis Aubl. Euphorbiaceae. "Shiringa", "Shiringa amarilla", "Jebe entre fino". Latex for rubber (RVM). Some Indians drink the latex to make them strong (SAR). "Waorani" use to treat infections following bites of the warble fly (SAR). Cooked seed edible (VAG).

Hevea pauciflora (Spruce ex Benth.). Muell.-Arg. var. *coriacea* Ducke. Euphorbiaceae. "Shiringa", "Shiringa maposa". Latex harvested, mainly to bulk with the more desirable latices of other species. Cooked seed edible (VAG).

♀ *Hibiscus mutabilis* L. Malvaceae. "Flor variable", "**Cotton rose**". Cultivated ornamental (RVM). Elsewhere, considered alexiteric, anodyne, demulcent, emollient, expectorant, and pectoral; used for abscesses, burns, cancer, cough, dysuria, fever, fistulae, inflammation, menorrhagia, snakebite, sores, swellings, and tumors (DAW).

♀ *Hibiscus rosa-sinensis* L. Malvaceae. "Rosa china", "**Chinese rose**", "Cucarda". Cultivated ornamental. The main "Créole" remedy for pulmonary diseases. The flowers, with leaves of *Lantana camara, Justicia pectoralis, Ocimum micranthum*, stems and leaves from *Macfadenya ungis-cati*, plus some fat, sugar, and a spoonful of rum, makes an antitussive tea for flu (GMJ). Flower extracts said to have antifertility effects (SAR).

Hibiscus sabdariffa L. Malvaceae. "Rosella", **Roselle**". Brazilians poultice the leaves, mashed in salt and alcohol, onto wounds, especially streptococcus-infected wounds (erysipelas), which they call "isipla" on the Rio Tapajos (BDS). (They also apply the red-spotted tree frog to such wounds.) (Fig. 119)

Hibiscus schizopetalus (Masters) Hook.f. Malvaceae. "Campanilla", "**Coral hibiscus**", "**Japanese lantern**". Cultivated ornamental (RVM). Elsewhere used for colds, coughs and eye ailments (DAW).

Hieronima alchornoides F.Allen. Euphorbiaceae. "Palo de sangre", "Piñaquiro colorado". Wood for lumber, use for construction in general, and for parquets (RVM). Seed oil vermifuge (SAR).

Hieronima sp. Euphorbiaceae. "Mojarra caspi". Wood for lumber, used for construction of interiors.

Himatanthus lancifolia (Muell.-Arg) Woods. Apocynaceae. "Socoba". Bark decoction used for fever near Pucallpa (VDF).

Himatanthus sucuuba (Spruce) Woods. Apocynaceae. "Bellaco caspi". Latex poulticed onto hernias and lumbar pains; also used to treat tumors; bark for gastric ulcers (RVM). "Tikuna" apply fresh latex to wounds (SAR). "Karijona" apply powdered bark to recalcitrant sores (SAR). "Waorani" rub the latex over larvae of botfly infections (SAR). Brazilians use bark tea for asthma, coughs, tuberculosis; latex for worms (BDS), fever, rheumatism (RAR). (Fig. 120)

Homolepis aturensis (HBK) Chase. Poaceae. "Torurco", "Colchón quihua". Forage grass common in pastures (RAF).

Huberodendron swietenioides (Gleason) Ducke. Bombacaceae. "Aguano masha". Wood for lumber, interior decorations, veneer.

Humiria balsamifera (Aubl.) St. Hill. Humiriaceae. "Chamisa", "Parinarillo", "Quinilla negra", "Loro shungo". The wood is used for heavy construction; dormers, jam posts for bridges, posts, forked poles (most used (after "balatillo") in house construction around J. Herrera zone, along the Rio Pucallpa) (RVM). Cultivated by the "Karaja" for its edible fruit. Bark infusion used for amebic dysentery (RVM). "Palikur" soak cotton in bark decoction for toothache; decoction also used for erysilepas (GMJ). Fruits eaten by Indians (SAR). "Barasana" apply powdered bark to cuts and wounds (SAR). Expectorant, used for gonorrhea and worms (RAR). Contains the antiinflammatory bergenin (JBH). (Fig. 121)

Humiriastrum cuspidatum (Benth.) Cuatr. Humiriaceae. "Manchari blanco". Wood for construction of rural houses; decks, and columns.

Humiriastrum excelsum (Ducke) Cuatr. Humiriaceae. "Ucho mullaca", "Manchari". The wood is used in heavy contruction, jam posts for bridges, dormers, etc.

Hura crepitans L. Euphorbiaceae. "Catahua", "Catahua blanca", "Catahua amarilla". Wood is used for interior decorations, plywood veneers, small boats, and shafts for floating houses. Latex caustic, sometimes used for fish POISON, even to POISON anacondas (AYA) and insects (DAW). In order to cut the tree, it is necessary to peel the bark ahead of time. It is said that in the old days, they used to dry the small reservoirs by pouring some liters of latex in the water. Some natives use the seeds as a laxative. The "Palikur", and "Wayãpi" use latex as POISON (GMJ). (Fig. 122)

Hydrocotyle ranunculoides L.f. Apiaceae. "Oreja sacha". Sometimes used as an ornamental in ponds.

Hydrocotyle umbellulata L. Apiaceae. "Matteclu". Leaf decoction cholagogue, depurative, for fever and headache, externally applied as ocular decongestant (FEO).

Hymenachne amplexicaulis (Rudge) Nees. Poaceae. "Gramalote negro". Forage (for small "antelopes").

Hymenachne donacifolia (Raddi) Chase. Poaceae. "Gramalote". Forage.

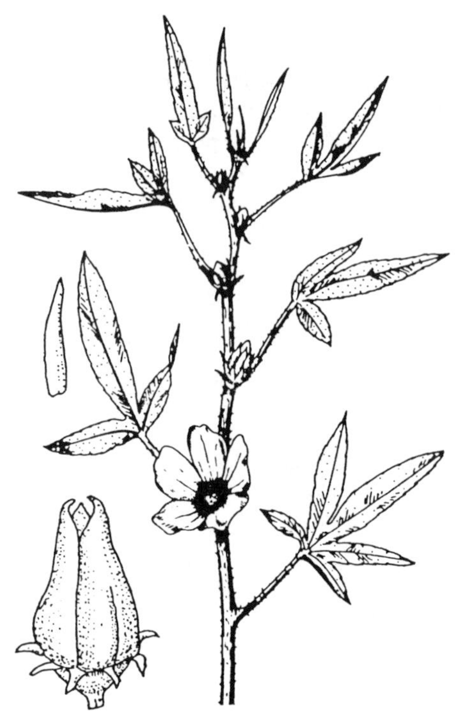

Fig. 119. *Hibiscus sabdariffa* (IIC)

Fig, 120. *Himatanthus sucuuba* (GAV)

Fig. 121. *Humiria balsamifera* (RVM)

Fig. 122. *Hura crepitans* (GAV)

Fig. 123. *Hymenaea oblongifolia* (GAV)

Hymenaea courbaril L. Fabaceae. "Algarrobo", "Azucar huayo". Brazilians drink the sap for cough (BDS). Reportedly useful for cystitis, hepatitis, prostatitis, and tuberculosis (RAR). Bark tea used for athlete's foot or foot fungus (BDS). "Karaja", like "Créoles" take macerated bark for diarrhea (RVM, MJP). Resin in old stumps used for tinder (MJP).

Hymenaea oblongifolia Huber. Fabaceae. "Azúcar huayo". Wood used for lumber, general construction, decorative plaques. Bark decoction or tincture used for rheumatism, arthritis, and diarrhea. Powdery fruit pulp edible, made into beverages (RVM). "Yukuna" paint the resin on fungal infections (SAR). Resin used as incense, and in varnishes. Fruit resin used as purgative. The bark decoction is recommended as a vermifuge (RVM). (Fig. 123)

Hymenaea palustris (Ducke) Lee & Langenh. Fabaceae. "Amahuaca", "Azúcar huayo". Somewhat as the previous species. Resin used for torches. For headache (RAR).

Hymenaea reticulata Ducke. Fabaceae. "Azúcar huayo". Similar uses as *H. oblongifolia*

Hymenolobium sp. Fabaceae. "Mari mari". The wood is used for forked poles, posts, jam posts for bridges, parquets and handicrafts.

Hyparrhenia rufa (Nees) Stapf. Poaceae. "Pasto yaragua". Forage grass.

Hyptis brevipes Poit. Lamiaceae. "Tipo". Used for gastritis and inflammation around Pucallpa (VDF).

Hyptis capitata Jacq. Coll. Lamiaceae. "Cadillo cabezon". Used in Ecuador for fungal infections (in Taiwan for asthma, colds, fever), the aerial parts contain the antioxidant rosmarinic acid, oleanolic-acid, and ursolic acid, stigmasterol, 10-epi-olguine, and 2,3-di(3',4'-methylenedioxybenzyl)-2-buten-4-olide, a lignan, and apigenin-4',7'-dimethyl-ether. No alkaloids. Crude extracts showed little fungicidal or insecticidal activity. (PC 30(8):2753-6. 1991).

♀ *Hyptis mutabilis* (Rich) Brig. Lamiaceae. "Albaca cimarrona", "Barin rao", "Matapasto", "Ovena micuna", "Soro sacha". Infusion considered soporific and antipyretic (VDF). Also used to avoid bad luck, mixed with *Petiveria alliacea*. "Créoles" use this as a vermifuge for infants (GMJ). Argentinans use as abortive (EAP).

Hyptis recurvata Poit. Lamiaceae. "Albaquilla". The leaves are used in baths by the "Boras" to reduce fever (DAT).

- I -

Ichnanthus pallens (Sw.) Munro ex Benth. Poaceae. "Nudillo". Forage grass.

Ichthyothere terminalis (Spreng.) Malme. Asteraceae. "Dictamo real", Galicosa", "Jarilla". Juice or solar tea of leaves used to wash sores (BDS).

♀ *Ilex guayusa* Loes. Aquifoliaceae. "Guayusa". Cultivated. In Piura the leaf decoction, considered antipyretic, antirheumatic, antiseptic, and cholagogue, is used to treat venereal diseases and female sterility (FEO). Leaf infusion used by the "Achuales" as an emetic. Women get up early in the morning and prepare the infusion in the biggest pot available; then everyone, including the children, drinks as much as they can, and minutes later they all start vomiting. They do this to clean body and spirit; bad things they have consumed the day before are eliminated, to start a new day with clean body and renewed spirit (RVM). Amazonian Ecuadorians drink guayusa to settle nerves and to prevent the ayahuasca hangover. Also believed useful in aphrodisia, dysmenorrhea, fever, hepatosis, malaria, pregnancy, stomach problems, syphilis, and perhaps other venereal diseases (SAR). (Fig. 124)

♀ *Impatiens balsamina* L. Balsaminaceae. "Trujillo", "Trujillo amarillo". Ornamental. Said to be POISONOUS for pigs (RVM). Elsewhere, considered antidotal, astringent, cathartic, cyanogenetic, diuretic, emetic, fungicidal, laxative, refrigerant, resolvent, stomachic and vulnerary; used for cancer, caries, dysmenorrhea, dysphagia, headache, inflammation, labor, lumbago, neuralgia, POISON, polyps, snakebite, sores, stasis and wounds (DAW, RAR).

Imperata tenuis Hack. Poaceae. "Grama dulce", "Colchón quihua". For forage. When young, this plant is used to stuff mattresses. Rhizome infusion is diuretic (RVM).

Indigofera suffruticosa Mill. Fabaceae. "Añil", **"Indigo"**. Root decoction used to clean infected wounds. Seeds used by the "Aztecas" to treat urinal problems and ulcers; leaves poulticed on forehead for fever. Plant used for syphilis. Also said to be antipyretic, vulnerary, purgative, antispasmodic, diuretic, for upset stomach; favorite local remedy for epilepsy (PCS). In Ambo cream of indigo mixed with vinegar for scorpion bites (SOU).

Inga alba (Sw.) Willd. Fabaceae. "Shimbillo". Fruits edible. Good firewood.

Inga altissima Ducke. Fabaceae. "Shimbillo". Wood for rural construction.

Inga aria Benth. Fabaceae. "Shimbillo". Fruit edible; wood for firewood.

Inga bourgonii (A.) DC. Fabaceae. "Shimbillo". Fruit edible (VAG).

Inga brachyrachis Harms. Fabaceae. "Shimbillo". Wood for rural construction.

Inga ciliata C.Prest. Fabaceae. "Pairajo de altura". Fruit edible (VAG).

Inga cinnamomea Sp. ex Benth. Fabaceae. "Vaca paleta". Fruit edible (VAG).

Inga coruscans H.&B. ex Will. Fabaceae. "Shimbillo". Fruit edible (VAG).

Inga dumosa Benth. Fabaceae. "Shimbillo". As *I. alba.*

Inga edulis Mart. Fabaceae. "Guaba", "**Ice cream bean**". Cultivated fruit tree, the white pulp around the seeds eaten. (Fig. 125)

Inga feullei DC. Fabaceae. "Pacay". Cultivated fruit tree.

Inga gracilifolia Ducke. Fabaceae. "Shimbillo". Fruit edible (VAG).

Inga heterophylla Willd. Fabaceae. "Shimbillo". Fruit edible (VAG).

Inga ingoides (Rich) Willd. Fabaceae. "Guabilla". Fruit edible (VAG).

Inga killipiana F.Mcbr. Fabaceae. "Shimbillo". Fruit edible (VAG).

Inga klugii Sty. F.Mcbr. Fabaceae. "Shimbillo". Fruit edible (VAG).

Inga lallensis Spr. ex Benth. Fabaceae. "Shimbillo". Fruit edible (VAG).

Inga lateriflora Miq. Fabaceae. "Shimbillo". Fruit edible. (VAG).

Inga leiocalycina Benth. Fabaceae. "Rosario shimbillo". Fruit edible (VAG).

Inga lineata Benth. Fabaceae. "Shimbillo". Fruit edible (VAG).

Inga longipes Benth. Fabaceae. "Rosca pacae". Fruit edible (VAG).

Inga lopadenia Harms. Fabaceae. "Shimbillo". Fruit edible (VAG).

Inga macrophylla H.B.W. Fabaceae. "Pacae", "Pacay". Fruit edible (VAG).

Inga marginata Willd. Fabaceae. "Shimbillo". Fruit edible (VAG).

Inga mathewsiana P.&E. Fabaceae. "Shimbillo". Fruit edible (VAG).

Inga minutula (S.) T.E. Fabaceae. "Guabilla". Fruit edible (VAG).

Inga multijuga Benth. Fabaceae. "Tabla shimbillo". Fruit edible (VAG).

Inga myriantha P.&E. Fabaceae. "Shimbillo". Fruit edible (VAG).

Inga nobilis Willd. Fabaceae. "Yacu shimbillo". Fruit edible (VAG).

Inga obidensis Ducke. Fabaceae. "Shimbillo". Fruit edible (VAG).

Inga oerstediana Benth. Fabaceae. "Guabilla". Fruit edible (VAG).

Inga peladenia Harm. Fabaceae. "Shimbillo". Wood for rural construction.

Inga pilosula (Rich) Mcbr. Fabaceae. "Purma shimbillo". Fruit edible (VAG).

Inga plumifera Spr. ex Benth. Fabaceae. "Coto shimbillo". Fruit edible (VAG).

Inga poeppigiana Benth. Fabaceae. "Shimbillo". Fruit edible (VAG).

Inga pruriens Poepp. Fabaceae. "Huapo shimbillo". Fruit edible (VAG).

Inga punctata Willd. Fabaceae. "Shimbillo". Fruit edible (VAG).

Inga quaternata Poepp. Fabaceae. "Pairajo". Fruit edible (VAG).

Inga ruiziana G.Don. Fabaceae. "Rufindi". Fruit edible (VAG).

Inga salzmanniana Benth. Fabaceae. "Shimbillo". Fruit edible (VAG).

Inga santaremnensis Ducke. Fabaceae. "Shimbillo". Fruit edible (VAG).

Inga semialata Martius. Fabaceae. "Poroto shimbillo". Fruit edible (VAG).

Inga spectabilis (Vh) Willd. Fabaceae. "Pacae colombiano". Fruit edible (VAG).

Inga splendens Wills. Fabaceae. "Shimbillo". Fruit edible (VAG).

Inga stenocarpa Benth. Fabaceae. "Shimbillo". Fruit edible (VAG).

Inga strigillosa Spr. ex Benth. Fabaceae. "Shimbillo". Fruit edible (VAG).

Inga thibaudiana DC. Fabaceae. "Rufinde", "Shimbillo". Fruit edible (VAG).

Inga tocacheana D.Simp. Fabaceae. "Shimbillo". Fruit edible (VAG).

Inga tomentosa Benth. Fabaceae. "Shimbillo". Fruit edible (VAG).

Inga tessmannii Harms. Fabaceae. "Shimbillo". Fruit edible (VAG).

Inga umbellifera (M.Vhl) Std. Fabaceae. "Shimbillo H.menuda". Fruit edible (VAG).

Inga villosissima Benth. Fabaceae. "Shimbillo". Fruit edible (VAG).

Inga virescens Benth. Fabaceae. "Shimbillo". Fruit edible (VAG).

Ipomoea batatas (L.) Lam. Convolvulaceae. "Camote", **"Sweet potato"**. Cultivated. The tubers are edible. Various cultivars can be distinguished by: color of tubers, flowers, or buds, and shape of leaves (RVM). Elsewhere considered alterative, aphrodisiac, astringent, bactericide, demulcent, fungicide, and laxative; used for asthma, bugbites, burns, catarrh, ciguatera, diarrhea, fever, kidneys, nausea, scorpion stings, splenitis, stomach problems, and tumors (DAW).

Ipomoea carnea Jacq. spp. *fistulosa* (Roem. & Schult.) Austin. Convolvulaceae. ""Algodon bravo", Campanilla morada", "Camote caspi". Cultivated ornamental. Purgative (RAR).

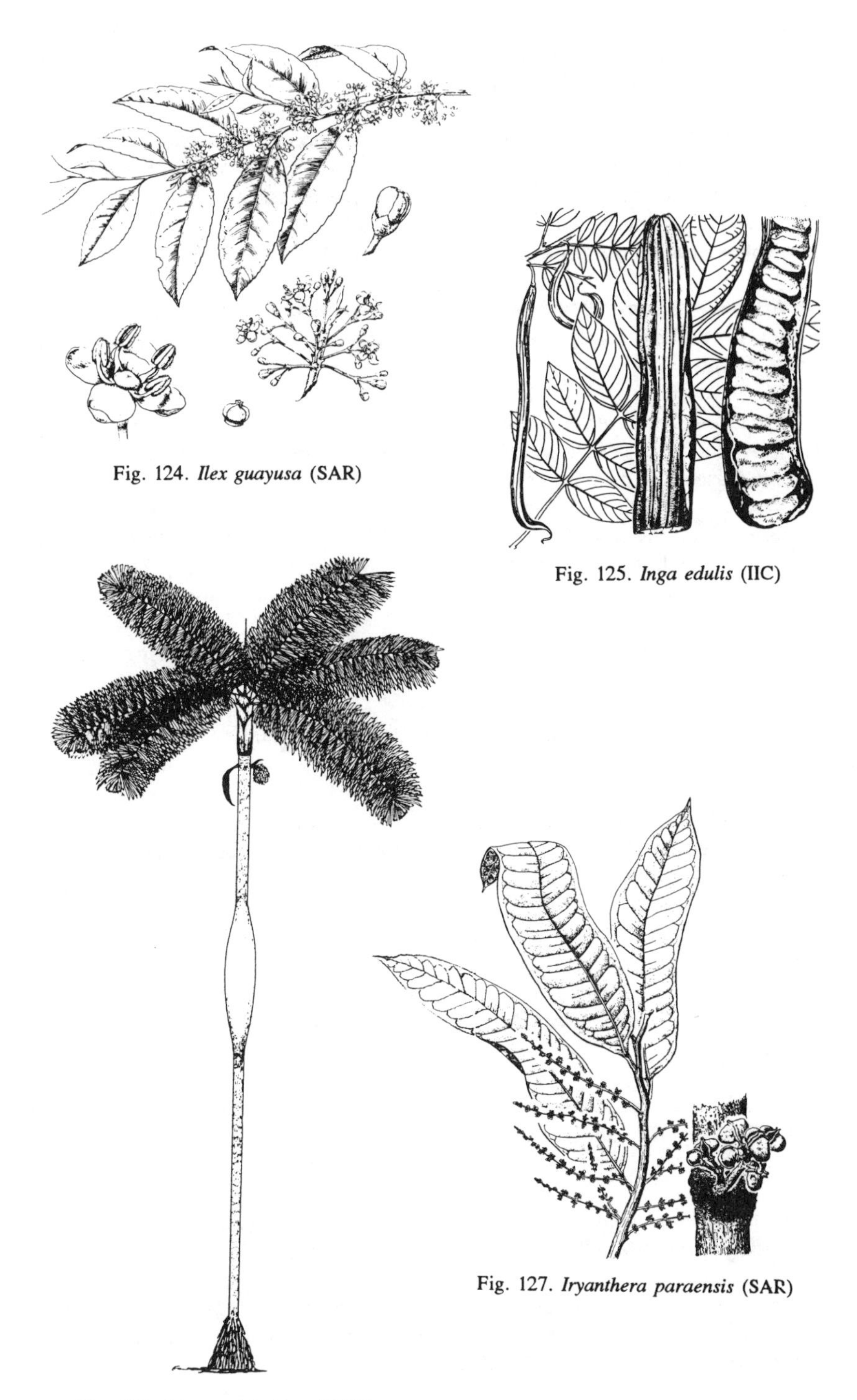

Fig. 124. *Ilex guayusa* (SAR)

Fig. 125. *Inga edulis* (IIC)

Fig. 127. *Iryanthera paraensis* (SAR)

Fig. 126. *Iriartea deltoidea* (GAV)

Ipomoea quamoclit L. Convolvulaceae. "Enredadera". Cultivated ornamental (RVM). Elsewhere, considered anodyne, cyanogenic, detergent, purgative and sternutatory, used for carbuncles, catarrh, piles, snakebites and sores (DAW).

Iriartea deltoidea R.& P. Arecaceae. "Huacrapona", "Barrigón", "Camona", "Cosho", **"Stilt palm"**. The shaft, much heavier than that of *Socratea*, is slatted in strips for the construction of floors of rural houses; for parquets, and for gutters in the sugar mills; also used as an improvised canoe (RVM). Used in some types of "Rompe Calzon", a popular aphrodisiac. (Fig. 126 and Fig. 2-Introduction)

Iriartella setigera (Mart.) H. Wendl. Arecaceae. "Ponilla". Stems used in construction as dividers, also to build movable platforms, etc. From the stems they make blowguns (MJB).

Irlbachia alatus (Aubl.) Maas. Gentianaceae. "Amaraguna", "Campanita del campo", "Tabaco bravo", "Uña de tigre". "Kubeos" take tea of the roots and leaves for stomachache caused by tainted food (SAR). "Witotos" use with *Senna* for fungal infections, sometimes adding pulverized root to coca ash. Pulverized leaves and flowers placed in clothes and bedding to discourage insects (SAR). "Chami" chew and swallow the leaves as a pain killer. Leaf decoction with lemon used for fever and colics (RVM). The "Palikur" prepare a bile remedy from the salty decoction, known for its bitterness (GMJ).

Iryanthera elliptica Ducke. Myristicaceae. "Cumala colorada". Aril consumed cooked (VAG).

Iryanthera grandis Ducke. Myristicaceae. "Cumala colorada". Wood for interiors (RVM).

Iryanthera juruensis Warb. Myristicaceae. "Cumala colorada". Wood for construction of interiors (RVM). "Puinave" and "Waorani" rub the inner bark and/or resin onto fungal infections (SAR). Aril eaten cooked (VAG).

Iryanthera laevis Markgraf. Myristicaceae. "Cumala colorada". Wood for interior construction.

Iryanthera lancifolia Ducke. Myristicaceae. "Cumala colorada". Wood for timber; arils edible.

Iryanthera paraensis Huber. Myristicaceae. "Cumala colorada". The fruit has edible arils (RVM). "Waorani" rub the inner bark and/or resin onto fungal infections and mites (SAR). (Fig. 127)

Iryanthera tessmannii Markgraf. Myristicaceae. "Cumala colorada". Wood for interior construction (RVM). Bark used for diarrhea around Iquitos (SAR).

Iryanthera tricornis Ducke. Myristicaceae. "Cumala colorada", "Pucuna caspi", "Cumala capirona". Wood for lumber. Good for firewood. From the branches they make blowguns as with other *Iryanthera* (RVM).

Iryanthera ulei Warb. Myristicaceae. "Cumala colorada". Fruit edible (RVM). Resin applied to roof of mouth for "patco" (=thrush according to some people, not thrush;

more like a sticky sore throat according to others) (SAR). "Secoya" used bark as aromatic arm bands (SAR). "Taiwano" mix bark ash with clay in pottery making (SAR).

Ischnosiphon arouna (Aubl.) Koern. Marantaceae. "Bijao". Stems used in making baskets and handicrafts, leaves used to wrap food.

Ischnosiphon obliquus (Rudge) Koern. Marantaceae. "Bijao", "Bombonaje sacha". Stems used by the "Boras" to press cassava; also used in making baskets and handicrafts, especially flower vases and in flower arrangements; leaves used to wrap food (RVM).

Ischnosiphon puberulus var. *verruculosus* (Macbr.) Anderss. Marantaceae. "Huasca bijao". Stems used in making rustic baskets (RVM).

♀ *Isertia hypoleuca* Benth. Rubiaceae. "Azar quiro". Some ranchers let it grow in the pastures to provide shade for cattle (RVM). Around Iquitos, leaf tea, with papaya leaves, used for dysmenorrhea (SAR). "Tikuna" use the bark for malaria. "Taiwano" drink hot diaphoretic leaf tea for fever (SAR). Hartwell mentions its use for tumors (JLH). Leaves contain alpha-amyrin, sitosterol and taraxasterol (SAR). (Fig. 128)

Isotoma longiflora (L.) T. Vimm. Campanulaceae. "Flor de sapo", **"Frog's flower"**. Cultivated ornamental. Considered POISONOUS for domestic animals (RVM). Mexicans use for asthma, bronchitis, epilepsy, rheumatism and venereal ailments, Chinese for cancer and snakebite (DAW).

♀ *Ixora chinensis* Lam. Rubiaceae. "Bouquet de novia", **Bride's bouquet"**. Cultivated ornamental. Chinese use for bruises, extravasation, pregnancy and stomachache (DAW).

♀ *Ixora coccinea* L. Rubiaceae. "Bouquet de novia", **"Jungle flame"**. Cultivated ornamental (RVM). Elswehere considered anodyne, antiseptic, apertif, astringent, sedative and stomachic; used for bronchitis, catarrh, diarrhea, dysentery, dysmenorrhea, fever, gonorrhea, headache, hemoptysis, hiccups, leucorrhea, and sores (DAW).

Ixora finlanysoniana Wall. Rubiaceae. "Bouquet de novia". As an ornamental.

Fig. 128. *Isertia hypoleuca* (SAR)

- J -

Jacaranda acutifolia Humb. & Bonpl. Bignoniaceae. "Arabisco", "Yaravisco". Leaf decoction, diuretic and vulnerary, used as urogenital antiseptic (FEO). Used for dermatosis, gangrene, sores (RAR).

Jacaranda copaia Aubl. ssp. *spectabilis* A. Gentry. Bignoniaceae. "Asphingo", "Chichicara caspi", "Huamanzamana", "Ishtapi", **"Jacaranda"**, "Mami rao", "Meneco", "Paravisco", "Soliman". Wood for light construction; to make furniture, pulp for paper, beams and decks (RVM). Pucallpa natives use the leaf decoction for bronchitis, fever, rheumatism (VDF). "Andoke" use crushed leaves as a cicatrizant on wounds (SAR). Rio Vaupes natives use shredded bark in teas for colds and pneumonia, the sap for skin infections (SAR). Elsewhere considered cathartic and emetic (DAW). Brazilians believe burning the leaves and bark will keep illness and mosquitoes away. Also used for sores, syphylis, and toothache (dental abscesses) (RAR). "Créoles" and "Maroons" use it for leishmaniasis (MJP). (Fig. 129)

Jacaranda macrocarpa Bur. & K. Schum. Bignoniaceae. "Solimán del monte", "Huamanzamana del varillal". Wood used for rural construction.

Jacaratia digitata (Poepp. & Endl.) Solms-Laubach. Caricaceae. "Papaya caspi", "Papaya del venado", "Shamburi", **"Tree papaya"**. Latex used as cicatrizant and vermifuge. Fruits edible cooked. Some farmers prune the plants for Coleoptera to lay eggs, so they can harvest the larvae. "Campas" use the tree as a blind, when they hunt large animals (RVM). (Fig. 130)

Jacqueshuberia loretensis Cowan. Fabaceae. "Pashaquilla colorada". Planted close to houses as an ornamental. (Fig. 131)

Jatropha curcas L. Euphorbiaceae. "Piñón", "Piñón blanco", **"Physic nut"**. Cultivated ornamental, with multiple uses among rural people. Leaf decoction piscicidal; roasted leaves poulticed on swollen infections; 1-4 raw seeds are used as a laxative, 5-10 seeds mixed with food for constipation. Some people warn that they are too POISONOUS for internal consumption (TRA). Keeping leaves in rooms is said to be healthy. Leaf decoction used to protect the color of stained wood (SOU). "Palikur" use the latex as a dental analgesic (GMJ). "Tikuna" use crushed leaves in febrifugal baths (SAR). Crushed leaves with those of *Petiveria* used to bathe aching heads (SAR). Juice from the petioles applied in pediatric gingivitis (SAR). Chopped leaves applied externally in rheumatism; latex used for earache (FEO). Brazilians take the leaf juice, mix sulfur in, and apply to streptococcus-infected wounds (erysipelas) (BDS). Seeds yield an excellent industrial oil. (Fig. 132)

Jatropha gossypifolia L. Euphorbiaceae. "Piñón negro", **"Black physic nut"**. Cultivated. Latex used as a cicatrizant for infected wounds and erysipelas (BDS). Seeds contain oil and have purgative and emetic properties. The leaf decoction is used for venereal diseases as blood purifier, and as an emetic for stomachache. The roots are used as antidote to *Hippomane mancinella* and *Guarea guara*. The latex is used for hemorrhoids and burns. The leaves are poulticed onto swellings (PEA, SOU). Leaf tea used in baths for flu in Brazil (BDS). Mashed leaves poulticed onto headache (RAR). "Créoles" use seed oil and leaf decoction as a purge; "Palikur" and "Wayãpi" use against witchcraft (GMJ).

Another example of a reputedly POISONOUS folk cancer remedy containing compounds with antitumor activity, e.g. jatrophone (CRC).

Jatropha macrantha M.Arg. Euphorbiaceae. "Huanarpo macho". Tincture of young male branches taken as aphrodisiac (FEO).

Jessenia bataua (Mart.) Burret ssp. *bataua.* Arecaceae. "Hungurahui", "Ungurahui". Fruit edible fresh or in the beverage "hungurahuina", popsicles, and lollypops. Leaves used to roof houses and to make baskets. Stem used for rural construction; cut stems provide the "suris" (or coleopterous larvae), valued as food by the Amazonians. Fruit oil used as hair tonic, liniment, and laxative (RVM). "Ashuar" use the aerial roots for hepatitis (NIC). Mesocarp fruit may contain >50% oil, physically and chemically similar to olive oil. The main component fatty acids are oleic 77.7%, palmitic 13.2%, stearic 3.2%, and linoleic 2.7% (RVM). The protein comes closer to animal protein than does soy (MJB). Blowgun darts are made from the petiole, arrowheads and bows from the trunk. "Boras" use leaves for baskets, roofs, room dividers, and chicken coops. "Guahibo" from Colombia and Venezuela use the oil for tuberculosis, asthma, cough, and other respiratory problems (RVM). "Waorani" use the adventitious roots for diarrhea, gastrosis, headache, and worms (SAR). (Fig. 133)

Juanulloa ochracea Cuatr. Solanaceae. "Azul", "Cuya cuya", "Pishco isma colorado". Mixed with other plants by "ayahuasqueros" (RVM). "Karijona" use the dried leaves of the epiphyte for earache and magic (SAR). "Siona" use to treat wounds (SAR).

Justicia brandegeana Wasshausen & L.B.Smith. Acanthaceae. "Camaroncillo". Cultivated ornamental.

Justicia pectoralis Jacq. Acanthaceae. "Lluichu lancetilla", "Pinipisa" "Yoman rao". "Boras" use it for antipyretic baths (DAT). Putumayo natives probably add this to the bark of *Virola*; the natives from Orinoco (Venezuela), and from Rio Negro (Brazil) use the leaves of var. *stenophylla* as an additive to powdered *Virola* for inhaling. (contains N, N-dimethyltryptamine, betaine, coumarin, umbelliferone) (SAR). "Créoles" apply leaf maceration externally to hematomas; leaf infusion bechic and pectoral. "Wayãpi" use decoction for stomachache (GMJ). Pucallpa natives use the decoction for fever, gastritis and inflammation (VDF). Amazonian Colombians value the decoction for pneumonia (SAR). Leaf and stem vapors inhaled for fever, headache, and pain (RAR). Listed as aromatic, aphrodisiac, narcotic (RAR). (Fig. 134)

Fig. 129. *Jacaranda copaia* (GAV)

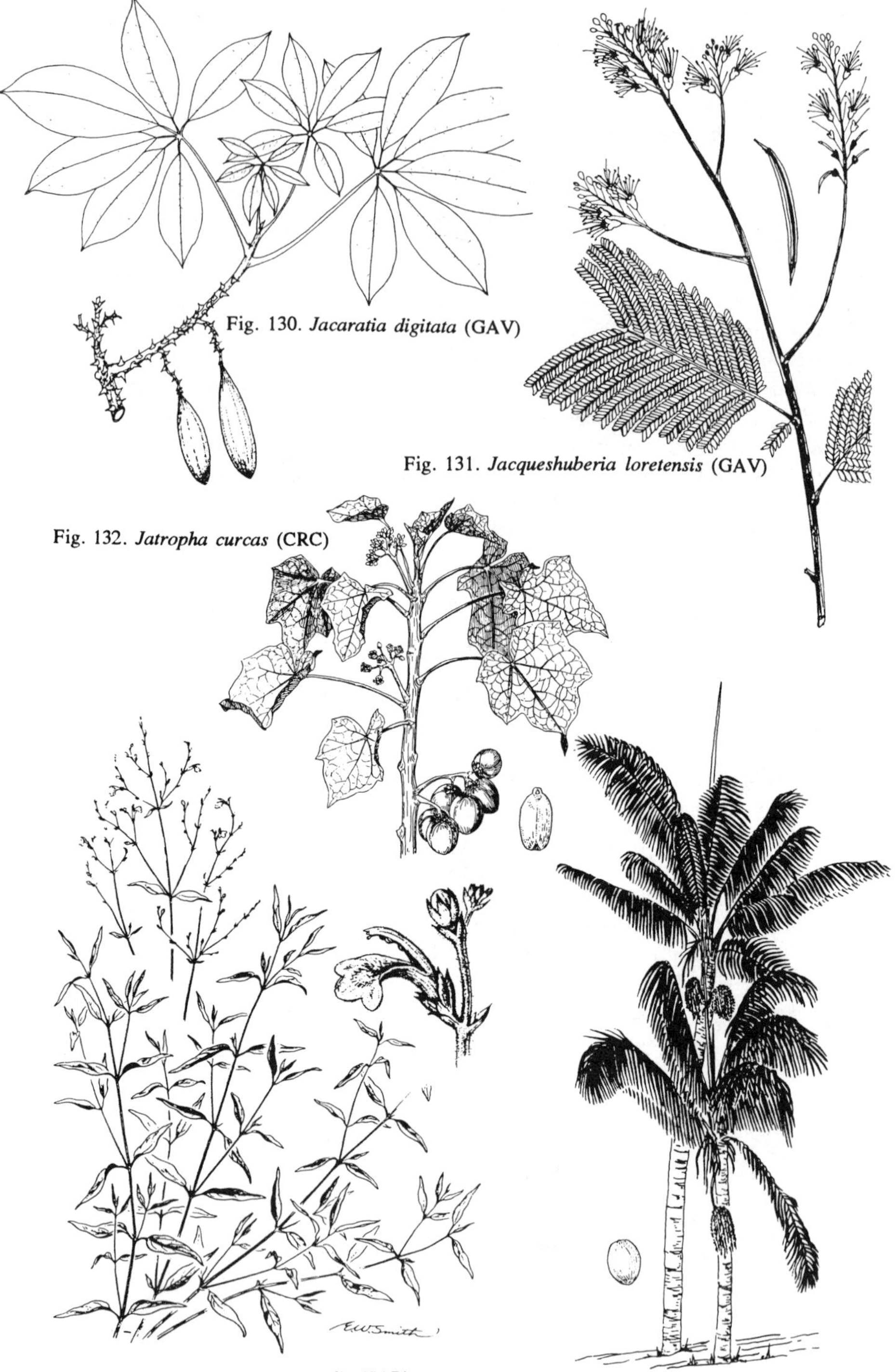

Fig. 130. *Jacaratia digitata* (GAV)

Fig. 131. *Jacqueshuberia loretensis* (GAV)

Fig. 132. *Jatropha curcas* (CRC)

Fig. 134. *Justicia pectoralis* (SAR)

Fig. 133. *Jessenia bataua* (CRC)

- K -

Kalanchoe pinnata (Lam.) Pers. Crassulaceae. "Hoja de aire", **"Air plant"**, "Paichecara". Cultivated ornamental and medicinal plant. Crushed leaves mixed with aguardiente for fever and headache. Crushed stems and leaves are soaked in water and left outside overnight; the next morning they drink this water for heartburn and internal fever. This same mix, with wet starch, is used for urinal tract inflammation (urethritis). A few drops of the extracted juice from the fresh leaves mixed with maternal milk is used for earache. "Créoles" use the lightly roasted leaves for mycosis and inflammations. The infusion of the fresh or dried plant is well known as an antipyretic. "Palikur" mix leaf juice with oil from coconut or *Carapa* to rub on head for migraines (GMJ). Leaves contain malic, citric, and isocitric acid, as well as rutin and quercetin. The leaf extract is active against bacteria (gram positive), because of bryophylline. Mashed and macerated leaves are poulticed onto headaches; the juice with a pinch of salt is used for bronchitis, and to cicatrize ulcers or sores, and to clear eye irritations (RVM). "Siona" apply heated leaves to boils (SAR). Rio Pastaza natives used leaf tea for broken bones and internal bruises (SAR). Peruvians drink the decoction for intestinal upsets (SAR). (Fig. 135)

Fig. 135. *Kalanchoe pinnata* (GAV)

- L -

Lablab purpureus (L.) Sweet. Fabaceae. "Zarandeja", **"Lablab bean"**. Seeds edible cooked (VAG).

Lacistema aggregatum (Berg) Rusby Flacourtiaceae. "Trompo huayo". Wood for house construction; for malaria, rheumatism (RAR).

Lacmellea arborescens (M.A.) Monachino. var. *peruviana* (V.Hev. & M.Arg.) Monachino. Apocynaceae. "Chicle huayo", **"Chicle"**. Fruit edible (RVM).

Lacmellea floribunda (Poepp.) Benth. Apocynaceae. "Chicle huayo", **"Chicle"**. Fruit edible (RVM).

Lacmellea lactescens (Kuhlmann) Markgraf. Apocynaceae. "Chicle huayo". Fruit edible (RVM). When lacking coca, "Bora" and "Witoto" mix the toasted powdered leaves with *Cecropia* ashes as a substitute (SAR). (Fig. 136)

Lactuca sativa L. Asteraceae. "Lechuga", **"Lettuce"**. Cultivated. As garden vegetable.

Lacunaria sp. Quiinaceae. "Sacha guayaba". Fruit edible (VAG).

Ladenbergia magnifolia (R.&P.) Kl. Rubiaceae. "Cascarilla", "Cascarilla verde". Bark infusion taken for malaria.

Lagenaria siceraria (Mol.) Standley. Cucurbitaceae. "Calabaza", "Poto-pate", **"White-flowered gourd"**. Cultivated. Fruit edible (RAR). The shell of the fruit is used as a container, and to make musical instruments (RVM). For nephritis (RAR). (Fig. 137)

Lagerstroemia indica L. Lythraceae. "Locura", **"Crepe myrtle"**. Cultivated ornamental.

♀ *Lantana camara* L. Verbenaceae. "Aya albaca", "Tunchi albaca", "Hierba de la maestranza", **"Yellow sage"**. The vapors from the decoction of the leaves mixed with *Ocimum micranthum* are used to clear respiratory passageways. The leaf decoction, mixed with onion, garlic, and bee honey, for cough and bronchitis. Dried leaves used to cleanse. Suggested for arterial hypertension. Root antiasthmatic and pectoral. (Decoction of 10 g root with 250 g water, or a mixture of 40 g root in 800 g water with enough sugar to obtain syrup.) Leaf decoction used to treat rheumatism and as a stomach tonic (SOU). Leaf decoction used for upset stomach and colds. In El Salvador, leaf decoction used to reduce fever, roots as blood purifiers, and for hepatoses. The flowers and roots are used as expectorant, and to treat bronchitis. The infusion of the entire plant is an emmenagogue, and an antiseptic (GAB, PCS). Used in Brazil as an antirheumatic in baths, emmenagogue, and diaphoretic. It is popular to rub anemic children with it. In Pedra Azul, Brazil, recommended for cough. A good antipyretic because it contains lantanine (considered a substitute for quinine), which is effective, especially when quinine does not work. Leaf decoction used for rheumatism. Vapors are useful with mycoses. People from Santa Catarina, Brazil, consider the leaf infusion tonic, sudorific, expectorant, and emollient, easing pain of the respiratory passageways. "Exumas" use boiled leaves to sooth the itch caused by measles and chickenpox; drinking the tea it said to remove spots from the skin. The "Cuna" use the entire plant in cold water baths for anemic children (RVM). "Créoles" use the leaf infusion, alone or mixed with other plants. "Wayãpi" use it for sedative baths, or to make tea; the "Palikur" mix it with the leaves of *Hedychium coronarium,* in baths, and to make tea. The decoction constitutes a specific antipyretic for infants (GMJ).

♀ *Lantana canescens* H.B.K. Verbenaceae. "Canirca". Peruvians consider digestive, emmenagogue (RAR); Brazilians use the leaf tea for stomachache (BDS).

♀ *Lantana trifolia* L. Verbenaceae. "Aya machana", "Lauraimena", "Siete colores","Tunchi albaca", "Yona rao". Though its fruits are very small, they are eaten by children. Leaf decoction used for colds and bronchial diseases. Infusion used in antipyretic baths around Pucallpa (VDF), and to relieve headaches; the alcoholic maceration of the leaves sweetened with bee honey is an expectorant. In Palmira (Valle), women use the decoction of the flowers and leaves to regulate menstrual periods, drinking one portion a day; roots famous as antibechics; flowers make a good uterine tonic (GAB). Natives of Colgedes use the infusion of the roots and flowers as cleansing baths for bad luck. When the body temperature drops suddenly, they leave the infusion outside overnight in order to obtain a stronger curative effect. "Cuna" use the infusion for mycosis between the toes, the root decoction for oral hemorrhages (RVM). "Jivaro" chew the leaves to blacken the teeth (SAR).

Laportea aestuans (L.) Chew. Urticaceae. "Ishanga blanca", **"White nettle"**. Commonly used to relieve rheumatic pains, and to whip children when they misbehave. Used by the "Créoles" as a diuretic (GMJ). Elsewhere used for burns, constipation, dysentery, rickets, and wounds (DAW).

Lecythis chartacea Berg. Lecythidaceae. "Cachimbo", "Machimango colorado". Wood used for posts, poles, and firewood.

Lecythis peruviana L.O.Williams. Lecythidaceae. "Machimango colorado". Lumber, fiber and ropes.

Lecythis pisonis Cambs. Lecythidaceae. "Castaña de monte". Wood for heavy construction, posts,and forked poles. Seeds edible. (Fig. 138)

Leonia crassa Smith & Fernandez. Violaceae. "Tamara blanca". Ripe fruits used as a fishbait.

Leonia glycicarpa R.& P. Violaceae. "Tamara", "Nina caspi". The fresh leaves left in the sun or lightly warmed, are poulticed, as an emollient, for abscesses, tumors, and phlegm. Fruit pulp used for hemorrhoids, seeds for pulmonary diseases (RVM).

Lepidocaryum tessmannii Burret. Arecaceae. "Irapay", **"Thatch palm"**. Leaves used to make "crisnejas", (a bunch of leaves interweaved and tied to a slat of *Socratea exorrhiza),* which are used to roof the houses; its use is so common as to be commercialized around Iquitos. Said to last 5-6 years (cf 1 year for *Carludovica*). Took two men ca two months to do the 500,000 fronds for the roof of the hammock house (upper) at Explorama. *L. gracile* is used by the "Boras" to clear eye infections; for this they roast the stems to soften, and the juice is squeezed in the eyes; as good as an antibiotic (MJB). (Fig.139 and Fig. 2-Introduction)

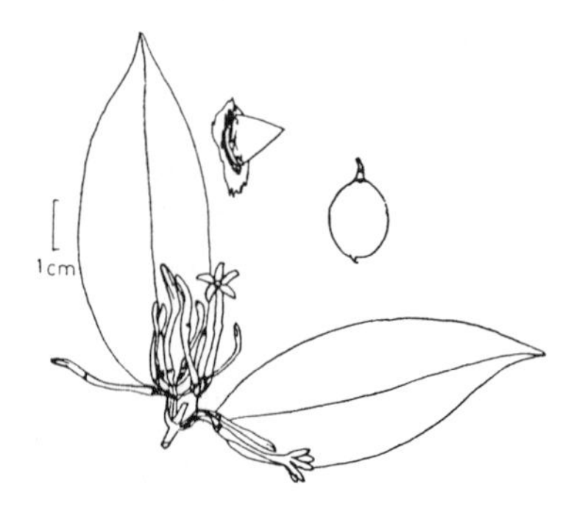

Fig. 136. *Lacmellea lactescens* (GAV)

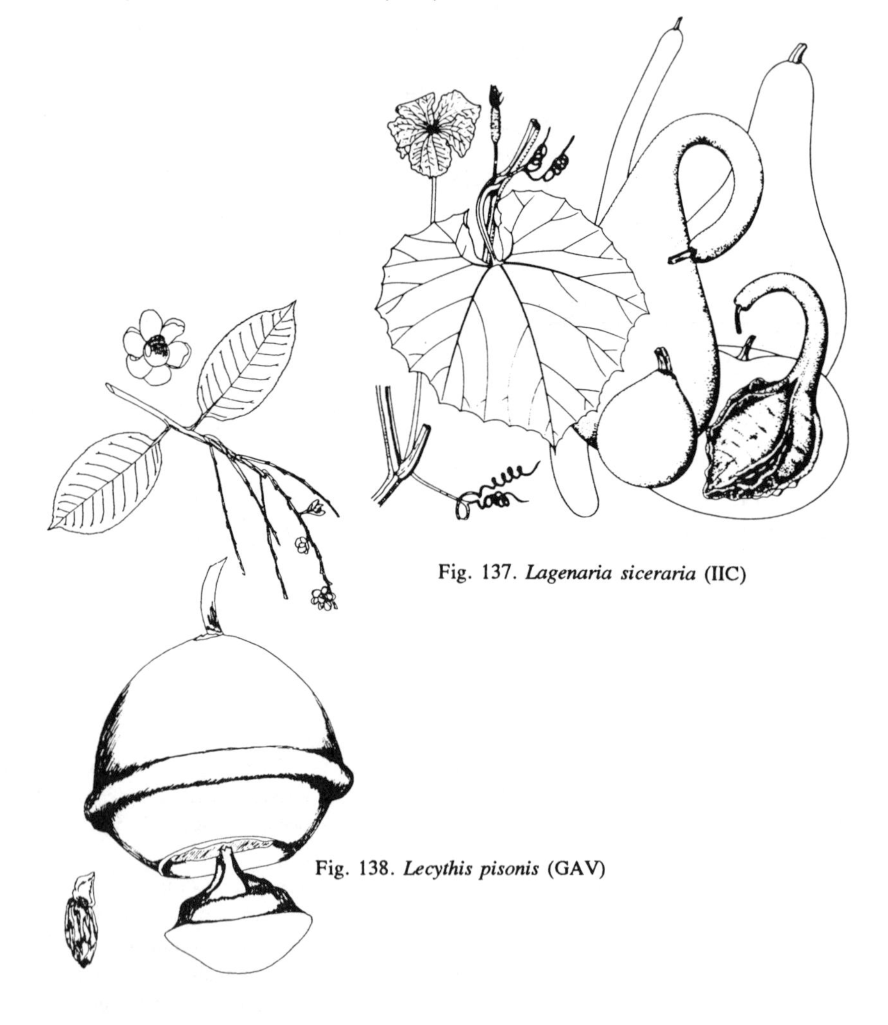

Fig. 137. *Lagenaria siceraria* (IIC)

Fig. 138. *Lecythis pisonis* (GAV)

Leptochloa dubia Nees. Poaceae. "Nudillo", **"Green sprangle top"**. Forage grass.

Leptochloa filiformis (Lam.) Beauv. Poaceae. "Grama dulce", "Nudillo", **"Red sprangle top"**. Forage grass. (Fig. 140)

Leptochloa panicoides (Presl) Hitchc. Poaceae. "Grama". Forage grass.

Leptochloa scabra Nees. Poaceae. "Ucsha gramalote". Forage grass.

Leptochloa uninervia (Presl) H.&C. Poaceae. "Nudillo". Forage grass.

Leucaena leucocephala (Lam.) de Wit. Fabaceae. "Leucaena", "Pashaquilla", **"White popinac"**. Cultivated ornamental, fixing nitrogen. Contains the depilatory mimosine. (Fig. 141)

Licania apetala (E. Meyer) Fritsch. Chrysobalanaceae. "Apacharama". Wood for posts, forked poles, and dormers.

Licania britteniana Fritsch. Chrysobalanaceae. "Apacharama", "Nina caspi". Wood for posts, forked poles, and dormers.

Licania caudata Prance. Chrysobalanaceae. "Parinari", "Apacharama". Wood for posts, forked poles, dormers, beams, and decks.

Licania heteromorpha Benth. var. *heteromorpha.* Chrysobalanaceae. "Apacharama", "Casharana". Wood for forked poles, beams, and decks.

Licania lata Macb. Chrysobalanaceae. "Apacharama", "Tinaja caspi". Wood for forked poles, beams, decks; the ashes of the bark are added to clays used in ceramics (RVM).

Licania macrocarpa Cuatr. Chrysobalanaceae. "Parinari". Fruit edible.

Licania unguiculata Prance. Chrysobalanaceae. "Apacharama", "Parinari colorado". Wood for jam posts for bridges, dormers, parquets, pillars, and forked poles (RVM).

Licaria canella (Meiss.) Kost. Lauraceae. "Muena", "Moena". Wood for carpentry.

Licaria triandra (Sw) Kost. Lauraceae. "Canela moena". Wood for carpentry and canoes.

Limnobium laevigatum (Willd.) Heine. Hydrocharitaceae. "Amanso". As an ornamental plant.

Limnocharis flava Buch. Butomaceae. "Cuchara panga". Sometimes cultivated in ponds as an ornamental.

Lindackeria paludosa (Benth.) Gilg. Flacourtiaceae. "Huacapusillo", "Casha huayo", "Iluicho caspi". Wood for forked poles in rural construction.

♀ *Lindernia crustacea* (L.) Muell. Scrophulariaceae. "Aretillo", "Llama plata". The "Cuna" boil the roots, drinking the decoction twice a day to eliminate intestinal parasites, especially worms. Leaf infusion used for rashes in children (RVM). "Créoles" prepare an infusion for albuminuria; the "Palikur" use the plant decoction as an antipyretic (GMJ). Considered anthelmintic, antibilious, diuretic, emetic, emmenagogue (RAR).

Lindsaea divaricata Kl. Dennstaedtiaceae. "Cilantrilla". Ornamental.

Lindsaea lancea (L.) Bedd. var. *falcata* (Dry) Rosent. Dennstaedtiaceae. "Yarinilla", "Culantrillo". Planted as an ornamental. Mashed leaves rubbed on faces of snakebite victims (RVM).

Lindsaea latifronds Kramer. Dennstaedtiaceae. "Culantrilla". Planted as an ornamental.

Lippia alba (Mill.) N.E.Br. Verbenaceae. "Pampa orégano". Curanderos mix with other plants to bathe patients during magic rituals; also used to relieve vomiting and upset stomach (RVM). "Créoles" use the leaf infusion with sugar to sooth cardiac pain. The leaf decoction is relaxant and soporific (GMJ). "Tikuna" wash headache with the crushed leaves in water (SAR). Mixed with *Mentha*, leaves are used for diarrhea (SAR). Brazilians use the leaf tea for stomachache (BDS).

Livistona chinensis (Jacq.) R.Br. Arecaceae. "**Fountain palm**". Cultivated ornamental.

Lonchocarpus nicou (Aubl.) DC. Fabaceae. "Barbasco", "Cubé", "**Rotenone**". Semicultivated. Even though fishing with barbasco or other ichthyotoxics is forbidden, this plant is still being used in places (RVM). Brazil's "Timbo", at 3 ppm, eliminates piranha and their eggs in 15 minutes (MJB). "Ketchwa" and "Shuar" use in arrow POISONs (SAR). Brazilians use *L. urucu* to kill leaf cutters (SAR).

Lonicera japonica Thunb. Caprifoliaceae. "Madre selva". Cultivated ornamental.

Loreya arborescens (Aubl.) DC. Melastomataceae. "Sacha nisperillo". Fruit edible (VAG).

Lucuma macrocarpa Hub. Sapotaceae. "Lucma", "Caimito brasilero". Cultivated fruit tree.

Ludwigia hyssopifolia (G.Don) Exell. Onagraceae. "Chirapa sacha". "Palikur" use for fever (GMJ).

♀ *Luffa operculata* (L.) Cogn. Cucurbitaceae. "Espongilla", "**Sponge gourd**". Dry fruit "skeleton", with a sponge consistency, used for cellulitis, etc. Fruit mixed with *Jatropha curcas* for sinusitis (RVM). Brazilians use the purgative fruit pulp for dropsy (SAR), massaging rheumatism with bits of fruit in andiroba oil. Fruit tea somewhat POISONOUS, ingested for rheumatism (BDS). Considered abortifacient (RAR). Contains luffanine (SAR).

Luziola subintegra Sw. Poaceae. "Grama". Forage grass especially for water buffalos.

Lycopersicum esculentum Mill. Solanaceae. "Tomate", "**Tomato**". Cultivated vegetable.

♀ *Lycopodium cernuum* L. Lycopodiaceae. "Shapumba", "Helecho", "Licopodio", "**Club moss**". Used with *Selaginella* as greenery for Christmas nativity displays. Some pharmacologists recommend it for bladder diseases (SOU). "Créoles" use it against bad spirits; "Palikur" use the decoction for antipyretic baths, and poisonous spider bites. In Surinam, lycopods are used as a substitute for tobacco (GMJ). (Some contain nicotine JAD). In some places they use this plant to stuff pillows. The abundant spores are used in pharmacy to coat pills and condoms, and can be used as baby powders. Used as a diuretic for gonorrhea and leucorrhea. The decoction is used for dysentery, and in baths for arthritic or gouty tumors. Pollen is used as carminative (RVM).

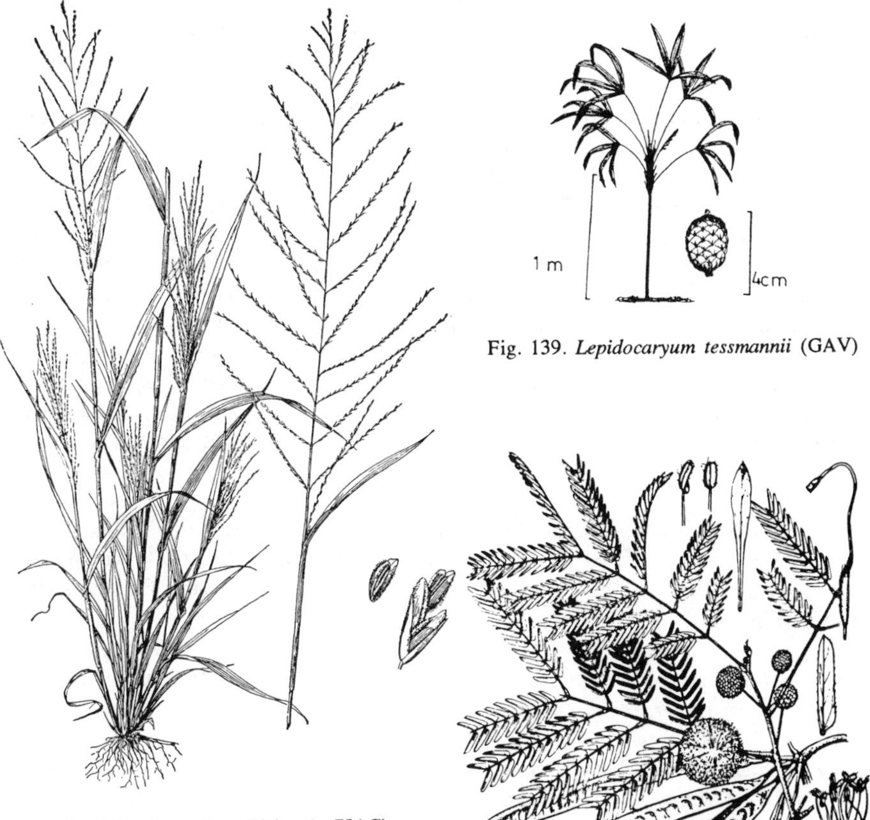

Fig. 139. *Lepidocaryum tessmannii* (GAV)

Fig. 140. *Leptochloa filiformis* (HAC)

Fig. 141. *Leucaena leucocephala* (JAD)

- M -

Mabea nitida Spruce ex Benth. Euphorbiaceae. "Shiringuilla". Wood for beams, decks, columns, and charcoal (RVM). "Kubeo" rub seed oil into scalp to prevent hair loss (SAR).

Macfadyena uncata (Andrews) Sprague & Sandw. Bignoniaceae. "Uña de gato", **"Cat's claw"**. "Créoles" use leaf infusion as flu preventive (GMJ).

Macfadyena unguis-cati (L.) A. Gentry. Bignoniaceae. "Cashi tae", "Garra de Murcielago", "Uña de gato", **"Cat's claw"**. Around Pucallpa, natives apply leaves to temples for headache and the fruit infusion for arthritis and rheumatism (VDF). "Créoles" value the leaf infusion as flu preventive. "Wayãpi" use plant in antipyretic baths. "Palikur" use plant as cough suppressant preparing a decoction with the bark of *Tabebuia serratifolia*, and honey (GMJ). Seed oil contains 15% vaccenic acid (JBH).

Machaerium floribundum Benth. Fabaceae. "Uñe-gato". Stem decoction used for diarrhea; sap poulticed onto wounds and indolent sores (RVM).

Maclura tinctoria (L.) Gaud. Moraceae. "Insira", "Insira amarilla". Fruits edible. Wood occasionally used in carpentry. Cotton soaked in the latex is used to relieve toothaches. An olive green dye is derived from the plant. Because it contains phloroglucin and gallic acid, it is probably antiseptic and astringent. Moringin is also antiseptic (AYA). This species also works as diuretic and anti-venereal. Highly recommended for urinary infections like blennorrhea. Colombians soak latex in 'cotton' of *Ochroma pyramidale* or *Ceiba samauma*, using it as a filling. Latex removes teeth, whether carious or healthy, without pain and bleeding (NIC). Used by the "Chami" for lumber.Considered analgesic, diuretic, purgative; used for cough, gout, pharyngitis, rheumatism, sore throat, syphilis (RAR).

Macoubea guianensis Aubl. Apocynaceae. "Loro micuna", "Jarabe huayo". Timber. Tree used as a blind when hunting game birds (RVM). Fruit pulp edible (SAR). Latex serves as chewing gum (SAR). Amazonian Brazilians use the latex for lung ailments (SAR). (Fig. 142)

Macoubea witotorum R. E. Schultes. Apocynaceae. "Amapa". Fruit edible (RVM).

Macrolobium acaciaefolium (Benth.) Benth. Fabaceae. "Aripari", "Chavapallana" "Faveira" , "Pasha quilla", "Pashaco colorado", "Yacu pashaco", "Plata pashaco". Wood used for canoes, plywood. Bark sometimes used to treat diarrhea (RVM). "Tikuna" dust ulcerated wounds with powdered leaves (SAR).

Malachra alceifolia Jacq. Malvaceae. "Malva". Leaves used for nephritis and stomachache (RVM). (Fig. 143)

Malachra capitata L. Malvaceae. "Malva", "Marica". Cultivated. Crushed leaves, left in water outside overnight, used for stomachache and kidney inflammation (RVM). Used around Pucallpa for fever and headache (VDF). "Tikuna" use leaf decoction for colds, fever and stomachache (SAR).

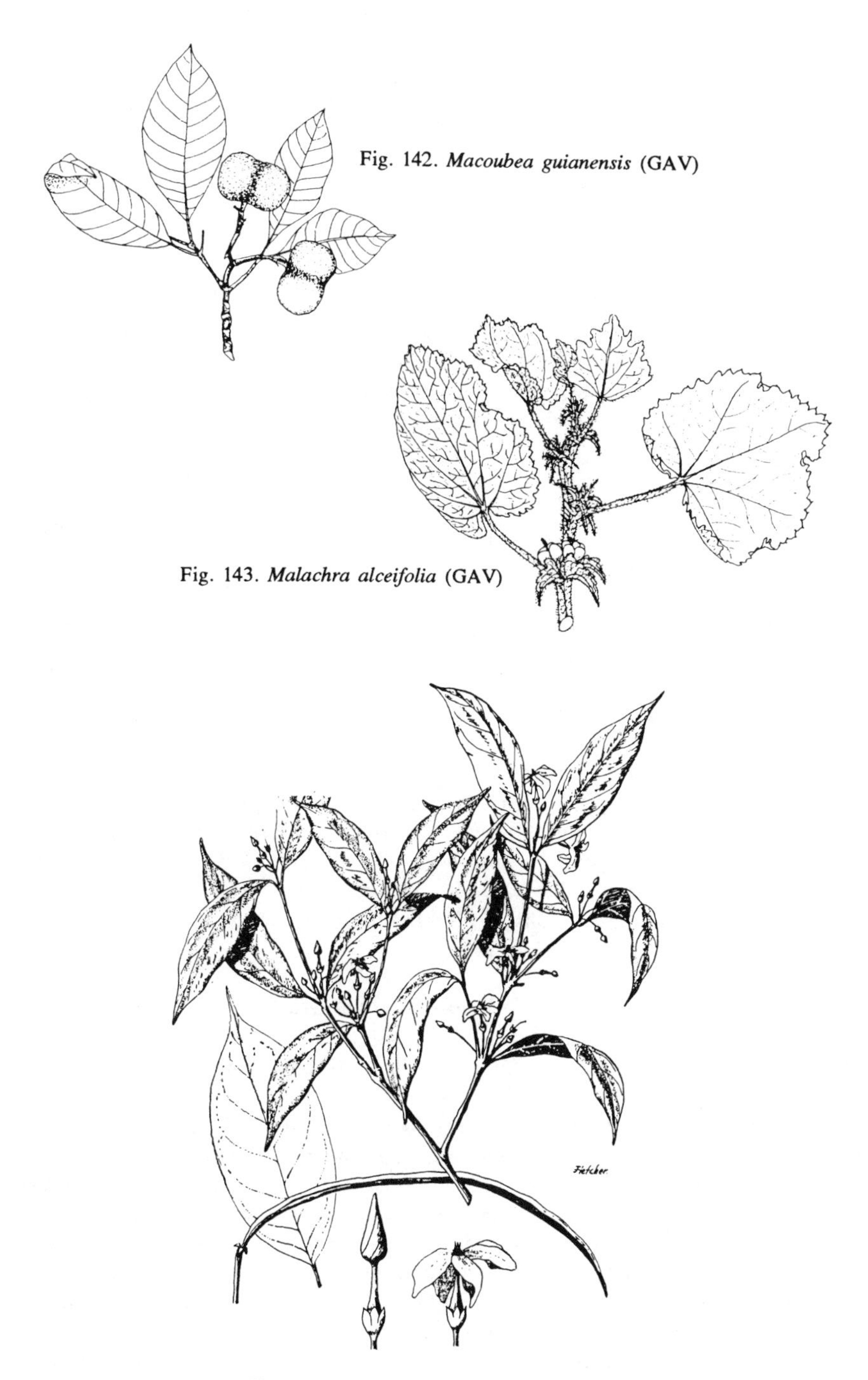

Fig. 142. *Macoubea guianensis* (GAV)

Fig. 143. *Malachra alceifolia* (GAV)

Fig. 144. *Malouetia tamaquarina* (SAR)

Malmea diclina R.E.Fr. Annonaceae. "Espintana". Wood used for beams and decks.

Malouetia tamaquarina (Aubl.) A. DC. Apocynaceae. "Chicle caspi", "**Chicle**". An additive to ayahuasca (RVM). "Kubeo", "Puinave" and "Tikuna" paint the vulnerary latex onto wounds (SAR). Fruits fed to certain birds ("pajuil") render the bones POISONOUS to dogs (SAR). (Fig. 144)

Malpighia glabra L. Malpighiaceae. "Acerola", Cereso", "**Barbados Cherry**". Cultivated fruit, very high in vitamin C.

Malvaviscus penduliflorus Cav. Malvaceae. "Cucarda caspi", "Malvavisco", "Piñón ceqeati", "**Pepper hibiscus**". Cultivated ornamental. Around Pucallpa, used for earache (VDF). In Brazil used to attenuate noise (RVM). Elsewhere, used for amygdalitis, aphthae, diarrhea, dysentery, fever and lung ailments (DAW).

Mammea americana L. Clusiaceae. "Mamey", "**Mamee apple**". Fruit edible (JAD). In Amazonian Brazil, latex, bark and/or fruit pulp are used for bugbites and parasitic infections (SAR). Seeds considered antieczemic, febrifuge, insecticide, parasiticide, vermifuge (RAR). (Fig. 145)

Manettia divaricata Wernham. Rubiaceae. "Yanamuco". "Jivaro" use leaves to coat their teeth to prevent cavities (SAR).

Manettia glandulosa P.& E. Rubiaceae. "Yanamuco". Pucallpa natives chew leaves to coat teeth and prevent cavities.

♀ *Mangifera indica* L. Anacardiaceae. "Magua", "**Mango**", "Mango inherto", "Mango chico-rico", "Mangua dulce". Cultivated ornamental fruit tree, providing shade (RVM). "Tikuna" take a cupful of leaf decoction on two successive days during menstruation as a contraceptive, three days as an abortifacient (SAR). Leaves reportedly antiviral (SAR). Flower infusion considered antiasthmatic, antitussive, and expectorant (FEO). Mangiferin is antiseptic (SAR). (Fig. 146)

Manicaria saccifera Gaestn. Arecaceae. "Yarinilla", "**Monkey-cap palm**". Immature seeds eaten. In Panama spathes used as hats; leaves for thatch. Fruit eaten by pigs and other frugivorous mammals. Juice used for colds and asthma (JAD). (Fig. 147)

♀ *Manihot esculenta* Crantz. Euphorbiaceae. "Cassava", "Mandioca", "Yuca". Cultivated. Many cultivars are morphologically different, and vary in cyanide content. Some are quite POISONOUS (JAD)! The edible roots yield farina, tapioca, and starch. Roots are used cooked, fried, roasted, and in other culinary applications. Also used to make the popular alcoholic refreshments, "mazatto", and "beshu", as well as a gelatinous beverage. Only "cassaba brava", is used to make farina. A poultice of cassava mixed with aguardiente, is used for chills and fever (RVM). "Créoles" apply to a child's body a mixture of starch and rum to relieve cutaneous eruptions. "Wayãpi" use leaves as a "remedy against the arrow", also in hemostatic poultice. They use root juice in ritual baths to treat sterility in women. The "Palikur" use the starch in poultice soaked in oil of *Carapa* sp. for tender muscles (GMJ). "Makuna" use the yuca water to treat scabies (SAR). "Witoto" used the leachings from cyanidiferous yuca as a fish POISON (SAR). A cupful of sweet squeezings is given for diarrhea (SAR).

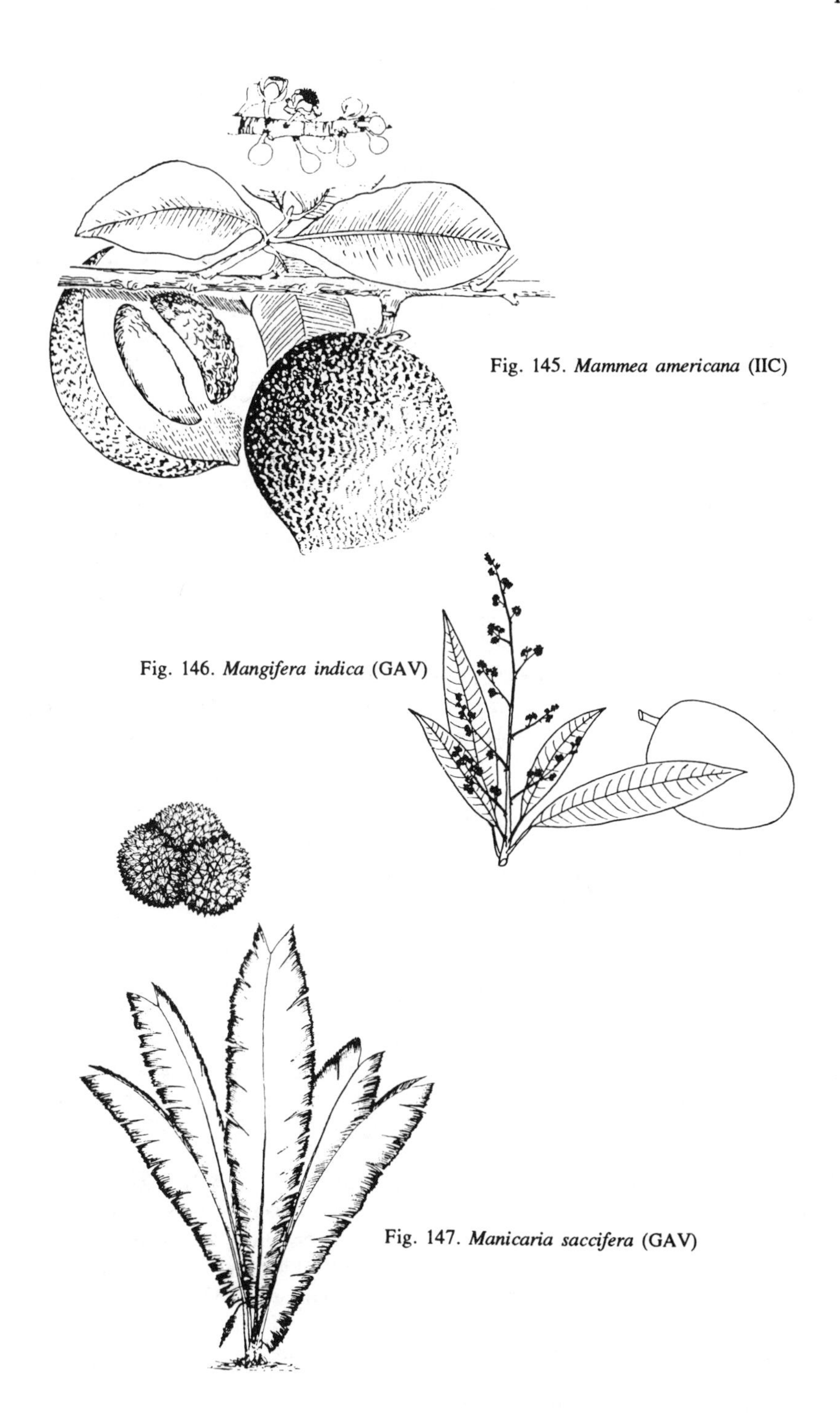

Fig. 145. *Mammea americana* (IIC)

Fig. 146. *Mangifera indica* (GAV)

Fig. 147. *Manicaria saccifera* (GAV)

Manilkara bidentata (A.DC.) A.Chew. Sapotaceae. "Quinilla". Wood for posts, forked poles, dormers, and parquets. The latex is used in handicrafts to make toys (RVM). Used for kidney stones (RAR).

Manilkara zapota (L.) Van Royen. Sapotaceae. "Caimito brasilero", "Sapotilla". Cultivated fruit tree. (Fig. 148)

Mansoa alliacea (Lam.) A.Gentry. Bignoniaceae. "Ajo sacha", "Boens", "Nia boens", **"Wild garlic"**. Alcoholic maceration of the stem and roots used for rheumatism; leaf infusion used in baths to relieve "manchiari" (a nervous state caused by terror or sudden shock), especially in children. Also used as cleansing baths for bad luck. "Achuales" use the roots as antirheumatic (RVM). "Créoles" use the stem decoction in baths, to relieve fatigue and small needle-like cramps. "Palikur" use it to protect themselves against the bad spirits (shades of Dracula?). "Wayãpi" use the decoction of leaves and stems as antipyretic baths (GMJ), Tapajos natives for body aches, flu (BDS). Contains alline, allicin, allyl-disulfoxide, diallyl sulfide, dimethyl sulfide, divinyl sulfide, propylallyl disulfide (AYA), and two cytotoxic naphthoquinones, 9-methoxy-alpha-lapachone and 4-hydroxy-9 methoxy-alpha-lapachone (Phytochemistry 31(3):1061. 1992).

Mansoa hymenaea (DC.) A.Gentry. Bignoniaceae. "Ajo sacha macho". As with *M. alliacea.* Tocache natives use infusions for tuberculosis and rheumatism.

Mansoa standleyi (Steyerm). A.Gentry. Bignoniaceae. "Ajo sacha". As *M. alliacea* (RVM). "Tikuna" use the emetic crushed leaf decoction for fever and headache, "Waorani" for arthritis, fever and myalgia (SAR). In Piura, the branch infusion used for inflammation and tumors (FEO).

Maprounea guianensis Aubl. Euphorbiaceae. "Airana", "Machinguilla". Decoction of leaves and bark used by "Créoles" for leg rash. "Wayãpi" prepare an antidiarrheic decoction. "Palikur" grind the bark and apply it to the umbilical cord of the newborn baby to hasten healing (GMJ).

Maquira calophylla (Poepp. & Endl.) C.C.Berg. Moraceae. "Chimicua colorada". Saw timber (RVM). Latex considered caustic and POISONOUS (SAR).

Maquira coriacea (Karst.) C.Berg. Moraceae. "Capinuri", "Capinuri del bajo". The wood is mainly used for plywood veneer, in lumbers. Latex poulticed onto hernias, luxations, lumbar pain, etc. Some of the spur branches, naturally pruned, are much more phallic in appearance than the one illustrated by Vasquez. (Fig. 149)

Maranta arundinacea L. Marantaceae. "Shimi pampana", **"Arrowroot"**. Cultivated. Rhizomes edible, used to make starch and love potions. Easy to digest, the starch is used in convalescent children's diet. Dominicans once used the mashed rhizome to treat wounds caused by arrows (SOU). Flour used in poultice, and for acid indigestion (GMJ). Used for asthenia, gall bladder, fever, sprains, urethritis (RAR). (Fig. 150)

Maranta ruiziana L. Marantaceae. "Maaihiiba". Cultivated. Rhizome edible.

Martinella obovata (HBK) Bur. & K. Schum. Bignoniaceae. "Yuquilla". Drops of root juice used for conjunctivitis, or other eye irritations. Also used to clean chronic wounds (RVM). The natives from Pichis-Palcazu use the roots as vermifuge. Its flocculant

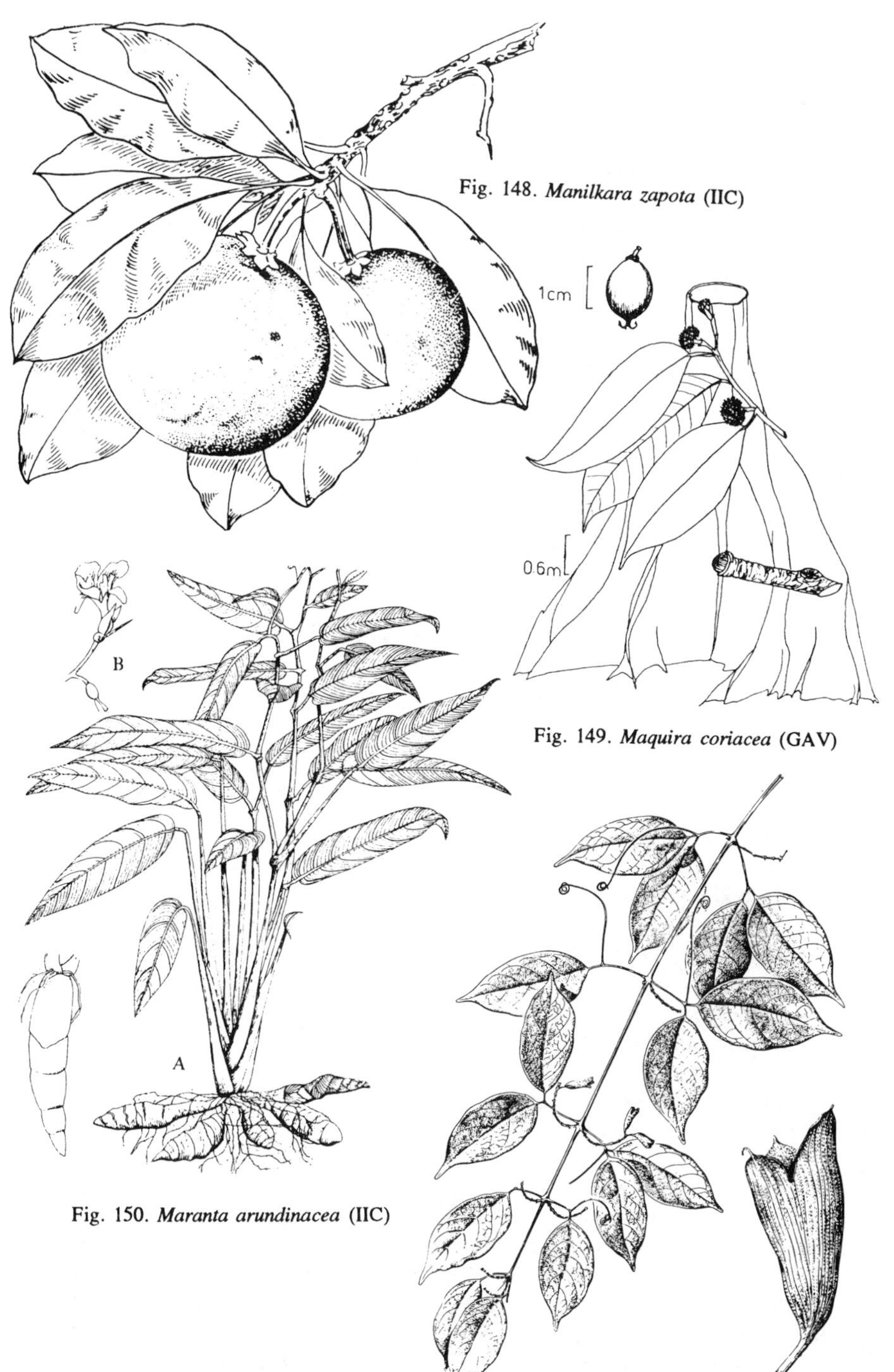

Fig. 148. *Manilkara zapota* (IIC)

Fig. 149. *Maquira coriacea* (GAV)

Fig. 150. *Maranta arundinacea* (IIC)

Fig. 151. *Martinella obovata* (SAR)

activity may help render water potable. Bark of this species with leaves of *Ambelania lopezii* and bark of *Distictella racemosa* are used by Colombia's "Barasana" Indians as ingredients in their arrow POISON. Bark infusion makes an effective but dangerous antipyretic (SAR). "Candoshi" use the root sap, Vaupes natives the fruits, for eye infections (SAR). (Fig. 151)

Mauritia carana Wallace. Arecaceae. "Aguaje del varillal". Fruit edible (RVM).

Mauritia flexuosa L.f. Arecaceae. "Aguaje", "Moriche", "Buriti", "**Moriche palm**". Unripe fruit edible fresh or in beverages (aguajina), popsicles (frozen sugared fruit juice), lollypops. From the petiole they make mats, lanterns, candles. Host for the "papaso" (*Rhynchophorus palmarum)* which lays its eggs on the palm; the larva is called "suri"; it is eaten raw, boiled or fried. (For more on the suri, see *Euterpe*. Leaves used for roofing. Stem used to make improvised bridges. Humboldt called it "The tree of life". The starch is collected and consumed as an important carbohydrate source. From the petiole they make fishing rafts; they also extract a high quality fiber (MJB). (Fig. 152)

Mauritiella aculeata (H.B.K.) Burret, Arecaceae. "Aguajillo". Fruit edible; stems used for floors and room dividers. (Fig. 153)

Maximiliana maripa (Mart.) Drude. Arecaceae. "Inayuga", "Shapajilla". Cooked pulp and seed edible (RVM). Petioles used in "Waorani" darts. Infusion taken for colds (SAR).

Maximiliana stenocarpa Burret. Arecaceae. "Inayuga". Buds edible; fruits yield oil.

Maximiliana venatorum (Poepp.) Wendl. Arecaceae. "Inayuga". Natives use petiole in blowgun darts. Nut contains 60-67% fat, while the mesocarp has 42.1% fat. Fruit an excellent food source (MJB).

Mayna amazonica (Mart.) Macbr. Flacourtiaceae. "Sapote yacu", "Shambo huayo". "Kofán" use in "curaré" (SAR). Leaf bath used for leg cramps (SAR). "Siona" use leaf/wood decoction for aching legs and prickly sensations (SAR). Used for leprosy, POISON for armadillo (RAR).

Maytenus guyanensis Klotzch. Celastraceae. "Chuchuhuasi", "Tonipulmon". Decoction of branches considered stimulant and tonic (FEO).

♀ *Maytenus macrocarpa* (R.&P.) Briq. Celastraceae. "Chuchuasi", "Chuchasha", "Chuchuhuasi". Bark maceration considered antidiarrheic, antiarthritic, used to regulate menstrual periods, for upset stomach. Its main use is in a cordial! Bark decoction used for dysentery. The wood is used for lumber (RVM). A shot of chuchuhuasi with aguardiente and honey was given many ecotourists on departure from the Iquitos airport in 1991 (JAD). Aril of a brazilian species contained 8,500 ppm caffeine (SAR). "Siona" boil stems in water for arthritis and rheumatism (SAR, under *M. laevis*). Under the name *M. ebenifolia*, Maxwell mentions the "chuchuhuasi" as an effective insect repellent. "Chuchuhuasi" is "probably the best known of all jungle remedies, in Colombia as well as Peru. Aphrodisiac...best of all antirheumatic medicines" (NIC).

♀ *Melia azedarach* L. Meliaceae. "Cinamono", "Paraiso", "**China berry**". Cultivated ornamental; some authors say fruit edible, others POISONOUS. Leaves used as an insecticide. Bark and roots used in Colombia as an emetic, anthelmintic, and quinine substitute. Used for intermittent fevers, the leaves are antipyretic and emetic (PEA). In Santa Catarina, Brazil, fruit and rootbark anthelmintic. Containing abundant bitter substances, bark can be used as an abortive, emetic, or toxic, externally to clean ulcers, especially syphilitic ulcers. Bahamans use it for colds and flu. "Cuna" put the bark in cold water bath to regain strength and vitality.

Melicocca bijuga L. Sapindaceae. "Pitomba", "**Genip**". Cultivated fruit. Seed also edible toasted (VAG).

Melinis minutiflora Beav. Poaceae. "Gordura", "**Molasses grass**". Cultivated forage.

Melissa officinalis L. Lamiaceae. "Toronjil", "**Lemon balm**". Cultivated. Leaf infusion used as a sedative (contains at least 5 sedative compounds), antiflatulent, and antispasmodic.

Mezilaurus itauba (Meissn.) Taub. ex Mez. Lauraceae. "Itauba". Wood used for tables, boats, canoes, posts, decorative plaques, etc. (Fig. 154)

Mezilaurus opaca Kubit. & V.D.Werff. Lauraceae. "Itauba". Wood for lumber.

♀ *Miconia impetiolaris* (Sw) D.Don. Melastomataceae. "Rifari". The "Cuna" use the pulverized bark in poultice to treat sores on the breasts (FOR).

Miconia poeppigii Triana. Melastomataceae. "Rifari". Wood for columns, decks and beams.

Micrandra spruceana (Baill.) R.E.Schult. Euphorbiaceae. "Conoco", "Shiringa masha". Wood used for interior decorations. The latex is used to bulk with that of *Hevea* (RVM). POISONOUS seeds processed to make food by Vaupes natives (SAR). "Bora" and "Witoto" use as a hemostat following umbilical separation and dab latex on sore mouth and gums (SAR).

Micropholis egensis (A.DC.) Pierre. Sapotaceae. "Quinilla negra", "Caimitillo". Wood used for house construction. Fruit edible (VAG).

Micropholis guyanensis (A.DC.) Pierre, ssp. *duckeana* (Baehni) Penn. "Balata rosada". Wood for house construction. Latex extracted for gum "Balata".

Micropholis guyanensis (A.DC.) Pierre, ssp. *guyanensis*. Sapotaceae. "Balata", "Quinilla". Wood for house construction.

Micropholis porphyrocarpa (Baehni) Monach. Sapotaceae. "Quinilla". Wood for house construction.

Micropholis venulosa (Mart. & Endl.) Pierre. Sapotaceae. "Caimitillo", "Quinilla", "Balatilla". Fruit edible. Wood for house construction.

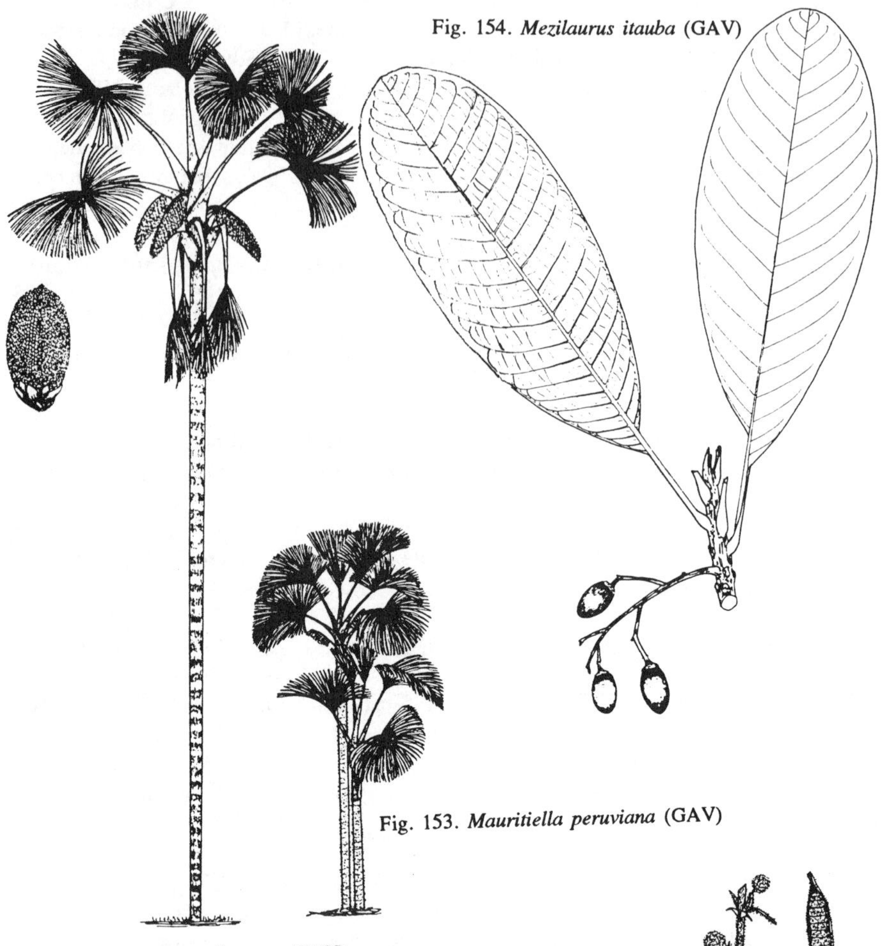

Fig. 154. *Mezilaurus itauba* (GAV)

Fig. 153. *Mauritiella peruviana* (GAV)

Fig. 152. *Mauritia flexuosa* (GAV)

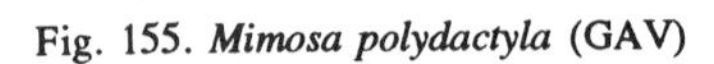

Fig. 155. *Mimosa polydactyla* (GAV)

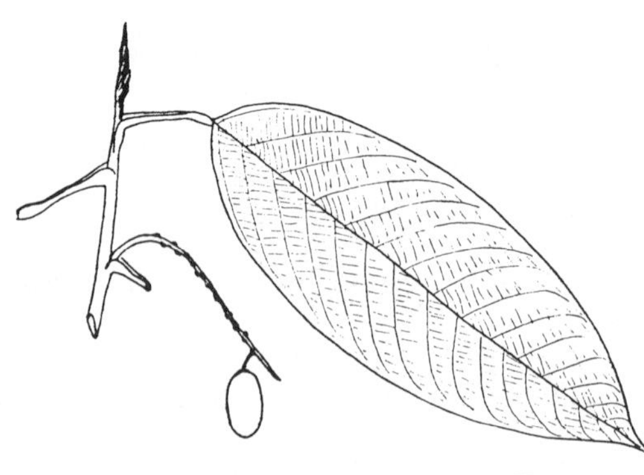

Fig. 156. *Minquartia guianensis* (GAV)

Microtea debilis Sw. Phytolaccaceae. "Capushi". Juice, extracted by crushing leaves in water, used for acid indigestion; leaves used to soothe swellings and burns. Infusion used by "Créoles" as a hypotensive diuretic (GMJ).

Mikania congesta DC. Asteraceae. "Sanquillo". "Créoles" use the juice of the bruised leaves as an aperitive tonic, taking it 3 times a day. The leaf decoction is used for malaria, and as a laxative. "Wayãpi" use leaf decoction in an antipyretic bath. "Palikur" take the leaf decoction to stimulate bile secretion (GMJ).

Mikania micrantha HBK. Asteraceae. "Playa huasca". As the preceeding (GMJ).

Mikania guaco HBK. Asteraceae. "Guaco". Crushed stem applied to snakebite; then they drink the decoction. Infusion is stomachic and antirheumatic; also used as an antipyretic (SOU). "Créoles" warm the leaves to apply to itching eruptions (GMJ).

Mimosa polydactyla Humb. & Bonpl. ex Willd. Fabaceae. "Amor dormido" **"Sleeping love"**, "Vergonsosa". Floral infusion sedative for insomnia and nerves. (Fig. 155)

Mimosa pudica L. Fabaceae. **"Sensitive plant"**. "Chami" dry and pulverize leaves, to mix a drink for insomniacs. Strong doses may cause madness (RVM). "Palikur" make a decoction mixed with pieces of *Scoparia dulcis,* to bathe irritable people (GMJ). Contains noradrenalin (JBH).

Minquartia guianensis Aubl. Olacaceae. "Huacapú", "Fierro caspi", **"Ironwood"**. Wood one of the best for posts, beams, dormers, bridges, supports; also used for parquets and handicrafts. High quality wood said to last 30 years, even in contact with the ground. Fruit edible (RVM). "Ketchwa" and "Waorani" use pounded bark as fish POISON (SAR). (Fig. 156)

Mirabilis jalapa L. Nyctaginaceae. "Isabelita", "Clavelilla", **"Four o'clock"**. Cultivated ornamental, root decoction used as a diuretic (RVM). Elsewhere, considered alterative, carminative, cathartic, hydragogue, purgative, stomachic, tonic and vermifuge; used for abscesses, boils, bruises, colic, diabetes, dropsy, hepatitis, herpes, hypochondria, pimples, sores, splenitis, strains, tumors, urticaria, and wounds (DAW).

Mollia lepidota Spruce ex Benth. Tiliaceae. "Achote vara". Wood used for house construction; columns, decks, beams. Bark used as rope (RVM). Rio Apapori Indians use bark tea for stomach problems following food poisoning (SAR).

Momordica balsimina L. Cucurbitaceae. "Balsamina", **"Balsam apple"**. Fruit tincture antiecchymotic, decongestant, vulnerary; decoction purgative (FEO). Used for bruises (RAR).

♀ *Momordica charantia* L. Cucurbitaceae. "Papailla", **"Balsam pear"**. Fruit edible cooked. Plant decoction used for colic, and worms; infusion of fruit and flowers used for hepatitis. Seed pulp mixed with lard as a suppurative (SOU). Considered vermicide, stomachic, emmenagogue, and very effective in the expulsion of *Trichocephalos.* Fruit decoction used as febrifuge and emetic (PEA). Leaf decoction used by the "Cuna" for measles (RVM), by Brazilians for fever, itch, and sores (BDS). Seeds and pericarp contain saponin glycosides which produce elaterin and alkaloids, which causes vomiting and diarrhea (LAE). Leaf infusion a common folk remedy for diabetes around Iquitos (AYA). TRAMIL cites it as relatively POISONOUS (TRA). On the patent for Compound Q for AIDS, as a source of momocharin. Also contains rosmarinic acid, with antiviral activity and calceolarioside and verbascoside. (Fig. 157)

Monstera dilacerata C. Koch. Araceae. "Costilla de adan". Fruit edible (RVM).

Monstera expilata Schott. Araceae. "Boa". Decoction given to infants for convulsions (GMJ).

Monstera obliqua Miq. Araceae. "Costilla de Adan", "**Adam's rib**". "Palikur" use for leishmaniasis sores (GMJ).

Montrichardia arborescens (L.) Schotto. Araceae. "Castaña", "Raya balsa", "**Water chestnut**". Toasted seeds edible. Surinamese use the sap to stop bleeding in fresh wounds (MJP). (Fig. 158)

Moronobea coccinea Aubl. Clusiaceae. "Lagartillo", "Azufre caspi". Wood used for construction of boats and canoes. "Créoles" use the latex for various dermatoses; they also use the latex to reinforce the ligature on their arrows (GMJ). (Fig. 159)

Mouriri acutiflora Naud. Melastomataceae. "Lanza caspi", "Guayavilla". Wood for house construction; beams, decks, jam posts. Fruit edible.

Mouriri apiranga Spr. Melastomataceae. "Apiranga". Bark decoction applied to sores on Rio Tapajos (BDS).

Mouriri grandiflora DC. Melastomataceae. "Lanza caspi". Fruit edible (VAG).

Mouriri oligantha Pilger. Melastomataceae. "Lanza caspi". Fruit edible (VAG).

Moutabea aculeata (R.&P.) P.&E. Polygalaceae. "Coto huayo". Fruit edible.

Moutabea sp. Polygalaceae. "Huasca caimito". Fruit edible (VAG).

Mucoa duckei (Markgraf) Zarruchi. Apocynaceae. "Yahuarhuayo blanco". Fruits edible (RVM).

Mucuna huberi Ducke. Fabaceae. "Mucuna", "Téwatacaá". Seeds used in making necklaces.

Mucuna pruriens (L.) DC. Fabaceae. "Nescafé", "Nescao". Cultivated. Toasted ground seeds are used as a coffee substitute (RVM). Elsewhere regarded as anodyne, antidotal, aphrodisiac, diuretic, nervine, resolvent, rubefacient, and vermifuge; used for anasarca, asthma, cancer, cholera, cough, diarrhea, dogbite, dropsy, dysuria, insanity, mumps, pleuritis, ringworm, snakebite, sores, syphilis, tumors, and worms (DAW). Interesting that this reputedly aphrodisiac plant should contain l-dopa, side effects of which include priapism (JAD).

Mucuna rostrata Benth. Fabaceae. "Vaca ñahui", "Corpus sacha". Seed infusion used as a diuretic, as an antidote and antihemorrhoidal. They carry seeds in case they are

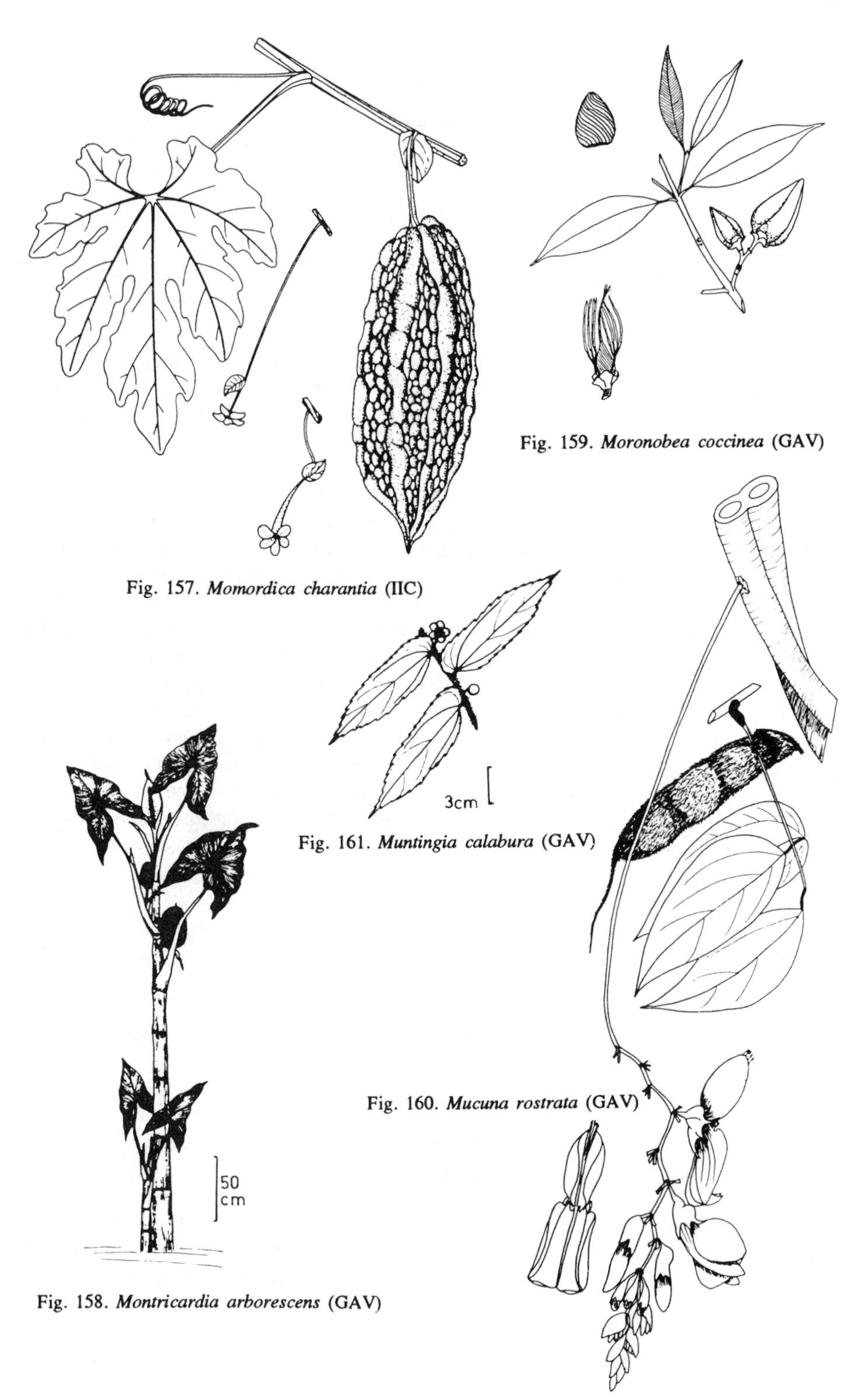

Fig. 157. *Momordica charantia* (IIC)

Fig. 158. *Montricardia arborescens* (GAV)

Fig. 159. *Moronobea coccinea* (GAV)

Fig. 160. *Mucuna rostrata* (GAV)

Fig. 161. *Muntingia calabura* (GAV)

bitten by a spider or a snake. The hair from the fruits is used as a mechanical vermifuge (SOU). (Fig. 160)

Mucuna urens (L.) Medik. Fabaceae. "Vaca ñahui". "Tiriyo" use to treat gonorrhea and migraine (RVM). Used as a vermifuge because of the mechanical action of the stinging hairs mixed with honey (GMJ).

Muntingia calabura L. Elaeocarpaceae. "Yumanasa", "Bolaina", "Cereso caspi", **"Calabur"**. Fruit edible; bark used as rope. Flower infusion used as antispasmodic (SOU). Cuna macerate the plant without roots in cold water for malignant and cronic ulcers (FOR). Used elsewhere for colds, headache, nerves, and spasms (DAW). (Fig. 161)

♀ *Murraya paniculata* (L.) Jack. Rutaceae. "Naranjilla", **"Orange jasmine"**. Cultivated ornamental. Elsewhere considered astringent, bactericide, cosmetic, and stimulant; used for cough, dentifrice, diarrhea, dysentery, eyes, hysteria, rheumatism, stomachache, and wounds.

Musa spp. Musaceae. "Plátano", "Banano", "Guineo", **"Banana"**, **"Plantain"**. Around Iquitos, there are plantains, with angular fruits and hard pulp (common plantain, bellaco, preto, coto, sapucha, and isleno) and bananas, with cylindrical fruits and soft pulp (banano, guineo, seda, capirona, guineo morado, pumerillo, or muquichi, viejillo or enano, and manzano). Recently they are called *M. sapientum* for "banana", and *M. paradisiaca* for "plantain". Most common dish is "inguiri", (boiled with water and salt). Green fruits together with the bark are boiled until half of the original liquid is gone; this astringent liquid is given to people recovering from tuberculosis; they are to drink it every day for more than six months. Some people recommend the water that accumulates in the cut stem of the "manzano" for rapid healing of pulmonary scars. The green fruit is boiled, the resulting juice used to treat diarrhea in infants (AYA). The plantain is used to treat gout, earache, warts, teeth, and freckles (RVM). Brazilians make a cough syrup boiling the exudate from cut stems with sugar (BDS). Considered useful for bronchitis, cough, diarrhea, fever, tuberculosis, and urticaria (RAR). Heated banana leaves seem to help leishmaniasis (JAD). (Fig. 162)

Mussatia hyacinthina (Standl.) Sandw. Bignoniaceae. "Chamairo". Stem bark mixed with coca leaves to sweeten and improve the taste, used by "Campa", "Machinguenga", and "Chimane" (RVM).

Myrcia fallax (Rich) DC. Myrtaceae. "Rupiña". Fruit edible (VAG).

Myrcia sp. Myrtaceae. "Guayabilla". Fruit edible (VAG).

Myrcianthes quinquefolia (McVaugh) McVaugh. Myrtaceae. "Lanche", "Lanchi". Aromatic cholagogue leaves decocted for jaundice (FEO).

Myrciaria dubia (HBK) McVaugh. Myrtaceae. "Camu-camu". The edible fruit is the highest known source of ascorbic acid, with 2-3 g/kg (RVM). Provisionally, Plotkin notes that "a forest stand righ in Camu-Camu is worth twice the amount to be gained from cutting down the forest and replacing it with cattle." (MJP). (Fig. 163)

Myrciaria floribunda (West. ex Willd.) Ber. Myrtaceae. "Camu-camu", "Camu-camu arbol", "Camu-camu negro". Edible fruit used in beverages; wood used for house construction, and forked poles.

Myristica fragrans Houtt. Myristicaceae. "Nuez moscada", **"Nutmeg"**. Introduced cultivated spice. Chopped seeds used for hexes and for paralysis and parasites (FEO).

Myroxylon balsamum (L.). Harms. Fabaceae. "Balsamo", "Estoraque", **"Balsam of Peru"**. For parquets, dormers, posts, jam poles, handicrafts, keel plates for boats. Resin from trunks believed antipyretic and cicatrizant. Resin used for colds, lung ailments (SAR), abscesses, asthma, bronchitis, catarrh, headache, rheumatism, sores, sprains, tuberculosis, venereal diseases, and wounds (DAW, RAR). Powdered bark used as incense (SOU). (Fig. 164)

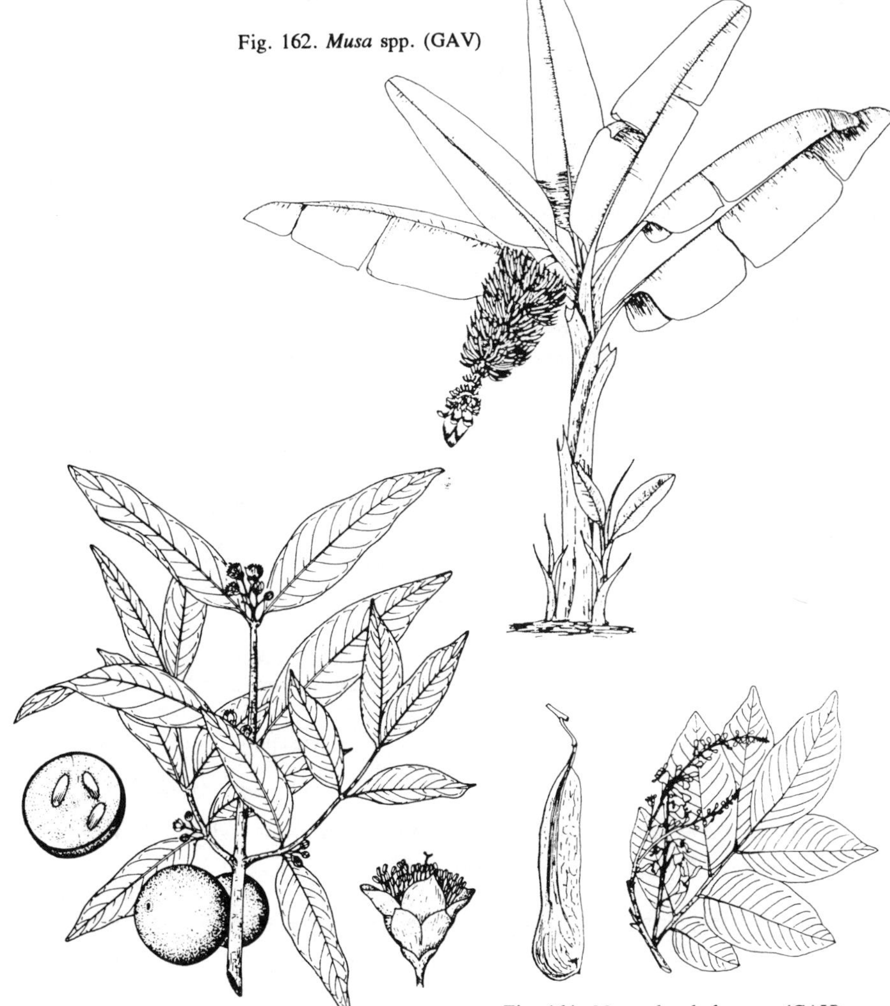

Fig. 162. *Musa* spp. (GAV)

Fig. 164. *Myroxylon balsamum* (GAV)

Fig. 163. *Myrciaria dubia* (IIC)

- N -

Naucleopsis concinna (Standley) C.C.Berg. Moraceae. "Llanchama", "Capinurí de altura". Fruit edible.

Naucleopsis mello-barretoi (Standley) C.C.Berg. Moraceae. "Llanchamillo". Fruit edible. Brazilian "Maku" use the latex with antiarigenin- and strophanthidin-based cardiac glycosides (JE 37:129. 1992).

Naucleopsis pseudo-naga C.C.Berg. Moraceae. "Puma chaqui". Fruit edible (VAG).

Naucleopsis ternstroemiiflora C.C.Berg. Moraceae. "Motelo chaqui". Fruit edible (VAG).

Nealchornea yapurensis Huber. Euphorbiaceae. "Huira caspi". The wood could be used for beams, joists, doors, windows, and drawers (RVM). "Taiwano" use crushed leaves as fish POISON (SAR).

Nectandra globosa (Aubl.) Mez. Lauraceae. "Moena amarilla". Wood for lumber RVM. Rio Loretoyacu natives use the bark tea for fever (SAR).

Nectandra kunthiana (Nees) Koster. Lauraceae. "Moena". Wood for lumber.

Nectandra membranacea (Sw.) Griseb. Lauraceae. "Pukeri". Seed decoction considered astringent, febrifuge, and tonic iin Piura (FEO). Branches in decoction used as gastric sedative (FEO).

Nectandra reticulata (R.&P.) Mez. Lauraceae. "Moena". Wood for lumber.

Nectandra woodsoniana Allen. Lauraceae. "Moena". Wood for lumber.

Neea divaricata Poepp. & Endl. Nyctaginaceae. "Piosha". "Achuales" chew the fresh leaves to coat their teeth and protect them from cavities (LAE).

Neea floribunda Poepp. & Endl. Nyctaginaceae. "Yanamuco". "Achuales" chew the fresh leaves to coat their teeth to protect them from cavities (LAE).

Neea laxa Poepp. & Endl. Nyctaginaceae. "Mesque", "Puca huayo". Leaf decoction used in bath or poultice for gastritis (VDF).

Neea macrophylla Poepp. & Endl. Nyctaginaceae. "Tupamaqui". "Yaguas" extract the purple-violet tint to dye their hammocks, handbags, and garments.

Neea parviflora Poepp. & Endl. Nyctaginaceae. "Palometa huayo", "Tupamaqui". "Yaguas" use it as fishbait.

Nephelea cuspidata (Kze) Tryon. Cyatheaceae. "Helecho arborescente". Buds yield a mucilaginous substance, rich in tannin, used as a vulnerary on small wounds. Trunks used as pillars for houses. The durable wood is nicely marked by the frond scars (RVM).

Nephrolepis biseriata (Sw.) Schott. Davalliaceae. "Plumilla". Ornamental.

Nephrolepis cordifolia (L.) Presl. Davalliaceae. "Plumilla". Ornamental.

Nephrolepis rivularis (Vahl) C. Chr. Davalliaceae. "Plumilla", "Serpentina". Ornamental.

Nerium oleander L. Apocynaceae. "Adelfa", **"Oleander"**. Cultivated ornamental plant, the latex POISONOUS.

Nicolaia elatior Zingiberaceae. "Bastón del emperador", **"Philippine waxflower"**, **"Torch ginger"**. Cultivated ornamental.

Nicotiana tabacum L. Solanaceae. "Tabaco", **"Tobacco"**. Cultivated. The black tobacco "mapacho or siricaipe", is smoked during the ayahuasca, witchcraft, healing, and cleansing rituals; the pitch left from the smoke is picked up on a piece of paper and applied on the skin to kill worms. Powdered tobacco is mixed with aguardiente and given to dogs to make them better hunters. "Créoles" mixed the dried leaves with *Scoparia dulcis* leaves, while the "Wayãpi" use the pitch, to suffocate the larvae of the worm "macaco", *Dermatobia hominis* (Euterebrides), parasites which live in the skin of humans and dogs. "Palikur" poultice it onto migraine headaches; it is also used as a cholagogue to treat liver diseases. One drop of tobacco juice makes a strong collyrium (GMJ). "Bora" and "Witoto" poultice fresh leaves onto boils and infected wounds (SAR). "Jivaro" take tobacco juice for chills, indisposition and snakebite (SAR). "Tukanoan" rub the leaf decoction onto bruise and sprains (SAR). Many Indian groups used it for lung ailments (SAR). In Piura the leaf decoction is applied externally for parasites and rheumatism.

Nymphaea ampla (Salisb.) DC. Nymphaeaceae. "Loto azul", "Flor de agua", **"Water lily"**. Planted in ponds. Flower infusion reputedly aphrodisiac.

- O -

Ochroma pyramidale (Cav. ex Lam.) Urban. Bombacaceae. "**Balsa**", "Topa", "Palo de balsa". Wood used for rafts, wooden buoys, net floats, fishing hooks and sounders. Bark used for cordage and belts. The cotton from the seeds, called "flor de topa", is used for stuffing toys (RVM).

Ocimum basilicum L. Lamiaceae. "Albaca", "**Basil**". Cultivated. Used as spice and medicine.

Ocimum gratissimum L. Lamiaceae. "Albaca". Cultivated. Used as spice (RVM).

Ocimum micranthum Willd. Lamiaceae. "Albaca", "Iroro", "Pichana albaca", "Pichana blanca", "**Wild basil**". Used for fever and headache around Pucallpa (VDF). Said to be hallucinogenic (RVM). "Créoles" prepare a collyrium from the flowers; with the decoction they make a tea to treat flu. The maceration is used by the "Wayãpi" in antipyretic baths, and in massage to relieve colic (GMJ). Leaves are used to relieve gastric pains (RVM). "Tikuna" wash the head with leaf macerations for fever (SAR). Leaf juice dropped into eyes for conjunctivitis (SAR). Sometimes used as spice and perfume (SAR). Tapajos residents use the plant on bugbites and stings.

Ocotea aciphylla (Nees) Mez. Lauraceae. "Canela muena", "Moena negra". Wood used to make furniture, keel plates for boats and canoes.

Ocotea argyrophylla Ducke. Lauraceae. "Puspo muena". Timber used for beams, decks, and columns.

Ocotea fragantissima Ducke. Lauraceae. "Anis muena". Wood for furniture.

Ocotea jelskii Mez. Lauraceae. "Hispingo". Seed decoction considered antiecchymotic and decongestant (FEO).

Ocotea marmellensis. Mez. Lauraceae. "Moena negra". Wood for furniture.

Ocotea oblonga ssp. *cuprea* (Meissn.) Rohwer. Lauraceae. "Sicshi muena". Wood used for furniture, canoes, and general construction. (Fig. 165)

Ocotea petalanthera (Meissn.) Mez. Lauraceae. "Moena amarilla". Wood for furniture and general construction.

Odontadenia macrantha (R.&S.) Margkraff. Apocynaceae. "Sapo huasca". "Wayãpi" consider the nectaries found near the base of the ovary a tonic (GMJ).

Odontocarya tripetala Diels Menispermaceae. Bark boiled with *Matisia cordata* and hot peppers to expel worms (SAR). "Tikuna" rub the leaves on body aches (SAR) .

Oenocarpus mapora Karst. Arecaceae. "Cinamillo", "Cinamo", "Bacaba". Warming the fruit in sun or hot water loosens the mesocarp, used to prepare a refreshing drink. Terminal buds edible; leaves used to roof houses; trunk cut in strips for room dividers. The whole plant is use to prepare "umshias" used during carnaval. Crushed stems yield a maroon or white dye (DAT). Green fruit mashed for diarrhea, malaria, and nausea.

Oenocarpus minor Mart. Arecaceae. "Cinamillo" "Bacaba". Similar to the previous species.

Oenocarpus multicaulis Spruce. Arecaceae. "Cinamillo", "Bacaba". Similar to *O. mapora*.

Oenothera rosea Ait. Onagraceae. "Chupa sangre", **Pink primrose**". Chopped shoots used for ecchymoses and fractures externally; leaf decoction used for respiratory problems and worms (FEO).

Olyra latifolia L. Poaceae. "Carricillo", "Huasca maronilla". "Cuna" use the leaves in cold water infusion for mild dermal mycosis; used preferably hot; also for sore throat (FOR). Hollow stems used to make flutes (DAT). (Fig. 166)

Omphalea diandra L. Euphorbiaceae. "Sapo huasca", "**Toad vine**". "Wayãpi" apply stem sap to the forehead for headache. The leaves, warmed by the fire, are applied over mycotic areas and wasp stings (just after the sting). Leaf decoction used for bugbites and sores (GMJ). Purgative (RAR). Cooked seeds consumed (VAG).

Opuntia ficus-indica Mill. Cactaceae. "Cacto", "Tuna", "Opuntia", "**Prickly pear**". Cultivated ornamental. In Loreto it is used in dentistry. In Piura, the fruit juice is drunk for whooping cough (Mexicans recommend it for diabetes, Israelis for prostate); chopped fruits are applied as a sedative to rheumatic pain and to stop nosebleed (epistaxis).

Orbignya polysticha Burret. Arecaceae. "Catirina", "Shapaja". Seeds and terminal buds edible. Leaves used for roofing. The liquid that comes out when the *Orbignya* is cut, fermented, and drunk (MJB).

Ormosia amazonica Ducke. Fabaceae. "Huayruro". Wood used for heavy construction, dormers, parquets, decorative plaques. Seeds used in handicrafts, necklaces, chaquiras, curtains (RVM). Contains the alkaloid amazonine (SAR).

Ormosia bopiensis Pierre ex Macbr. Fabaceae. "Huayruro". See preceeding. (Fig. 168)

Ormosia coccinea (Aubl.) Jacks. var. *subsimplex* (Spruce ex Benth.) Rudd. Fabaceae. "Huayruro colorado". See preceeding (RVM). "Witoto" consider seed POISONOUS (SAR).

Ormosia macrocalyx Ducke. Fabaceae. "Huayruro". Similar to *O. amazonica*.

Ormosia nobilis Tul. var. *santaremensis* (Ducke) Rudd. Fabaceae. "Huayruro". See preceeding.

Orthoclada laxa (L. Rich.) Beauv. Poaceae. "Ocajiníímune". White dye (DAT).

Orthomene schomburgkii (Miers) Barneby & Krukoff. Menispermaceae. "Coto huayo". Additive in curaré (RVM). "Kofán" take the leaf tea for insomnia (SAR). "Maku" bathe indolent ulcers with twig decoction (SAR).

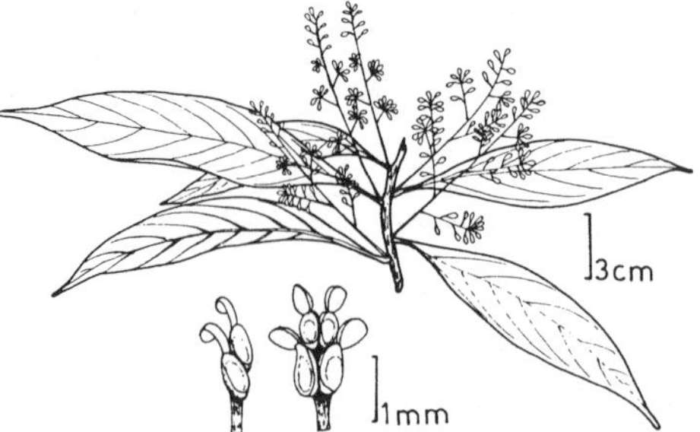

Fig. 165. *Ocotea cuprea* (GAV)

Fig. 166. *Olyra latifolia* (HAC)

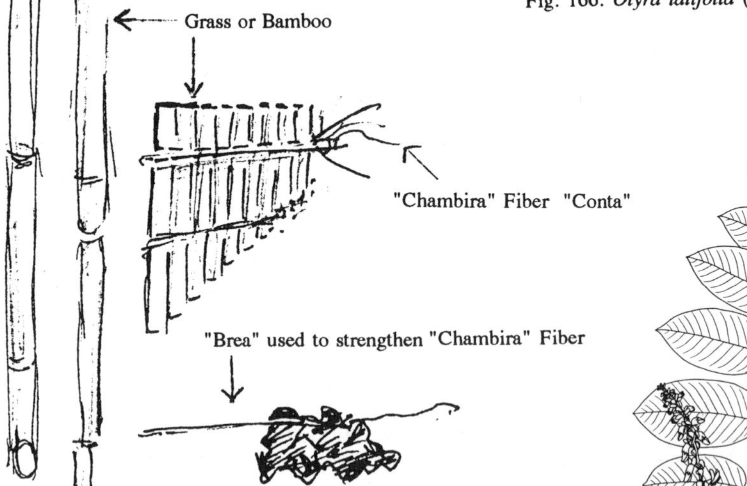

Fig. 167. Pan pipe (Propst)

Fig. 168. *Ormosia bopiensis* (GAV)

♀ *Oryctanthus alveolatus* (HBK) Eich. Loranthaceae. "Suelda con suelda", "Pishco isma". Depending on the host, this parasite may cure fast or slow. The most popular host is lime; used for fractures, dislocations, and cuts. Mashed leaves are applied over the affected area, before it is splinted. To hasten healing, they drink a cup of the decoction a day. They mix a leaf with foliar buds of *Psidium guayaba* and bark of *Spondias mombin* to give to women after childbirth, two cups a day, morning and afternoon. This helps them heal faster, enabling them to meet their marital duties, sooner than normally expected (RVM). Around Pucallpa, the leaf maceration is applied in bruises and fractures (VDF).

Oryctanthus florulentus (Rich.) Vant. Loranthaceae. "Suelda con suelda", "Pishco isma". "Palikur" poultice onto fractures and then splint with strips of *Gynerium sagitatum* (GMJ).

Oryza grandiglumis (Ducke) Presl. Poaceae. "Arroz bravo", "Grama playa". Forage.

Oryza sativa L. Poaceae. "Arroz" "**Rice**". Cultivated. Rainfed and paddy rice often grown in Amazonia.

Osteophloeum platyspermun (A.DC.) Warb. Myristicaceae. "Cumala blanca", "Favorito". Wood for carpentry, furniture and interior decorations. Manaus natives smoke the leaves for asthma (SAR). "Maku" drink the sap for colds and cough (SAR).

Otoba glycicarpa (Ducke) Rodr. Myristicaceae. "Aguanillo", "Cumala colorada". Wood for carpentry and general construction.

Otoba parvifolia (Markg) Gentry. Myristicaceae. "Aguanillo", "Cumala colorada", "Mamilla". Fruits edible. Used by the "Campas" as a blind when hunting partridges (RVM). "Waorani" crush the bark and rub the resin on fungal infections and mite bites (SAR). Wood used for carpentry.

Oxandra euneura Diels. Annonaceae. "Yahuarachi caspi", "Espintana". Wood used for house construction; beams and decks.

Oxandra spp. Diels. Annonaceae. "Espintana". Similar to the previous species. (Fig. 169)

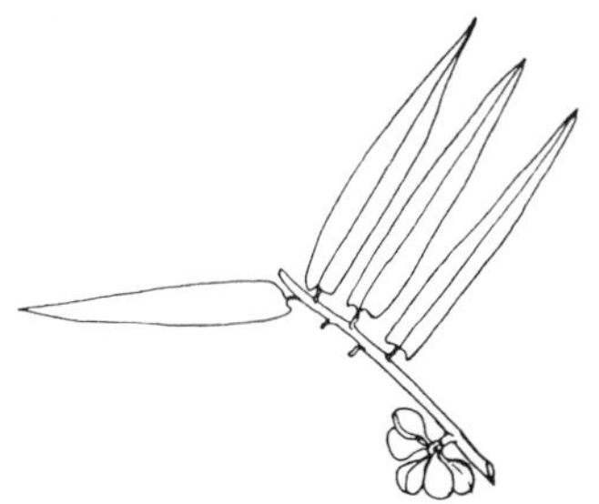

Fig. 169. *Oxandra* spp. (GAV)

- P -

Pachira aquatica Aubl. Bombacaceae. "Sacha pandisho", "Punga", "**Provision tree**". Seeds edible (RVM). Wood used for buoys; bark used for cordage, handicrafts, and kitchen utensils. Elsewhere used for diabetes (DAW).

Pachira insignis Sav. Bombacaceae. "Punga de altura". Wood for plaques, buoys; bark for cordage.

Pachyrrhizus tuberous (Lam.) Spreng. Fabaceae. "Ashipa", "Ajipa", "Nupe", "**Yam bean**". Cultivated. The roots are edible; the seeds are piscicidal (RAR).

Palicourea condensata. Standl. Rubiaceae. "Purma sisa", "Sacha huito". Semicultivated ornamental (RVM). Used to clear the sight on the R. Loretoyacu (SAR). Mocoa natives use the emetic decoction (SAR). "Siona" employ as fish POISON (SAR).

Palicourea puinapensis Aubl. Rubiaceae. "Purma sisa". Semicultivated ornamental.

Palicourea triphylla DC. Rubiaceae. "Huitillo". Mashed leaves mixed with water to stain the body black (DAT). So regarded by "Taiwano" that they won't collect it with a botanist (SAR). "Kuripako" use leaves as fish POISON (SAR).

Panicum dichotomiflorum Michx. Poaceae. "Gramalote", "**Fall panicum**". Forage.

Panicum hirsutum Swartz. Poaceae. "Grama". Forage grass.

Panicum laxum Poaceae. "Grama nudillo". Forage grass.

Panicum maximum Jacq. Poaceae. "Pasto guinea", "**Guinea grass**". Invasive grass, native of Africa, used as forage. Roots used for flu (FOR). (Fig. 170)

Panicum pilosum Sw. Pocaeae. "Torurco". Forage grass. "Cuna" give the root decoction to children when they lose their appetite, and don't want to drink water (FOR).

Panicum rudgei R.&S. Poaceae. "Grama". Forage grass.

Parahancornia peruviana Monachino. Apocynaceae. "Naranjo podrido". Fruit edible (RVM). (Fig. 171)

Pariana sp. Poaceae. "Shacapa". Curanderos use it rythmically in their rituals and chants. Hollow stems for flutes (DAT).

Parkia igneiflora Ducke. Fabaceae. "Pashaco", "Goma huayo", "Goma pashaco". Wood for lumber. Seeds used in handicrafts. (Fig. 172)

Parkia nitida Miq. Fabaceae. "Pashaco". Wood for lumber. Bark tea for dysentery (MJP).

Parkia panurensis Benth. ex M.C.Hopkins. Fabaceae. "Pashaco", "Goma pashaco". Wood for lumber; seeds for handicrafts.

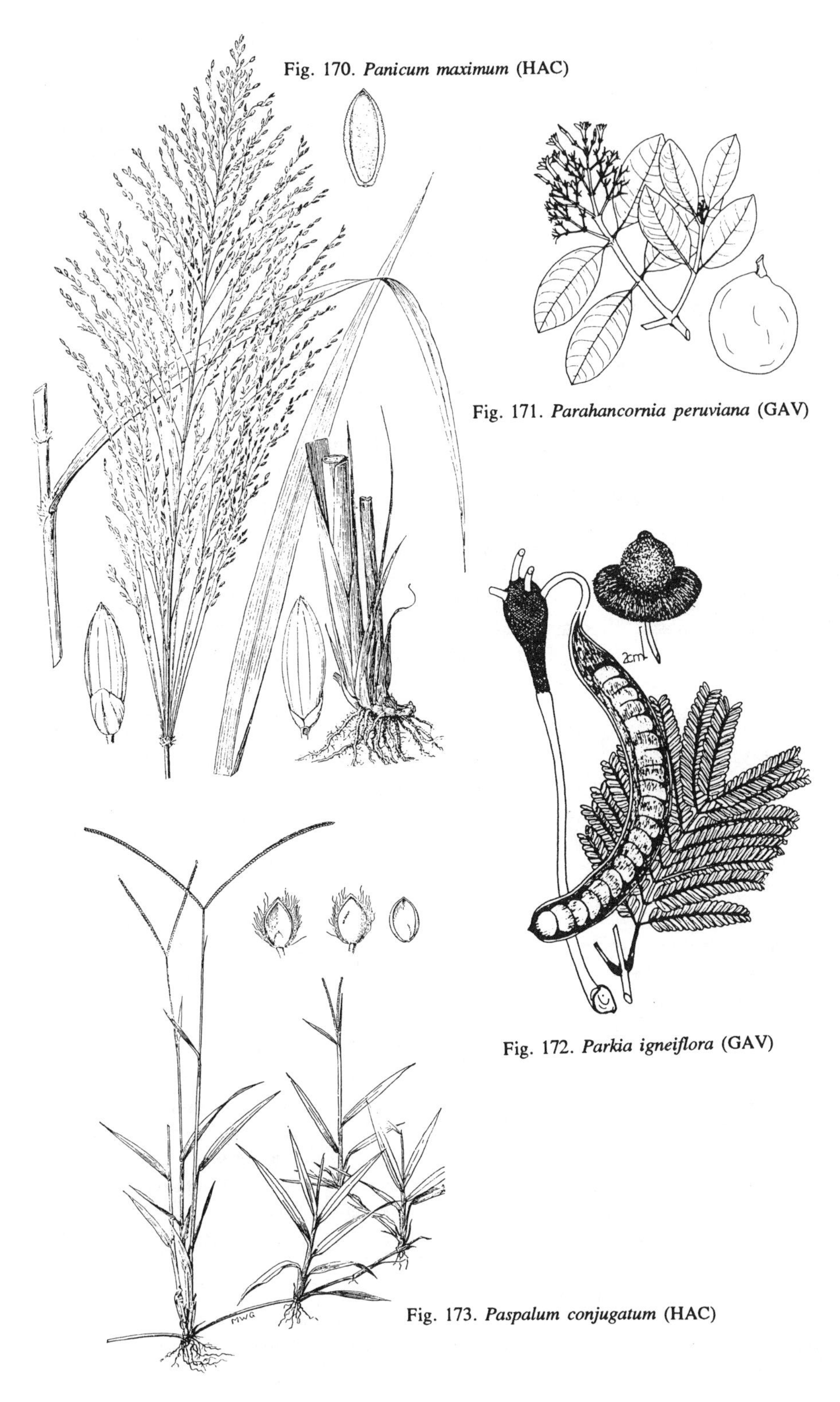

Fig. 170. *Panicum maximum* (HAC)

Fig. 171. *Parahancornia peruviana* (GAV)

Fig. 172. *Parkia igneiflora* (GAV)

Fig. 173. *Paspalum conjugatum* (HAC)

Parkia velutina Benoist. Fabaceae. "Cutana pashaco", "Pashaco curtidor". Wood for lumber; bark for tannery.

Paspalum conjugatum P. Bergius. Poaceae. "Catahua", "Grama", "Horquetilla", "Torurco". Forage grass (RAF). "Achuales" use 2 drops of sap for conjunctivitis. "Palikur" use with other plants to prepare the hunting dogs (RVM). Juice used to cure infected cornea and pterygium, often reventing blindness (NIC). (Fig. 173)

Paspalum spp. Poaceae. "Grama". Forage grass.

Paspalum fasciculatum Willd. ex Fluegge. Poaceae. "Gramalote". Forage grass.

Paspalum notatum Fluegge. Poaceae. "Grama", "Cañamazo", **"Bahia grass"**. Cultivated grass for sport fields and lawns.

Paspalum repens P.Bergius. Poaceae. "Gramalote negro". Forage grass.

Paspalum virgatum L. Poaceae. "Gramalote", "Remolina". Forage grass.

Passiflora alata Ait. Passifloraceae. "Granadilla". Fruit edible (VAG).

Passiflora coccinea Aubl. Passifloraceae. "Costada sacha", "Granadilla agria", "Granadilla venenosa", **"Red granadilla"**. Decoction taken 3 times a day for fever (VDF). Fruit and flower edible. A collyrium for conjunctivitis is extracted from macerations (GMJ).

Passiflora edulis Sims. Passifloraceae. "Maracuyá", **"Purple granadilla"**. Cultivated. Fruits edible. Brazilians on Rio Tapajos drink the pure fruit juice for the heart (BDS), using the leaf tea as a sedative. (Fig. 174)

Passiflora foetida Cas. Passifloraceae. "Bedoca", "Granadilla", "Ñorbo cimarrón", "Puro puro". Fruit edible (VAG).

Passiflora ligularis Juss. Passifloraceae. "Granadilla", "Tumbo", **"Sweet granadilla"**. In Piura, the leaf decoction is considered antimalarial, antipyretic, mucolytic, and stomachic. Fruits eaten or boiled in decoction to <u>prevent</u> gallstones, rabies, ulcers, and yellow fever (FEO).

Passiflora nitida HBK. Passifloraceae. "Granadilla". Fruits edible.

♀ *Passiflora quadrangularis* L. Passifloraceae. "Tumbo", **"Giant granadilla"**. Cultivated. Fruits edible; stems are considered POISON; the leaves, roots and flowers abortifacient. "Chami" make an infusion to treat fractures and bruises (RVM). Elsewhere considered calmant, CNS-depressant, cardiodepressant, decongestant, depurative, emollient, narcotic, sedative; used for arthritis, diabetes, hoarseness, hypertension, inflammation, liver ailments, neuralgia, sorethroat, and uvulitis (DAW). Contains noradrenalin (JBH). (Fig. 175)

Passiflora vespertilio L. Passifloraceae. "Cheshteya", "Granadilla", "Yacu granadilla". Fruits edible (RVM). Vapors from leaf decoction inhaled for rheumatic pain (VDF).

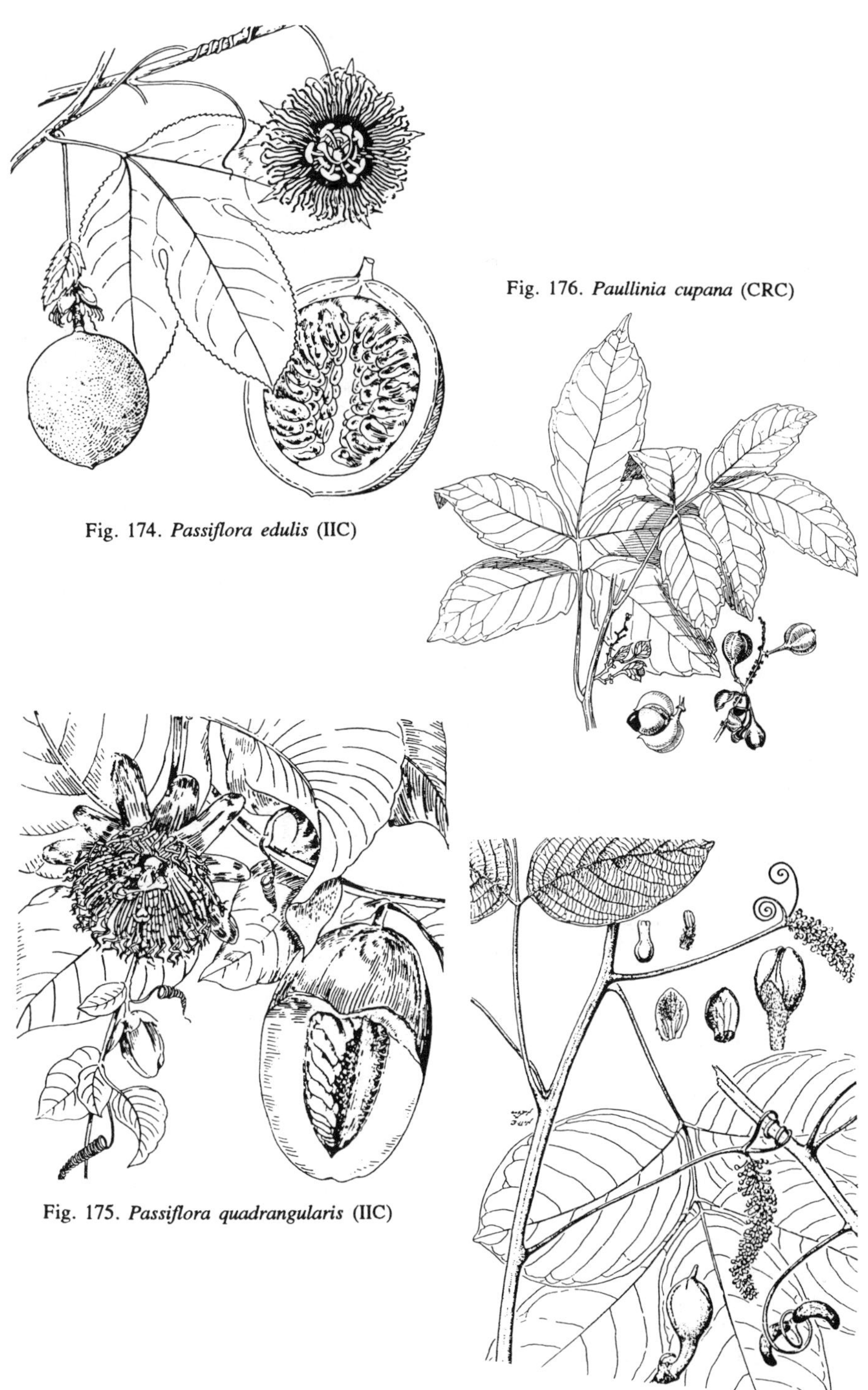

Fig. 174. *Passiflora edulis* (IIC)

Fig. 175. *Passiflora quadrangularis* (IIC)

Fig. 176. *Paullinia cupana* (CRC)

Fig. 177. *Paullinia yoco* (SAR)

Paullinia cupana HBK. Sapindaceae. "Guaraná", **"Cupana"**. Cultivated. Seed decoction an astringent, bitter, nervine tonic (FEO). From the seeds is prepared commercial guarana. Considered a preventive for arteriosclerosis, and an effective cardiovascular drug; also used to treat chronic diarrhea. Considered analgesic (MJB), aphrodisiac, astringent, febrifuge, intoxicant, piscicide, stimulant, and tonic; used for diarrhea, dysentery, hypertension, migraine, neuralgia (DAW, RAR). Seeds contains >5% caffeine, cf tea with 2.2%, and toasted coffee with 0.8%, green coffee with 2.2% , and cacao with 1.1% (RVM). Traces of theobromine and theophyllline also occur (Int. J. Pharmacogn. 31(3):174. 1993). (Fig. 176)

Paullinia cf *pinnata* L. Sapindaceae. "Sapo huasca", "Timbó". Considered POISONOUS; used as an intoxicant. The root bark is a narcotic toxin, used in Brazil by some natives to prepare the POISON "lanto" (SOU). Used for jaundice (RAR).

Paullinia yoco Schultes & Killip. Sapindaceae. "Yoco blanco", "Huarmi yoco". Cultivated. For fever, gall baldder, and dysentery (DAW, RAR). The bark contains 2.7% caffeine (SAR); from the stem is extracted in cold water a stimulant which quells the pangs of hunger; it contains hallucinogenic compounds. Schultes says you can feel the stimulant ca 10 minutes after taking one cup, a tingling on the fingers and a nice sensation of well-being (RVM). Used also for fever (SAR). Taken for bilious conditions following malaria (SAR). Unlike Vasquez, Schultes says it is never cultivated and becoming scarcer near the dwellings, perhaps endangered. "Every Indian household keeps a supply of yoco stems" (SAR). (Fig. 177)

♀ *Pavonia leucantha* Garcke. Malvaceae. "Mashushingo". Infusion used to strengthen the uterus to avoid an abortion.

Pavonia fruticosa (Willd.) Fawc. & Rendle. Malvaceae. "Hierba del conejo", "Pega-pega". "Chocó" use as a cough suppressant (JAD). "Cuna" put roots in cold water over night, and drink for headache (FOR).

Pavonia paniculata (Cav.) Mon. Malvaceae. "Pega-pega". "Chake" use it to treat female disorders (RVM). "Waunana" use it as an ornamental plant (FOR).

Paypayrola grandiflora Tul. Violaceae. "Umari del monte" Colombian Amazonians use the flower decoction for anemia (SAR).

Pedilanthus retusus Benth. Euphorbiaceae. "Planta plástica", "Zapatíto de Jesús". Cultivated ornamental.

Pennisetum nervosum (Nees) Trin. Poaceae. "Grama peinata". Forage grass.

Pennisetum purpureum Schum. Poaceae. "Pasto elefante", **"Elephant grass"**, **"Napier grass"**. Cultivated. Forage grass.

Pentaclethra macroloba (Willd.) Kuntze. Fabaceae. "Pashaco". Wood for lumber, bark for handicrafts and tanning.

Pentagonia cf *velutina* Standl. Rubiaceae. "Aromuhe". Ecuadorian "Waorani" use the plant for sting ray punctures, drinking the fruit juice (SAR). Colombian "Putumayo" decoct the leaves with those of a *Piper* to rub on swollen joints and purify the blood (SAR).

Peperomia flavamenta Trel. Piperaceae. "Congona". Shoot decoction considered vulnerary (FEO).

Peperomia galioides HBK. Piperaceae. "Congona". Shoot decoction considered vulnerary. Leaves applied externally against alopecia, burns, hemorrhoids, and otitis, the tea drunk for scurvy and hysteria.

♀ *Peperomia pellucida* HBK. Piperaceae. "Meralla", "Sacha yuyu". Brazilians pour hot water over the whole plant and drink the tea for metrorrhagia. Surinamese use for everything from athlete's foot to high blood pressure, gum problems, headaches, even to exorcise demons (MJP).

Peperomia rubea Trel. Piperaceae. "Lancetilla". Leaf juice or decoction applied for earaches and inflammations, drunk for tertian fever (FEO).

Perebea spp. Moraceae. "Chimicua". Wood for lumber; dormers.

♀ *Persea americana* Mill. Lauraceae. "Palta", "Huira palta", "**Avocado**". Cultivated fruit tree. Fruit juice considered aphrodisiac, used against dandruff and alopecia (FEO). Leaves well known as stomachic, emmenagogue, and resolvent. Seed decoction is an antidiarrheic, also used as an abortive. Used to treat amebic dysentery, diabetes, and snakebite (SOU). Also well known as antidiabetic (RVM). It eliminates uric acid, is a reconstituent tonic, antianemic, diuretic, antiinflammatory for the liver, for renal calculus, to strengthen weak muscles, for dysentery, and it is a mild aphrodisiac (RVM). "Tikuna" drink a cup of avocado leaf tea before meals to clean the liver (SAR). "Ketchwa" crush seed with *Brownea* wood and *Rudgea* leaves and make a decoction, said to stop menstruation for 3-6 months (SAR). As contraceptive, the seed decoction is taken each month during menses (SAR). "Siona-Secoya" also use as contraceptive (SAR). Ecuadorian "Shuar" take crushed seed in aguardiente for snakebite (SAR). Monounsaturates like oleic-acid are the health food rage now; avocado proved highest among 1,200 species (JAD). (Fig. 178)

♀ *Petiveria alliacea* L. Phytolaccaceae. "Chanviro", "Micura", "Mocosa", "Mucura", "Sacha ajo". Reportedly abortive, antispasmodic, antirheumatic, antipyretic, diuretic, emmenagogue, sudorific; mostly used in magic rituals call "limpias" ("cleansing"). The curanderos bathe the patients in the liquid left from the infusion to cleanse them from the "salt" (bad luck); other people bathe with it on the first hour of the new year. Colombians chew the plant in order to coat their teeth and protect them from cavities (GAB). Also used in ritual amulets. Preclinical tests show depressive effects on the central nervous system (CNS), with anticonvulsive effects (RVM). "Créoles" use it to get rid of bad spirits; the roots are antispasmodic and antipyretic; the leaf decoction, sudorific and cough suppressant. "Palikur" use to protect their children against bad luck, and in baths for the vitamin deficiency called "coqueluche" (GMJ). "Tikuna" bathe feverish patients in the leaf infusion and wash headache with the decoction. For bronchitis and pneumonia, a drop of kerosene and lemon juice is added to a teaspoon of macerated leaves (SAR). Rutter mentions beriberi, cramps, nerves, paralysis, rheumatism, scabies, scorpion sting, spider bites, toothache, venereal diseases, and vision, calling the herb abortifacient, analgesic, contraceptive, diuretic, emmenagogue, vermifuge, and insecticide (RAR).

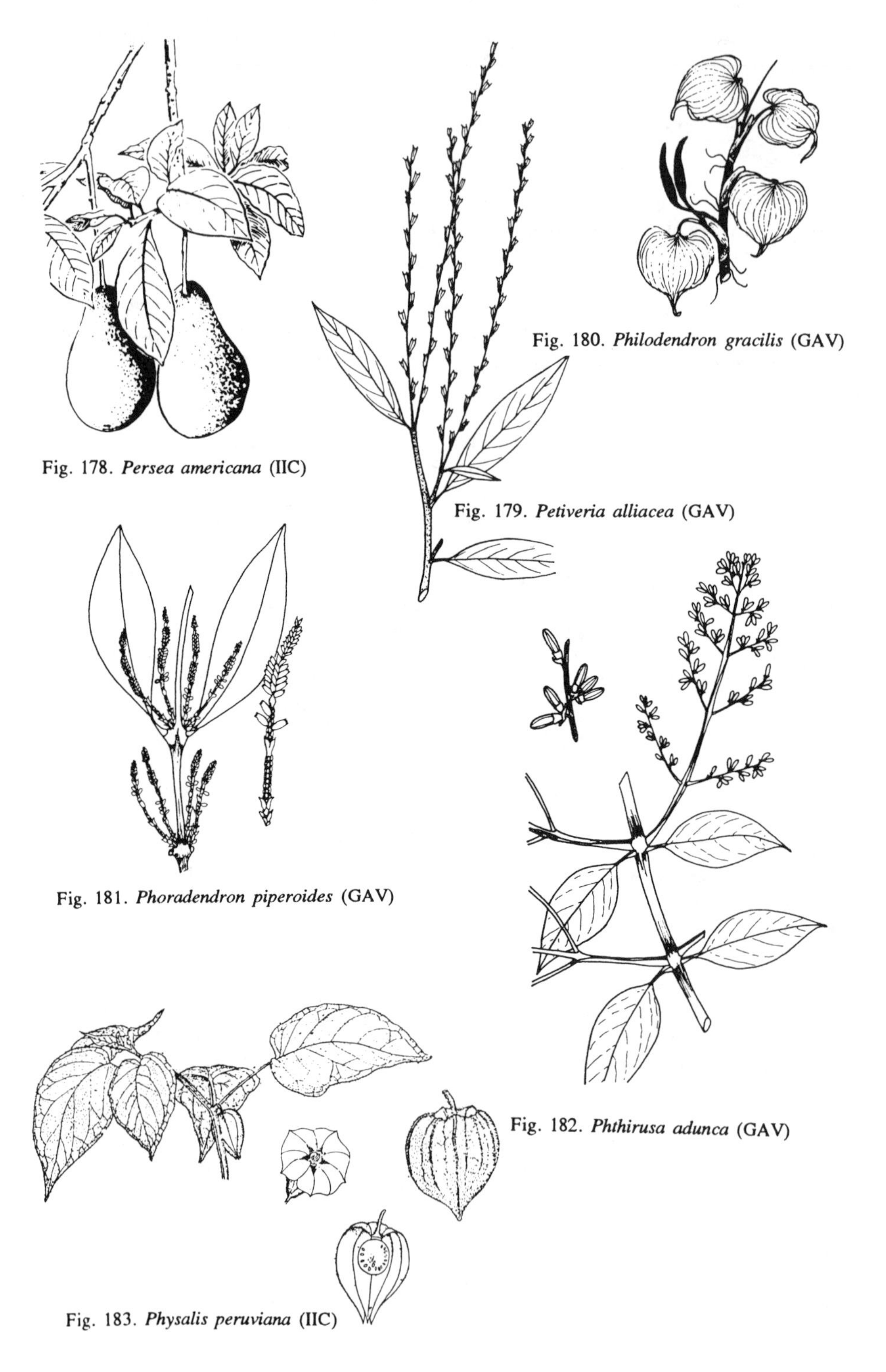

Fig. 178. *Persea americana* (IIC)

Fig. 179. *Petiveria alliacea* (GAV)

Fig. 180. *Philodendron gracilis* (GAV)

Fig. 181. *Phoradendron piperoides* (GAV)

Fig. 182. *Phthirusa adunca* (GAV)

Fig. 183. *Physalis peruviana* (IIC)

Independently, two different sources, one Venezuelan, one Colombian, related anecdotes about "curing" pancreatic cancer with *Petiveria* (JAD). Tramil all but endorses inhalation of the aroma for migraine and sinusitis, and using as a mouthwash for toothache (TRA). (Fig. 179)

Pharus latifolia L. Poaceae. "Puma barba", "Paujil chaqui". "Cuna" cook the roots for a long time and drink for diarrhea, taking 2 small cups a day (FOR).

Phaseolus vulgaris L. Fabaceae. "Frejol" "**Bean**", "Frijol", "Poroto", "**Green bean**". Cultivated. Seeds edible, served daily at Explorama as black or brown beans. Probably as good as soybean at preventing cancer (JAD).

Phenakospermum guyannense (L.C.Rich.) Endl. Strelitziaceae. "Abacá", "Platanillo". Fruits edible. Fiber from buds used for cordage; leaves used to wrap food (RVM). "Andoke" use sap for liver ailments (SAR). Stem chewed to prevent caries (SAR).

Philodendron cuneatum Engl. Araceae. "Itininga sacha". The hanging roots are used as ropes (RVM). "Taiwano" apply the crushed leaves in fat to dermatoses (SAR).

Philodendron deflexum Poepp. Araceae. "Tsutsihe". "Bora" use in remedy to keep children from eating soil DAT.

Philodendron gracilis Bunting. Araceae. "Boa sacha". As an ornamental. (Fig. 180)

Philodendron megalophyllum Schott. Araceae. "Itininga". The hanging roots are used as ropes. Juice of roots used for bugbites and snakebites.

Philodendron goeldii G.M.Barroso. Araceae. "Corona de Neron". Ornamental; sap used to extract worms from the skin.

Philodendron solimoesense A.C.Smith. Araceae. "Huambé". Roots are used for making baskets, handbags, handicrafts, and in finishing chairs and other furniture. The exudation of the leaves are used to extract worms from the skin (RVM).

Philodendron sp. Araceae. "Mai tanpeshco". Leaves poulticed onto toothache (VDF).

♀ *Phoradendron crassifolius* (DC.) Eichl. Loranthaceae. "Suelda con suelda", "Pishco isma", "**Mistletoe**". Depending on the host, this parasite cures fast or slow. If parasitic on lime, it is used for fractures, dislocations, and cuts. Mashed leaves are applied over the affected area, before it is splinted. To hasten healing, they drink a cup of the decoction a day. They mix a leaf with foliar buds of *Psidium guayaba* and bark of *Spondias mombin* for a mother after childbirth, two cups a day, morning and afternoon. This helps her heal faster, better able to meet her marital duties, sooner than normally expected.

Phoradendron huallagense Ule. Loranthaceae. "Beguefide". Leaves applied 3 times a day as a topical antiinflammatory (VDF).

♀ *Phoradendron piperoides* (HBK) Trelease. Loranthaceae. "Suelda con suelda", "Pishco isma". As *Phoradendron crassifolius*. Rubber workers drink the leaf tea for anemia (SAR). (Fig. 181)

♀ *Phthirusa adunca* (Meyer) Maguire. Loranthaceae. "Suelda con suelda", "Pishco isma". As *Phoradendron crassifolius*. "Karaja" use the leaf maceration for fever (RVM). (Fig. 182)

♀ *Phthirusa pyrifolia* (HBK) Eichl. Loranthaceae. "Suelda con suelda", "Pishco isma". As above.

♀ *Phyllanthus niruri* L. Euphorbiaceae. "Chanca piedra", "Sacha foster", "**Stone-breaker**". Like other species, quite effective in eliminating kidney- and gallstones (NIC). Considered anodyne, apertif, carminative, digestive, diuretic, emmenagogue, laxative, stomachic, tonic and vermifuge, used elsewhere for blennorrhagia, colic, diabetes, dropsy, dysentery, dyspepsia, fever, flu, gonorrhea, itch, jaundice, kidney ailments, malaria, proctitis, stomachache, tenesmus, tumors and vaginitis (DAW). Plant has proven antihepatotoxic, antispasmodic, antiviral, bactericidal, diuretic, febrifugal, and hypoglycemic activity (TRA).

Phyllanthus stipulatus (Raf.) Webster. Euphorbiaceae. "Chanca piedra". Well known for renal calculus and liver diseases.

Phyllanthus urinaria L. Euphorbiaceae. "Chanca piedra". As the previous species (RVM). Brazilian Indians use the stem decoction to prevent hair loss (SAR).

Physalis angulata L. Solanaceae. "Bolsa mullaca", "Capulí cimarrón", "Mullaca". Fruits edible; leaf infusion diuretic. Leaves and fruits used as narcotics, the decoction as an antiinflamatory, and disinfectant for skin diseases (GAB). Leaf juice used for worms (RAR). Also used for earache, liver, malaria (RAF), and rheumatism. "Cuna" often drink the leaf infusion for asthma (FOR). Root infusion used for hepatitis (AYA). Brazilians use the sap for earache (SAR), the roots boiled with *Bixa* and *Euterpe* for jaundice (BDS).

Physalis peruviana L. Solanaceae. "Aguaymanto". "**Cape gooseberry**". Fruit edible. Fruit juice for pharyngitis and stomatitis, the infusion as an ocular decongestant, the diuretic leaf infusion for cough and jaundice (FEO). (Fig. 183)

Phytelephas macrocarpa R.& P. Arecaceae. "Yarina", "Pelo ponto", "Cabeza de negro", "**Ivory palm**". The leaves are used to roof houses; terminal buds and immature seeds are edible. Ripe fruits, called "marfil vegetal" ("vegetable ivory "), used to make buttons and miniature handicrafts. The mesocarp of the ripe fruit is edible cooked. Diuretic (RAR). (Fig. 184 and Fig. 2-Introduction)

Phytelephas sp. Areaceae. "Piasaba". Produces a fiber for brushes and brooms. Leaves used in roofing; seeds edible.

Phytolacca rivinoides Kunth & Bouche. Phytolaccaceae. "Poe-hoe", "Airambo", "Apacas", "Jaboncillo", "Nipirihe", "**Pokeberry**". Saponins make this a soap substitute, occasionally used to wash rashes and pruritus. Ferreyra says young shoots are cooked as a vegetable (RAF). It is bacteriostatic and disinfectant (RVM). Ecuadorians use the leaf poultice for tumors (SAR). "Andokes" use the warm leaf infusion as an antiseptic for inflamed wounds (SAR). "Tanimuka" apply as an antidote to hot peppers (SAR). (Fig. 185)

Picramnia lineata Macbr. Simaroubaceae. "Amii", "Sani panga". Leaf cataplasm used as antiecchymotic and resolvent (VDF).

Picramnia magnifolia Macbr. Simaroubaceae. "Amiicuche". "Boras" poultice it onto sores (DAT).

Picramnia sprucei Hook.f. Simaroubaceae. "Ooniyatso", **"Sam Panga"***. Used by the "Boras" to treat skin irritations (DAT). Fruit soaked overnight for a purple dye (SAR).

Picrolemma pseudocoffea Ducke. Simaroubaceae. "Sacha café". Considered anthelmintic and febrifuge in Peru (RAR), leaf and root tea used for gastritis in Brazil (BDS).

Pilocarpus spicatus A.St.Hil. ex DC. Rutaceae. "Sapote yaru". Sedative (RAR).

Pinzona coriacea Mart. & Zucc. Dilleniaceae. "Paujil huasca", **"Watervine"**. Stems provide potable water.

♀ *Piper acutifolium* R.&P. Piperaceae. "Hierba del soldato", "Matico". Leaf decoction applied as topical and vaginal antiseptic; ingested for dyspepsia, dysmenorrhea, and gastrosis (FEO).

Piper aduncum L. Piperaceae. "Cordoncillo", "Matico". Leaves used in ritual baths for enteritis and upset stomach (RVM). "Karijona" use dried leaves as styptic (SAR).

Piper angustifolium R.&P. Piperaceae. "Cordoncillo", "Matico". Leaves applied externally as antiseptic vulnerary; the tea consumed for bronchitis, dysentery, gonorrhea, inflammation, and malaria (FEO, RAR). Infusion washed onto rheumatic areas around Pucallpa (VDF).

Piper angustum Rudge. Piperaceae. "Cordoncillo". The dried and burned leaves are applied on infected abscesses; the leaf infusion is antiseptic.

Piper arboreum L. Piperaceae. "Cordoncillo". Carminative, antirheumatic and emollient (RVM). Leaves given for debility, overeating and food poisoning (SAR).

Piper armatum Trecul & Yuncker. Piperaceae. "Niu weoko", "Cordoncillo". "Secoya" use to blacken their teeth (RVM).

♀ *Piper carpunya* R.&P. Piperaceae. "Carpunya". Leaf tea taken for bronchitis and dysmenorrhea (FEO). Analgesic (RAR).

Piper dumosum Rudge. Piperaceae. "Cordoncillo". "Orejones" once used it to coat their teeth to prevent cavities; also used for abdominal inflammation (RVM). "Tikuna" once used in curaré (SAR).

♀ *Piper hispidum* Sw. Piperaceae. "Cordoncillo", "Ungushurato", "Pióó". Leaf infusion used regularly for menstrual periods. "Achuales" on Rio Huasaga chew leaves to coat their teeth and prevent cavities (LAE). Leaves used by the "Boras" for mouth sores in children (DAT). Once used in "Tikuna" curaré (SAR). Leaf decoction used for malarial fever on the Rio Putumayo (SAR). Dried leaves used to delouse dogs (SAR). Mixed with *Phyllanthus* leaves as a fish POISON (SAR). Believed diuretic in La Pedrera (SAR).

♀ *Piper marginatum* Jacq. Piperaceae. "Cordoncillo", "Katio". Cultivated. Fresh leaf infusion used to relieve menstrual pains, and as an air refreshener. Colombians chew the plant to coat the teeth and protect them from cavities (GAB). Brazilians rub fat on swellings, then place the leaf over it (BDS). Regarded as anesthetic, carminative, digestive, antiinflammatory, antipyretic, antispasmodic, for hepatoses and blennorrhagia. Used in perfumery, dentifrices, aromatizers, synthesis of anisaldehyde, caramels, color photography, and microscopy (POV).

Piper nigrum L. Piperaceae. "Pimienta negra", "**Black Pepper**". Cultivated. As a digestive stimulant. Mixed with honey, copaiba, andiroba, and sugar, for coughs, bronchial affections, laryngitis and pharyngitis (RVM). Contains cineole.

♀ *Piper obliquum* R.&P. Piperaceae. "Cordoncillo". "Secoya" men and women chew it for 5 days after childbirth. "Wayãpi" use it for hernia (GMJ). "Ketchua" use for the teeth.

Piper peltatum L. Piperaceae. "Santa María". Leaves used as table cloths, to wrap food (RVM), and rubbed on the body as a tick repellent (DAW). Leaf decoction used as a diuretic, antipyretic, and emetic. The leaves passed over fire are applied directly on the head to relieve and reduce the swelling caused by trauma and hernias. Leaf poulticed onto sores (DAT). Believed anodyne, antiblennorrhagic, antiinflammatory, diuretic, lenitive, pediculicidal, piscicidal, resolvent, sudorific, vermifuge (JAD, RVM). "Créoles" use it as an antineuralgic, the leaf infusion as a sudorific (GMJ). Elsewhere used for abscesses, burns, colds, erysipelas, headache, hepatitis, leishmaniasis, swellings, toothache and urethritis (DAW). (Fig. 186)

Piper soledadense Trelease. Piperaceae. Amazonian Peruvians chew leaf and stem for oral sores (SAR).

Piptadenia suaveolens Miq. Fabaceae. "Pashaco". Wood for lumber.

Pistia stratiotes L. Araceae. "Huama", "Lechuga cimarrona", "**Water lettuce**". Vapors of the infusion used for mycosis (RVM). "Tikunas" mix crushed leaves with salt to remove warts (SAR). (Fig. 187)

Pitcairnia sprucei Bak. Bromeliaceae. "Bromilia". Flower shafts used as ornamental.

♀ *Pityrogramma calomelanos* (L.) Link. Adiantaceae. "Apaapalle", "Shapumba". "Huitotos" use as a preventive collyrium for cataracts. Natives from Trinidad use the leaf infusion for flu, amenorrhea, and hypertension (DAW). "Cuna" use the root infusion for stomachache (FOR). Plant infusion used as depurative, mild astringent, and pectoral; especially recommended in infusions for pulmonary affections. Frond applied to the gums for toothache (VDF). In Surinam, the whole plant with roots is decocted for bronchitis.

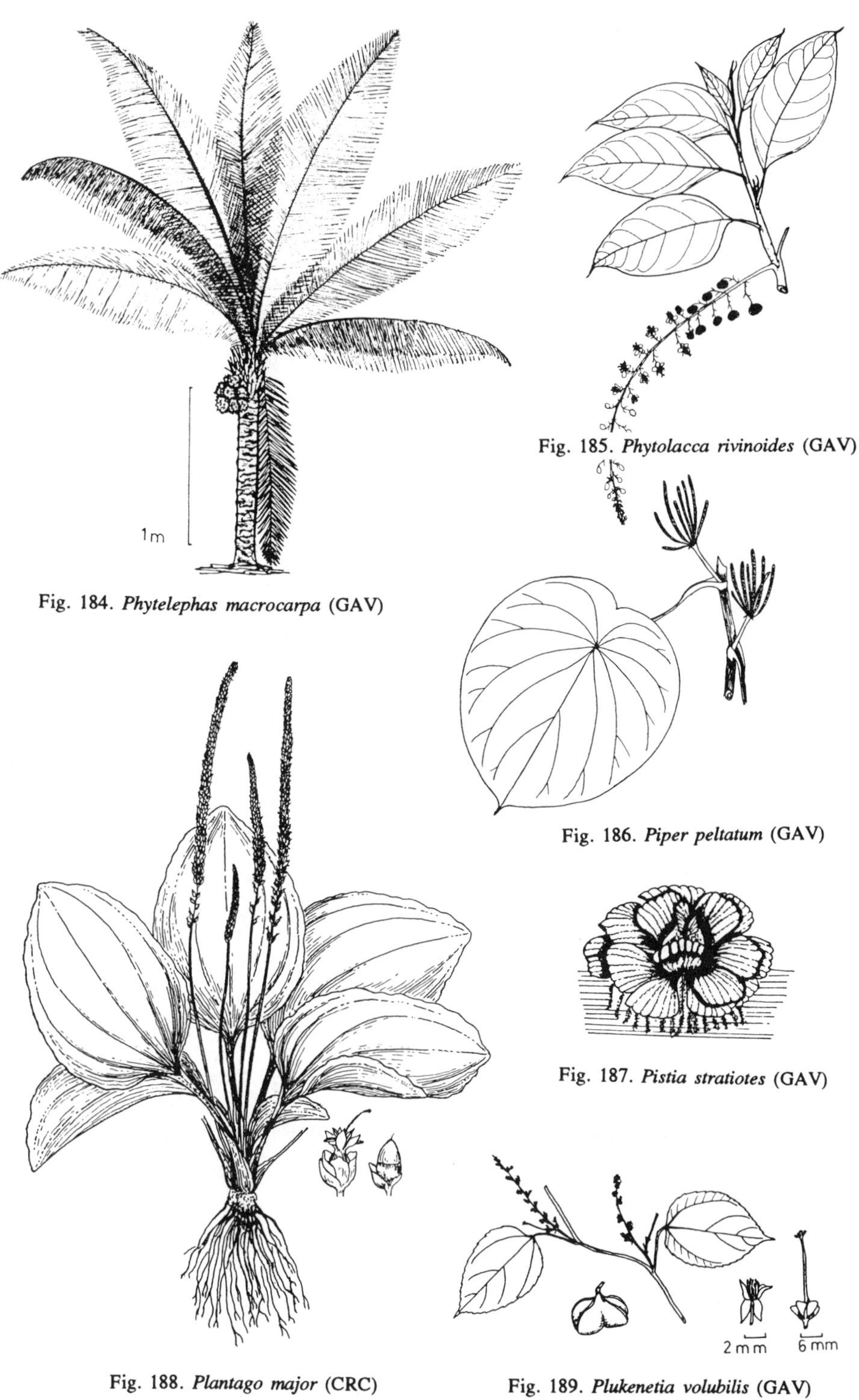

Fig. 184. *Phytelephas macrocarpa* (GAV)

Fig. 185. *Phytolacca rivinoides* (GAV)

Fig. 186. *Piper peltatum* (GAV)

Fig. 187. *Pistia stratiotes* (GAV)

Fig. 188. *Plantago major* (CRC)

Fig. 189. *Plukenetia volubilis* (GAV)

Leaves are applied to wounds and cuts to stop bleeding (JFM). Root infusion used as a bechic, for pulmonary ailments; also considered a talisman (GMJ).

♀ *Plantago major* L. Plantaginaceae. "Llanten", "**Plantain**". Cultivated. Decoction used as antiseptic for infected wounds. "Chami" use leaf maceration after childbirth; decoction used as liver analgesic (CAA). "Créoles" use for traumatic irritations and conjunctivitis (GMJ). "Tikuna" crush leaves with raw eggs for bronchitis and fever (SAR). For gastric ulcers (RAR). (Fig. 188)

Platymiscium spp. Fabaceae. "Añushi-cumaceba". Wood used for jam posts for bridges, dormers and posts.

Pleonotoma variabilis Miers. Bignoniaceae. "Nishi bata". Leaf infusion used internally around Pucallpa for gastritis (VDF).

Pleurothyrium densiflorum A.C. Smith. Lauraceae. "Pungar muena". Wood used for rafts, supports, decks, columns, lumber.

Plinia sp. Myrtaceae. "Anihuayo". Fruit edible (VAG).

Plukenetia volubilis L. Euphorbiaceae. "Maní del monte", "Correa", "Sacha inchi". Sometimes cultivated. Seeds are edible cooked or toasted. Fresh seeds mildly laxative. Among the "Witoto" who call it "amuebe", fresh leaves are poulticed to "heal extensive third-degree burns without scarring" (NIC). Leaves said to be edible (RAR). (Fig. 189)

Plumeria alba L. Apocynaceae. "Jasmin", "Suche rosado", "**Frangipani**". Cultivated ornamental. Elsewhere, regarded as bactericidal, cardiotonic, cathartic, caustic, emmenagogue, fungicidal, POISON, purgative, rubefacient, stimulant; used for abscesses, dermatoses, dropsy, fever, gingivitis, herpes, itch, rheumatism, scabies, toothache, tumors, venereal diseases and warts (DAW).

Podocarpus oleifolius D.Don ex Lamber. Podocarpaceae. "Aceitillo", "Pino regional", "**Podocarp**". Wood for house construction.

Poeppigia procera Presl. Fabaceae. "Cedro pashaco". Wood used for decorative plaques, bark for tannery. Astringent to heal sores (SOU).

Polianthes tuberosa L. Agavaceae. "Angélica", "**Tuberose**". Cultivated ornamental, grown elsewhere for perfumery (JAD). Elsewhere considered astringent, detergent, diuretic, emetic, resolvent and styptic; used for diarrhea, fever, gonorrhea, heat rash, pimples, and tumors (DAW).

Pollalesta discolor (HBK) Arist. Asteraceae. "Yanavara", "Ocuera negra". Wood is used for beams, columns, construction, decks, and fuel. (Fig. 190)

Polybotrya caudata Kunze. Aspleniaceae. "Shapumba huashu". The "Wayãpi" use it in rituals (GMJ).

Polygala acuminata Willd. Polygalaceae. "Irgapirina sacha", "Puru pagic sacha". Root used in Upper Huallaga Valley (UHV) for rheumatic ailments (RAF).

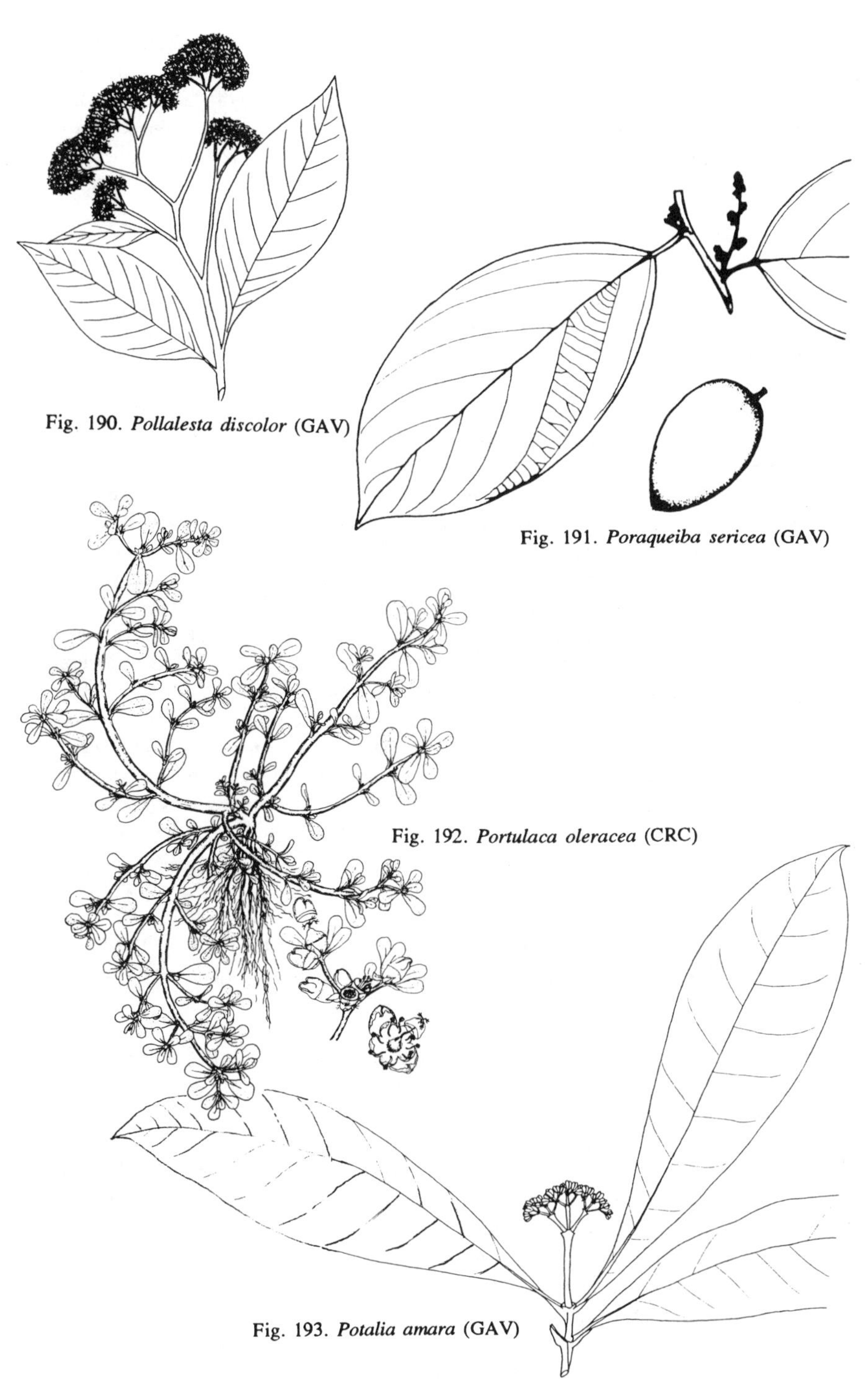

Fig. 190. *Pollalesta discolor* (GAV)

Fig. 191. *Poraqueiba sericea* (GAV)

Fig. 192. *Portulaca oleracea* (CRC)

Fig. 193. *Potalia amara* (GAV)

Polypodium decumanum Willd. Polypodiaceae. "Calaguala", "Huayhuashi-shupa". Rhizome maceration used for fever, whooping cough (grated or in infusion), and renal indispositions. From the leaves the "Boras" prepare a drink for coughs (DAT). Rhizome used to treat the pancreas (RVM). "Créoles" use the decoction in ritual baths for infants (GMJ). In Latin America, "calaguala", "llanten" and "matico" are among the first mentioned when the subject is medicinal plants, especially cancer (JAD).

♀ *Poraqueiba sericea* Tul. Icacinaceae. "Humarí". Cultivated. Mesocarp edible, frequently fermented; wood used for house construction, beams, charcoal, and decks (RVM). Used in a Witoto contraceptive (SAR). Astringent leaf tea valued for bacillary dysentery (SAR). Brazilians use the fruit oil for frying fish (SAR). Perhaps endangered or lost in the wild (SAR). (Fig. 191)

♀ *Portulaca oleracea* L. Portulacaceae. "Verdolaga", **"Purslane"**. Crushed plant used for fever, stings, and swellings. Containing noradenaline, purslane might logically be rubbed onto beestings and/or placed under the tongue, especially of allergic people (JAD). "Créoles" prepare an antidiabetic, digestive, and emollient tea. Used by the Palikur as a hypotensive (GMJ) (but contains hypertensive compounds JAD). Elsewhere considered alexeritic, alterative, aperient, astringent, bactericidal, cardiotonic, demulcent, detergent, diuretic, emmolient, fungicidal, hemostat, refrigerant, sedative, vermifugal and viricidal; used folklorically for anthrax, bladder ailments, blenorrhagia, boils, bugbites, burns, colds, colic, dermatitis, diarrhea, dysentery, dyspepsia, earache, eczema, edema, enterorrhagia, erysipeals, fever, gonorrhea, gravel, hematuria, hepatitis, herpes, hyperglycemia, hypotension, inflammation, insomnia, leucorrhea, nausea, nephritis, palpitations, piles, pleuritis, pruritis, snakebite, sores, splentitis, strangury, swellings, toothache, tumors,, warts and wounds DAW. A rather promising chemopreventive (="cancer-preventive") herb, loaded with antioxidants (JAD). Seeds of *P. peruviana* I.M. Johnston are considered emmenagogue and vermifuge. The shoot decoction, considered diuretic and cholagogue, is used for headache. Shoots are chopped and applied in pork fat to hemorrhoids (FEO). (Fig. 192)

♀ *Portulaca pilosa* L. Portulacaceae. "Flor de mediodia", "Flor de once". Considered digestive, diuretic, emmenagogue, vermifuge; used for burns, catarrh, colic, erythema, gallbladder, hemoptysis, hepatitis, nephritis (RAR). Brazilians drink the tea for sores.

Posoqueria latifolia (Rudge) Roem. & Sch. Rubiaceae. "Huitillo", "Sacha huito". Fruit edible (SOU). "Cuna" drink bark infusion for diarrhea (FOR). Powdered flowers used as a flea repellent. Febrifuge; tonic (RAR).

Potalia amara Aubl. Loganiaceae "Curarina", **"Sacha mangua"**. Important snakebite medicine, also famed for syphilis, urethritis, and other venereal diseases. The aqueous decoction or extract soothes body aches and pains. Natives boil the root bark to obtain a black substance which is filtered and taken orally. The "Boras" apply the maceration on snakebite, stingray, and ant bites (AYA, SAR). Shoots and leaves are considered antisyphilitic. Leaf decoction used for urethritis and ophthalmia. The strong tea is used to treat cassava poisoning. It is a cicatrizant, used in dental cavities (RVM). "Créoles", and "Wayãpi" use leaf decoction for fever. "Palikur" use tender leaves and small branches to treat the swellings caused by deep abscesses (GMJ). Colombian Rio Vaupes Indians regard the infusion as an emetic in food poisoning (SAR). Brazilians use the infusion for food poisoning, ophthalmia, and syphilis. Venezuelans consider it laxative (DAW). Contains squalene, saturated fat, methyl esters (SAR). (Fig. 193)

Poulsenia armata (Miq.) Standl. Moraceae. "Llanchama", "Yanchama". Bark used for cloth and handicrafts. Few groups use the bark for clothing, but it is still used for thin mattresses. (Fig. 194)

Pouraqueiba sericea Tul. Icacinaceae. "Humarí", "Umarí". Fruit edible (VAG).

Pouraqueiba paranesis Ducke. Icacinaceae. "Humarí", "Umarí". Fruit edible (VAG).

♀ *Pourouma cecropiaefolia* Mart. Moraceae. "Baacohe", "Ubilla", "**Grape tree**". Cultivated. Fruit edible, produced ca 3 years after planting (MJB). Wood used for paper pulp, and the toasted seeds as a substitute for coffee (RVM). Leaf ashes sometimes substituted for *Cecropia* as a coca additive. "Bara-Maku" use root scrapings to induce permanent sterility (SAR). "Cubeo" use as a masticatory, elsewhere considered intoxicant (DAW). (Fig. 195)

Pourouma guianensis Aubl. ssp. *guianensis*. Moraceae. "Sacha ubilla". Fruit edible.

Pourouma herrerense C.C.Berg. Moraceae. "Sacha ubilla". Fruit edible.

Pourouma minor Benoist. Moraceae. "Sacha ubilla". Fruit edible.

Pourouma ovata Trecul. Moraceae. "Chullachaqui blanco". Wood used in house construction.

Pouteria caimito (R.& P.) Radlk. Sapotaceae. "Caimito", "Tocino caimito", "Quinilla caimitillo". Cultivated as a fruit tree. Found wild with the name "quinilla caimito". Wood used for posts, forked poles, fence posts, dormers, bridges, parquet and handicrafts (RVM). "Witoto" apply toasted macerated leaves as a disinfectant to wounds (SAR). Contains alpha-amyrin, dammarenediol-II, erythrodiol, and lupeol (SAR). (Fig. 196)

Pouteria cladantha Sandwith. Sapotaceae. "Quinilla caimitillo". Fruit edible (VAG).

Pouteria cuspidata (A.DC.) Baehni. Sapotaceae. "Caimitillo". Fruit edible (VAG).

Pouteria gomphiaefolia (Mart. ex Miq.) Radlk. Sapotaceae. "Quinilla blanca del bajo". Wood for posts, forked poles, dormers, and fence posts.

Pouteria guianensis Aubl. Sapotaceae. "Caimitillo". Fruit edible (VAG).

Pouteria laevigata Radlk. Sapotaceae. "Caimitillo". Fruit edible (VAG).

Pouteria lucuma Kntz. Sapotaceae. "Lucuma". Fruit edible (VAG).

Pouteria macrophylla (Lam) Eyma. Sapotaceae. "Lucma". Cultivated. Fruit tree; wood for posts, forked poles, dormers and fence posts.

Pouteria multiflora (A.DC.) Eyma. Sapotaceae. "Caimitillo". Fruit edible (VAG).

Pouteria neglecta Cronq. Sapotaceae. "Quinilla negra". Wood for posts, forked poles, fence posts, and dormers.

Pouteria plicata Penn. Saptoaceae. "Caimitillo". Fruit edible. (VAG).

Pouteria procera (Mart.) Baehni. Sapotaceae. "Caimitillo", "Quinilla blanca". Fruit edible (VAG). Wood used like *P. neglecta*.

Pouteria reticulata (Engl.) Eyma. Sapotaceae. "Caimitillo". Fruit edible (VAG).

Pouteria simulans Monach. Sapotaceae. "Quinilla blanca". Wood used like the previous species.

Pouteria unilocularis (Donn. Sm.) Baehni. Sapotaceae. "Caimitillo". Fruit edible. Wood for forked poles, posts, dormers, and fence posts.

Pouteria sp. Sapotaceae. "Caimitillo", "Caracha quinilla". Wood for forked poles, posts, dormers, jam posts for bridges, parquets, and fence posts. Fruit edible (VAG).

♀ *Priva lappulacea* (L.) Pers. Verbenaceae. "Puspo quihua", "Bolsa quihua". Women drink an infusion of the leaves and roots during menstruation as a contraceptive. "Chocó" use it for whooping cough (JAD). "Cuna" drink root decoction with lemon juice for stomachache (FOR). "Créoles" poultice the maceration of the whole plant with salt onto sprains (GMJ). In Hispaniola used for colic, cough, diarrhea, gas, and gastritis (DAW).

♀ *Prosopis chilensis* (Mol.) Stuntz. Fabaceae. "Algarrobo". Unripe fruit considered astringent, lactagogue; unripe fruit applied to toothache. Seed infusion considered nutritious, tonic (FEO).

Protium altsonii Sandw. Burseraceae. "Copal". Wood for lumber. Latex once widely used for caulking boats; today more people use petroleum tar in caulking. For torches (shupihui), they put some coagulated latex on a palm leaf and then light it; bright, it gives a pleasant aroma. Latex also used for shining ceramics.

Protium divaricatum Engl. ssp. *divaricatum*. Burseraceae. "Copal colorado". For extraction of latex.

Protium grandifolium Engl. Burseraceae. "Copal", "Brea caspi". The latex is used for caulking boats. Aril of the seed edible.

Protium hebetatum Daly. Burseraceae. "Copal carana". Wood used for lumber. Latex gathered.

Protium nodulosum Swart. Burseraceae. "Copal". Latex extracted. Arils eaten (VAG).

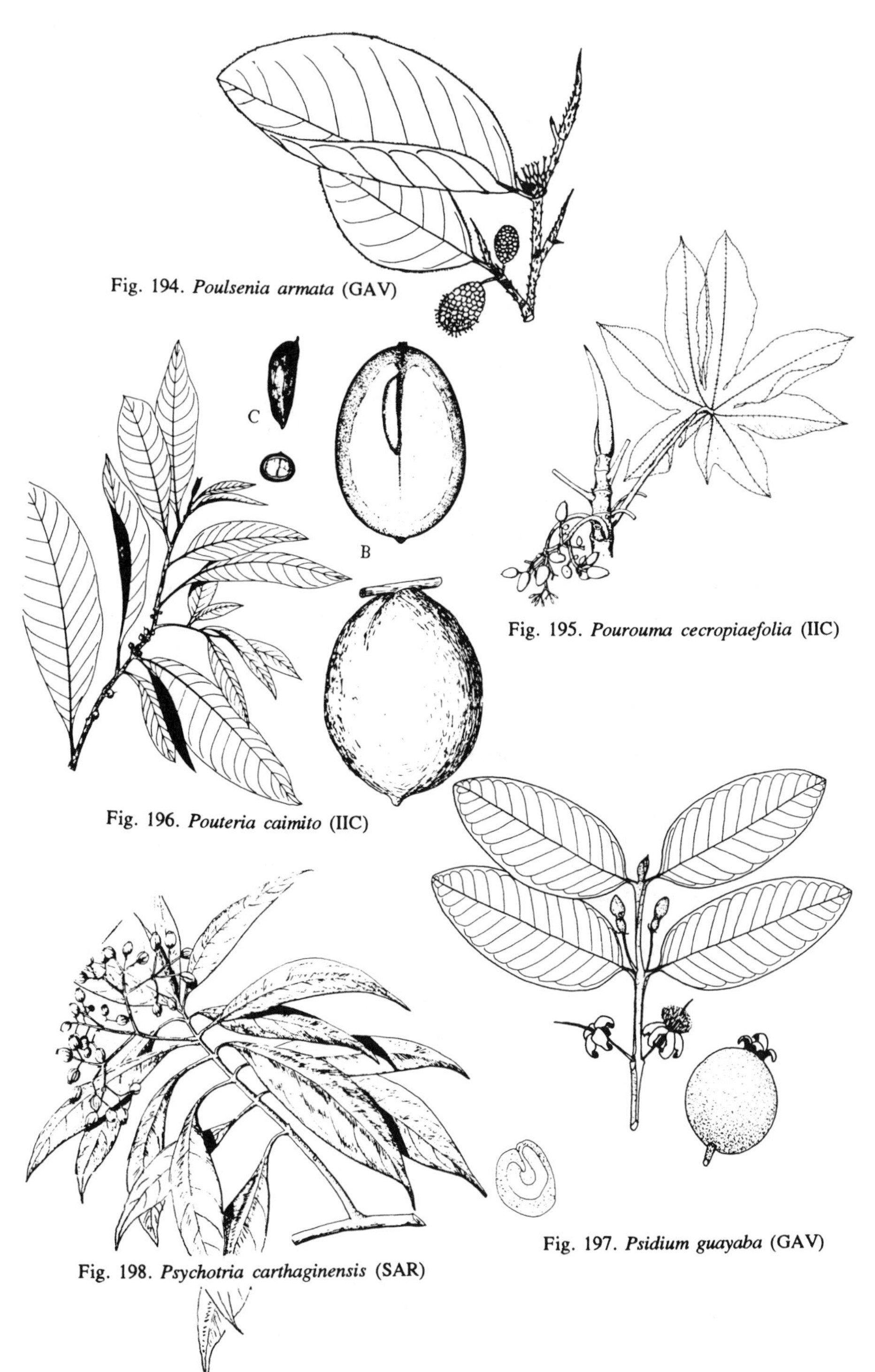

Fig. 194. *Poulsenia armata* (GAV)

Fig. 195. *Pourouma cecropiaefolia* (IIC)

Fig. 196. *Pouteria caimito* (IIC)

Fig. 197. *Psidium guayaba* (GAV)

Fig. 198. *Psychotria carthaginensis* (SAR)

Protium spruceanum Engl. Burseraceae. "Copal". Latex extracted. Arils eaten (VAG).

Protium subserratum (Engl.) Engl. Burseraceae. "Copal". Latex extracted. Arils from the fruit edible.

Prunus avium L. Rosaceae. "Cereso", **"Sweet cherry"**. Cultivated ornamental, also for fruit.

Prunus cerasus L. Rosaceae. "Guinda", **"Sour cherry"**. Cultivated ornamental, also for fruit.

Pseudobombax munguba (Mart. & Zucc.) Dugand. Bombacaceae. "Punga", "Pretino punga". Wood used for buoys; cotton from seeds is used with darts. "Boras" use the inner bark to make "tipití", an implement used to press cassava. Bark used for rope (RVM).

Pseudolmedia laevigata Trecul. Moraceae. "Chimicua", "Misho chaqui", "Motelo chaqui". Wood for lumber, posts, forked poles, dormers.

Pseudolmedia laevis (R.&P.) Macbr. Moraceae. "Chimicua", "Itauba amarilla". Wood for timber, stakes, and dormers. Fruit edible (VAG).

Pseudoxandra polyphleba (Diels) R.E. Fries. Annonaceae. "Siririca", "Espintana". Wood for house construction, and fishing poles.

Psidium acutangulum DC. Myrtaceae. "Guayabilla". Fruit edible (VAG).

♀ *Psidium guayaba* L. Myrtaceae. "Guayabo", "Guayabo blanco", **"Guava"**. Cultivated. Fruit is edible. Wood used to for tool handles, and for the "tramojo" (an implement put on pigs so they cannot walk easily). The infusion of foliar buds is used for diarrhea (especially that caused by bacteria, AYA). Also used for sanitary napkins; for dentition, and swellings of gout (VAM). "Exumas" use the leaves and roots for diarrhea. Natives of Cojeles (Venezuela) use the bark decoction for diarrhea, the floral infusion to regulate menstrual periods (FOR). "Créoles" and "Wayãpi" use decoction of bark, leaves, and shoots for diarrhea (GMJ). Tramil recommends the leaves for diarrhea, emotional shock, vertigo, and vomiting (TRA). (Fig. 197)

Psychotria acuminata Benth. Rubiaceae. "Sananguillo". "Cuna" use a daily portion of the cooked root for children who urinate too often (FOR).

Psychotria alba R.& P. Rubiaceae. "Tupamaqui", "Ucumi-micuna", "Yagé". Sometimes used with "ayahuasca".

Psychotria carthaginensis Jacq. Rubiaceae. "Sameruca" "Yagé". Added to "ayahuasca" (RVM). Considered TOXIC, the leaves contain tryptamines (SAR). (Fig. 198)

Psychotria deflexa DC. Rubiaceae. "Sananguillo". The "Cunas" put the leaves in cold water to bathe the children, twice a day, to relieve fever (FOR).

Psychotria horizontalis Sw. Rubiaceae. "Tupamaqui". Added to "ayahuasca".

Psychotria marginata Sw. Rubiaceae. "Yagé", "Sanaguillo". Added to "ayahuasca". "Cunas" use the hot leaf decoction to wash (many times a day) sore parts caused by blows (FOR).

Psychotria poepiggiana Muell.-Arg. Rubiaceae. "Boca pintada", "Oreja del diablo" **"Devil's ear"**, "Picho e mula". Ornamental. "Chami" use it in baths to treat hemorrhoids (CAA). "Wayãpi" use the flower bracts as an analgesic for the auricular pains; "Palikur" use the flower decoction as an antitussive (GMJ). Colombians take the hot root internally and massage it on the chest for lung ailments (SAR).

Psychotria stenostachya Standl. Rubiaceae. "Yagé", "Rumo sacha". Used in preparing "ayahuasca".

Psychotria viridis R.&P. Rubiaceae. "Chacruna", "Yagé", "Tupamaqui". Used in making "ayahuasca". Contains N,N-dimethyl tryptamine (SAR).

Pterocarpus amazonum (Mart.) Amshoff. Fabaceae. "Coshon tama", Jaguar caspi", "Maria buena", "Mututi", "Palo sangre blanca". Latex applied under the tongue three times a day for fever (VDF). Bites of inhabitant ants used for rheumatic pains (RVM).

Pterocarpus rufescens Benth. Fabaceae. "María buena", **"Good Mary"**. Wood for decorative plaques.

Pueraria phaseoloides (Roxb.) Benth. Fabaceae. "Kudzú". Cultivated. As a soil-ameliorating nitrogen-fixing forage, sometimes becoming undesirable (RVM). Elsewhere used for boils and ulcers (DAW).

♀ *Punica granatum* L. Punicaceae. "Granda", **"Pomegranate"**. Cultivated shrub, fruit edible. Bark taenicidal. Seeds, containing more than 15 ppm estrone, are reportedly estrogenic (JAD).

- Q -

Qualea paraensis Ducke. Vochysiaceae. "Quillo sisa", "Yesca caspi", "Moena sin olor". Wood for furniture, house construction.

♀ *Quararibea cordata.* (H.&B.) Vischer. Bombacaceae. "Numiallamihe", "Sapote", "Sapote de monte". Fruit is edible; wood for lumber. "Tikuna" use fruit pulp with *Pouzolzia* for dysmenorrhea (SAR). (Fig. 199)

Quararibea ochrocalyx (K.Schum.) Vischer. Bombaceae. "Machín sapote". Fruit edible (VAG).

Quassia amara L. Simaroubaceae. "Amargo", "Cuasia", **"Bitterwood"**. Insecticidal, tonic, for fever and hepatitis (RAR). Brazilians use the leaf tea in bathing for measles (BDS), a remedy that sounds a bit better than tea of ashes of dry white dog dung. Brazilians also wash the mouth with leaf tea after tooth extraction. Surinamese "Maroons" use the bark for fever and parasites (MJP). Potent aphidicide (MJP).

Fig. 199. *Quararibea cordata* (IIC)

- R -

Randia ruiziana DC. Rubiaceae. "Huitillo". Fruit edible (VAG).

Rauwolfia andina Markg. Apocynaceae. "Misho runto". Roots used as POISON. Reserpine extracted from *Rauwolfia* spp. is used for arterial hypertension and CNS problems.

Rauwolfia sprucei Muell.-Arg. Apocynaceae. "Sanango", "Misho runto". Source of reserpine. (Fig. 200)

Rauwolfia tetraphylla L. Apocynaceae. "Misho runto", "Pelilla", "Sanango", "Turcassa", "**Amazonian snakeroot**". Around Pucallpa, the leaf decoction is used for toothache (VDF). "Shipibos", "Yaguas", and "Achuales" use the roots as arrow POISON (AYA). Reserpine, tetraphylline, and tetraphyllicine are obtained from this species and from *R. sprucei* (LAE).

Remijia asperula Standl. Rubiaceae. "Padojcohe". Used in house construction (DAT).

Remijia peruviana Standl. Rubiaceae. "Chullachaqui caspi". Iquitos children drink it during the new moon to become strong (SAR). Febrifuge (RAR).

Renealmia alpina (Rottb.) Maas. Zingiberaceae. "Mishquipanga". Fruits yield a red-purple dye used for cloth and handicraft. Don Segundo suggests this plant as an ephemeral mosquito repellant; it seems to work, albeit briefly, on some of us.

Rhabdadenia biflora (Jacq.) Muell.-Arg. Apocynaceae. "Nea pono". Around Pucallpa, the leaf decoction is used as an antiseptic cicatrizant for abscesses. Latex used for toothache and gingirrhagia (VDF).

♀ *Rhacoma urbaniana* Loes. Celastraceae. "Ullucuy chuchuashi". Tocache natives use bark infusion as an abortive (RVM).

Rheedia gardneriana (Miers ex Planch.) Triana. Clusiaceae. "Charichuelo de hoja menuda". Fruit edible; wood for house construction.

Rhigospira quadrangularis Marcgraf. Apocynaceae. "Yahuarhuayo colorado". Fruit edible (RVM).

Rhodognaphalopsis brevipes A.Robyns. Bombacaceae. "Punguilla", "Punguilla del varilla". Wood used for house construction, beams, decks, and columns.

Rhodostemonodaphne grandis (Mez) Rohwer. Lauraceae. "Muena", "Garza muena". Wood for lumber, and boat contruction.

Ricinus communis L. Euphorbiaceae. "Higuerilla", "**Castor bean**". Cultivated. Seeds yield castor oil, used as liniment and laxative. Presently used industrially in paints, soaps, varnish, ink, and plastics (CRC). "Créoles" use as a purgative; "Palikur" use leaf decoction for fever. The oil is used as a liniment to soothe muscular pains (GMJ). The deadly POISON ricin can be put on "kamikaze" monoclonal antibodies and directed to tumors or viruses (JAD). (Fig. 201)

Rinorea racemosa (Mart.) Ktze. Violaceae. "Yutubanco", "Limóncillo", "Cafecillo". Wood used in home construction. "Tirio" use the crushed bark of a Surinamese species for wasp stings (MJP).

Rollinia cardiantha Diels. Annonaceae. "Anona", "Anonilla". Fruit edible (RVM).

Rollinia curvipetala R.E.Fries. Annonaceae. "Anonilla". Fruit edible (RVM).

Rollinia cuspidata Mart. Annonaceae. "Anonilla". Fruit edible (RVM).

Rollinia edulis Pl. & Tr. Annonaceae. "Anona", "Anonilla". Fruit edible (RVM).

Rollinia mucosa (Jaq.) Baill. Annonaceae. "Anona". Cultivated. Fruit edible. (Fig. 202)

Rollinia pittieri Safford. Annonaceae. "Sacha anona". Wood used for house construction, beams, decks.

Rosa canina L. Rosaceae. "Rosa", "**Rose**". Cultivated. Ornamental.

Rosa centifolia L. Rosaceae. "Rosa del remedio", "**Medicinal rose**". Ornamental.

Rosa gallica L. Rosaceae. "Rosa". Cultivated. Ornamental.

Rosa indica L. Rosaceae. "Rosa Alejandrina", "**Alexandran rose**". Ornamental.

Roucheria punctata Ducke. Linaceae. "Puma caspi". Wood used for construction, beams, decks, and columns (RVM). "Taiwano" believe the bark decoction of the related *R. calophylla* Planch. is a "sure cure" for malaria (SAR).

Ruitzerania trichanthera (Spruce ex Warm.) Marcano-Berti. Vochysiaceae. "Quillo sisa", "Moena sin olor". Wood for lumber, general construction.

Ruizodendron ovale (R.&P.) R.E.Fr. Annonaceae. "Espintana". Wood is used for interior decoration.

♀ *Ruta chalepensis* L. Rutaceae. "Ruda", "**Rue**". Cultivated. Around Iquitos, tied to many superstitions; a little branch of rue dispels bad spirits, and attracts clientele. Some merchants carry a rue leaf on their ears. Rue tea mixed with castor oil is well known as an oxytocic (SOU). Powdered leaves used to treat otitis, ophthalmia, and pediculosis. Leaf tea taken, perhaps dangerously, as antidysmenorrheic, antihysteric, cardiotonic, digestive, sedative, vermifuge. Vinegar decoction of leaves used for decongestant and sedative in myalgia (FEO). Brazilians use the tea for stroke (BDS). Psoralens in the plant may render it photoTOXIC (JAD).

Ryania speciosa Vahl var. *tomentosa* (Miq.) Monach. Flacourtiaceae. "Esponja huayo", "Espuma huayo". Highly toxic species used by the "Paumari" as fish POISON; used for making insecticides (RVM). On Rio Negro, they used the roots for rat POISON (SAR) "Maku" said to use the plant for euthanasia, homocide and suicide (DAW). (Fig. 203)

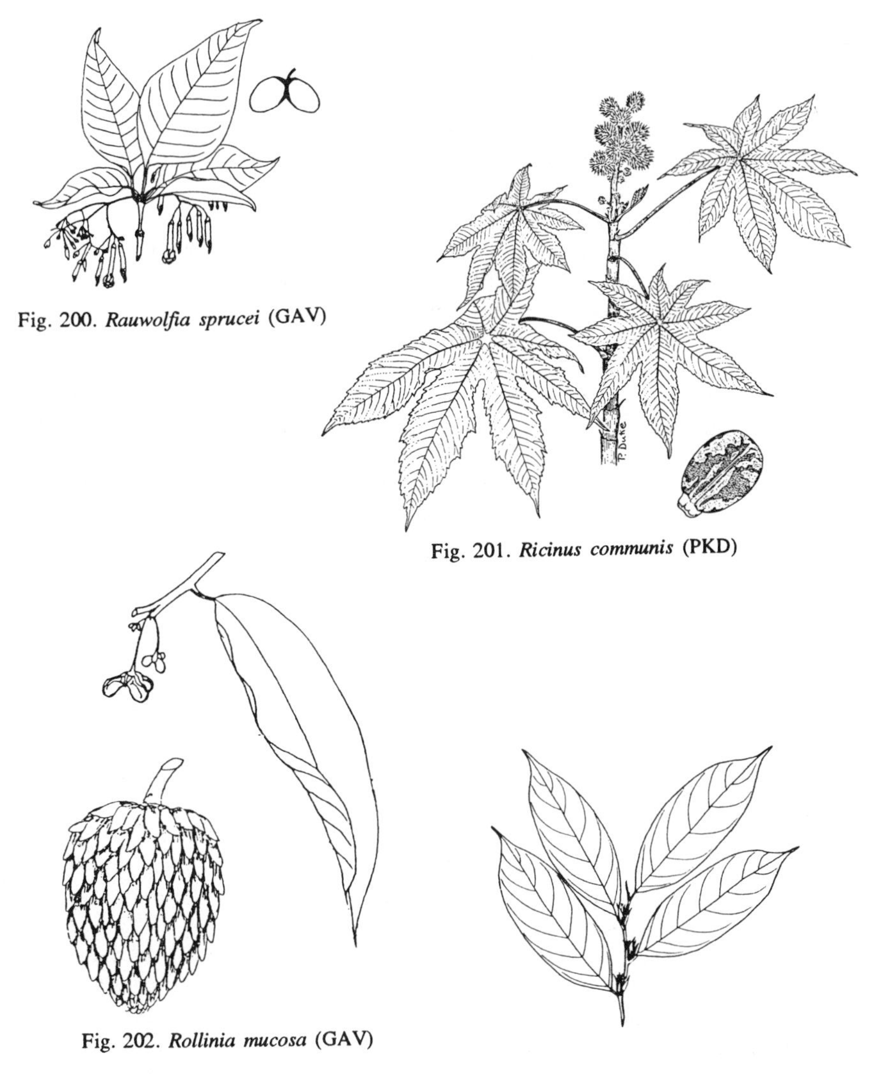

Fig. 200. *Rauwolfia sprucei* (GAV)

Fig. 201. *Ricinus communis* (PKD)

Fig. 202. *Rollinia mucosa* (GAV)

Fig. 203. *Ryania speciosa* (GAV)

- S -

Sabal palmetto (Walt.) Lodd. Aracaceae. "**Texas palmetto**". Cultivated ornamental.

Sabicea paraensis (Schum.) Vernh. Rubiaceae. "Huasca mullaca", "Ruichao". Edible fruits.

Sabicea villosa Roemer & Schultes. Rubiaceae. "Curihjau", "Huasca mullaca". Edible fruits. "Wayãpi" use the leaf decoction for dysentery, for cramps, colic, and abdominal pains (RVM). Around Rio Apaporis, the astringent leaf infusion is used for malaria (SAR).

Saccharum x *officinarum* L. Poaceae. "Caña de azúcar" "**Sugar cane**". Cultivated. Sugar cane juice used as a refreshing beverage, fermented into spiritous beverages. Known locally as "leva" or "guarapo", distilled to make "aguardiente". Used to sweeten foods and medicine.

Sacoglottis mattogrossensis Mal. Humiriaceae. "Loro shungo". Wood for construction, posts, dormers, bridges, and mortars for sugar mills.

Salix martiana Leyboyld. Salicaceae. "Sauco" "**Willow**". Bark and leaves used as sudorific antirrheumatics (RVM), both internally and as a wash (VDF). Also for gonorrhea, hemoptysis, and worms.

Salvia occidentalis Sw. Lamiaceae. "Salvia" "**Sage**". Used for colic, nausea, and flatulence.

Salvia splendens Ker-Gawl. Lamiaceae. "Flor de fuego", "**Scarlet sage**". Cultivated ornamental.

Salvinia minima Bab. Salviniaceae. "Lenteja de agua" "**Water lentil**". Aquatic ornamental.

♀ *Sambucus mexicana* Presl. var. *bipinnata* (S.&C.) Schw. Caprifoliaceae. "Sauco" "**Elder**". Cultivated ornamental. Flower decoction used as a sudorific, antipyretic, for stomachaches, and urinal problems (urethritis). Infusion of flowers and shoots drunk to relieve fever and flu (CAA). "Tikuna" take 3 cups leaf decoction a day for measles (SAR). Elsewhere, considered anodyne, aperient, cathartic, diaphoretic, diuretic, expectorant, and stimulant; used for amenorrhea, asthma, catarrh, cough, dropsy, epilepsy, fever, flu, headache, inflammation, rheumatism and syphilis (DAW). Fruit edible (RVM).

Sambucus peruviana HBK. Caprifoliaceae. "Rayan", "Sauco", "**Elderberry**". Chopped leaves applied as a "galactophorus application"; decoction used as an oral antiseptic. Flower decoction considered aphrodisiac, antirheumatic, antiseptic, depurative, and sudorific (FEO, BDS). Brazilians take flower tea of *Sambucus nigra* for chickenpox.

Sanchezia peruviana (Nees) Rusby. Acanthaceae. "Antara caspi". Planted ornamental.

Sanchezia tigrina Leon. Acanthaceae. "Matzi sanban", "Sanban". Leaves applied to rheumatic joints around Pucallpa (VDF).

Sansevieria thyrsiflora Thunb. Agavaceae. "Sanseveria", "**Bowstring hemp**". Cultivated ornamental.

Sapindus saponaria L. Sapindaceae. "Choloque", "Tingana", "**Soapberry**". Fruits once used for soap; now used in children's games ("canicas") like marbles.

Sapium glandulosum (L.) Morong. Euphorbiaceae. "Caucho masha", "Shiringarana". The latex is bulked with that of *Hevea*; used as gutta-percha. (Fig.204)

Sapium marmieri Huber. Euphorbiaceae. "Shiringarana", "**Gutapercha**". As the previous species.

Sauvagesia erecta L. Ochnaceae. "Hierba de San Martin", "Intimiracu". Decoction used as diuretic. "Créoles" use the leaf tea for fever (GMJ). "Siona" take the plant decoction for stomachache (SAR). Elsewhere used for snakebite (DAW); tuberculosis (RAR).

Scheelea basleriana Burret. Arecaceae. "Shebon". Fruit and/or seeds edible (RAR, RVM).

Scheelea cephalotes (Epp.) Karst. Arecaceae. "Shapaja". Seeds and terminal buds edible. Small edible grubs (suri) may eat the kernel (JAD). Used for construction.

Scheelea plowmanii Glass. Arecaceae. "Catirina". Seeds edible. Leaves used to roof houses.

Scheelea princeps Karst. Arecaceae. "Shapajilla". Seed edible cooked (VAG).

Scheelea salazerii Glassm. Arecaceae. "Shapaja". Seeds edible.

Scheelea tessmannii Burret. Arecaceae. "Shapaja". Seeds edible.

Schefflera morototoni (Aubl) Maguire, Steyerm. & Frodin. Araliaceae. "Moena sin olor", "Sacha uvilla". The clean wood is used for interior decorations. Young leaf rachis used by "Campa-ashaninca" for arrows (RVM). Considered anodyne, antirheumatic, antisciatic, aphrodisiac and tonic in Haiti (DAW). Cataplasm for "luxaciones" (RAR).

Schistostemon reticulatum (Ducke) Cuatr. Humiriaceae. "Parinari sacha". Fruit edible (RAR).

Schizaea elegans (Vahl) Sw. Schizaeaceae. "Señorita", "Paraguita". Ornamental. Leaf maceration with water for pimples. Cold water infusion is tonic (RVM).

Schizolobium amazonicum Huber ex Ducke. Fabaceae. "Pashaco". Wood for lumber, plaques, and plywood (RVM). "Tikuna" use leaf tea as febrifuge (SAR).

Schizolobium parahyba (Vell. Conc.) S.F.Blake. Fabaceae. "Pashaco". Wood for lumber, decorative plaques, and plywood. It can also be used as an ornamental.

Schoenobiblus peruvianus Standl. Thymelaeaceae. "Barbasco". Mashed roots a fish POISON (AYA). "Kofán" use in curaré (SAR). "Tikuna" poultice powdered leaves onto infected cuts and wounds (SAR). Ophthalmic analgesic (RAR).

Scleria flagellum-nigrorum Berg. Cyperaceae. "Cortadera", "Verdugo", "**Cutgrass**". The leaves and stems are put under the roofs to keep the bats away. (Fig. 205)

♀ *Scleria malaleuca* (Schlecht. & Cham.) Reichb. Cyperaceae. "Cortadera". Leaf decoction used in treating female sterility (FEO).

Scleria microcarpa (Liebm.) Steud. Cyperaceae. "Cortadera". Rhizome decoction for renal calculus (RVM).

Sclerolobium melinonii Harms. Fabaceae. "Tangarana de hoja menuda". Wood for lumber.

Sclerolobium rigidum Macbr. Fabaceae. "Tangarana de altura". Wood for lumber.

♀ *Scoparia dulcis* L. Scrophulariaceae. "Bati matsoti", "Escobilla", "Ñucñu-pichana", "Piqui pichana". Leaf infusion used for bronchitis, cough, diarrhea, fevers, kidney diseases, and hemorrhoids (RVM, VDF). Leaf infusion antidiarrheic and emetic (CAA). Antiseptic leaf decoction used for wounds; and fever. "Créoles" use the leaf decoction mixed with maternal milk as an antiemetic for infants. Dried leaves used by as a marihuana substitute. "Palikur" use the leaf decoction in antipyretic baths and in poultices for migraine headaches (GMJ). Ecuadorians take the tea for pain and swelling (SAR). "Tikuna" drink the tea, with or without "paico", three days during the menses as an abortifacient or contraceptive (SAR). Four to five plants tied together make the typical river-dweller's broom (RVM). Brazilians add the root to the bath when "cleaning their blood" (BDS). They apply strained leaf juice for eye ailments; and to infected wounds (erysipelas) (BDS).

Sechium edule (Jacq.) Sw. Cucurbitaceae. "Chayote". Fruit edible (VAG). Often cooked as a vegetable at Explorama.

Securidaca paniculata L.C. Rich. Polygalaceae. "Gallito". "Wayãpi" use the inner film of bark (cambium) to prepare a decoction used as a dental analgesic; "Palikur" use it to treat dermatosis (GMJ).

Selaginella exaltata (Kunze) Spring. Selaginellaceae. "Shapumba", "Helecho". Greenery used for Christmas decoration. "Cuna" boil the roots for a long time, and drink a cup a day of this tea for stomachache and pancreatic ailments (FOR).

Selaginella speciosa A.Br. Selaginellaceae. "Helecho", "Shapumba". Similar to *S. exaltata*.

Selaginella stellata Spring. Selaginellaceae. "Palillo", "Sapo magui". Brazilians bathe in the decoction as a flu treatment (BDS).

Senefeldera inclinata Muell.-Arg. Euphorbiaceae. "Tscaahe". Firewood (DAT). Mayna "Jivaro" apply scraped bark to toothache (SAR).

Sesamum indicum DC. Pedaliaceae. "Ajonjoli", "**Sesame**". Oil from macerated seeds used in massage for contortions (BDS). Sounds more pleasant than grated crocodile penis, drunk as a remedy, even better than crocodile oil for massage (BDS).

Sesuvium portulacastrum L. Aizoaceae. "Capin", "Litho". Decoction or infusion prescribed for bronchitis (VDF).

Setaria geniculata (Lam.) Beauv. Poaceae. "Grama-chilco", "**Knotroot-bristle-grass**". Forage grass. (Fig. 206)

♀ *Sicana odorifera* Naud. Cucurbitaceae. "Secana", Cassabanana". Young fruits cooked as vegetables, ripe fruits edible uncooked; taken for sore throat; leaf and flower decoction considered emmenagogue, laxative, vermifuge. Flowers may contain POISONOUS cyanide (JFM).

Sida acuta Burm. Malvaceae. "Jocuchuchupa", "Pichana". Plant used to make brooms. "Créoles" consider the leaf infusion diuretic. Macerated leaves soaked in water produce a mucilaginous solution valued as a shampoo to get rid of lice. "Wayãpi" use the decoction for fever. "Palikur" poultice the leaves onto migraine headaches (GMJ).

♀ *Sida rhombifolia* L. Malvaceae. "Ancusacha", "Pichana", Varilla". Considered analgesic, aphrodisiac, demulcent, diuretic, emmenagogue, emollient, lactagogue, and sedative; used for alopecia, antibiotic, bilious conditions, bladder ailments, boils, burns, conjunctivitis, dermatosis, diarrhea, dyspepsia, dyspnea, gastrosis, gonorrhea, impetigo, leucorrhea, lupus, piles, rheumatism, snakebite, sores, thrush, tuberculosis, tumors, ulcers, urethritis, and wounds (DAW, TRA, RAR).

Simarouba amara Aubl. Simaroubaceae. "Marupá". Wood for lumber, interior decorations, furniture, plywood veneer, paper pulp. Bark decoction for fever. "Créoles" mix macerated bark with rum as a tonic for malaria and dysentery (GMJ). Emetic, hemostat, purgative, tonic (RAR).

Simira rubescens (Benth.) Bremek. Rubiaceae. "Puca quiro". Wood for handicrafts and plaques. "Achuales" chew bark to prevent caries (AYA). "Boras" make a pink dye from the bark (SAR). "Taiwano" poultice the flowers onto skin infections (SAR).

Simira tinctoria (HBK) K.Schum. Rubiaceae. "Huacamayo caspi", "Puca quiro". The wood is used for handicrafts.

♀ *Siparuna guianensis* Aubl. Monimiaceae. "Isula huayo", "Picho huayo", "Asna huayo". Fruit used in fiestas, the leaf infusion believed aphrodisiac. Leaf decoction used in baths for mycosis. "Créoles" use the leaf tea as an abortive, oxytocic, and antipyretic; the alcoholic leaf maceration as vulnerary, and the salty leaf decoction as hypotensive. "Wayãpi" use the decoction of leaves and bark as a refreshment and antipyretic (GMJ). The tea of the leaves and flowers is used as a carminative, in dyspepsia, and painful spasms (RVM). Don Segundo informed one class that the aroma of this plant, applied to the skin to prevent hunted animals from smelling the hunter (by masking his body odor), was not only effective, but rendered the hunter all but irresistible to females. One of my taxonomic associates claims to have confirmed this empirically (JAD). "Tikuna" eat the fruits for dyspepsia (SAR). Elsewhere considered anodyne, insecticidal and stomachic; used folklorically for colds, colic, cramps, dermatosis, fever, headache, mange, rheumatism,

Fig. 204. *Sapium glandulosum* (GAV)

Fig. 205. *Scleria flagellum-nigrorum* (GAV)

Fig. 206. *Setaria geniculata* (HAC)

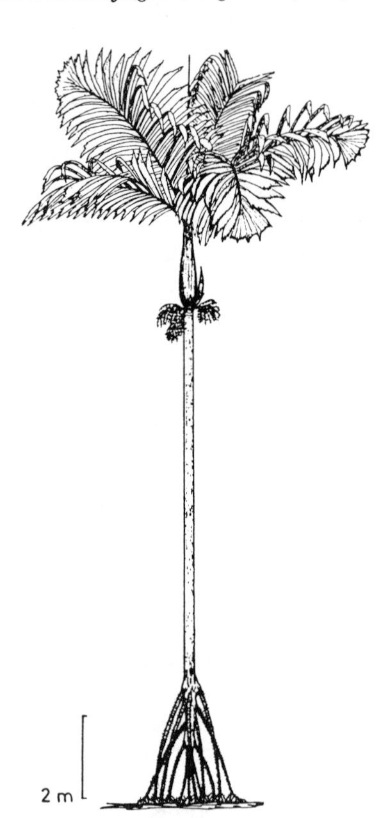

Fig. 207. *Socratea exorrhiza* (GAV)

Fig. 208. *Spigelia anthelmia* (GAV)

snakebite and wounds (DAW). Tapajos natives make solar tea from the leaves for bathing headache (BDS).

Sloanea laxiflora Spruce ex Benth. Elaeocarpaceae. "Cepanchina". Wood for lumber; buttress roots used to make oars, tables, doors, and handicrafts.

Sloanea terniflora (Moc. & Sesse). Standl. Elaeocarpaceae. "Cepanchina", "Yacu achotillo". Wood used for house construction; buttress roots used for doors, handicrafts, oars and tables.

Sloanea sp. Elaeocarpaceae. "Puzanga caspi". With dried leaves of *Jacaranda copaia,* the bark is used in love potions.

Smilax ruiziana Kunth. Smilacaceae. "Mai mosha". Leaves poulticed onto arthritic joints around Pucallpa (VDF).

Socratea exorrhiza (Mart.) H. Wendl. Arecaceae. "Pona", "Cashapona", "**Stilt palm**". Stem bark slatted and used for floors, walls, room dividers, supports for roofs etc. "Achuales" use the leaf decoction for hepatitis (AYA). Ashuar use the roots (NIC). "Palikur" consider the fruits TOXIC. "Wayãpi" use the grated stipes to poultice on umbilical cords of newborn babies (GMJ). (Fig. 207 and Fig. 2-Introduction)

Solanum americanum Mill. Solanaceae. "Ucsha-coconilla". Used to treat erysipelas and cough. "Cuna" use leaf infusion for mycoses, rubbing liquid on the affected areas, preferably hot (FOR).

Solanum barbeyanum Huber. Solanaceae. "Mashua". Leaves used to massage local inflammation around Pucallpa (VDF).

Solanum coconilla Huber. Solanaceae. "Coconilla colorada". Fruit edible.

Solanum grandiflorum R.&P. Solanaceae. "Poni ani mite", "San Pablo". Plant used internally as antiinflammatory resolvent around Pucallpa (VDF).

Solanum jamaicense Mill. Solanaceae. "Coconilla con espiñas". Fruit edible (RVM), but "Tikuna" regard the fruits as TOXIC (SAR). Leticia natives boil the fruits with yuca flour as an antidote to food-poisoning (SAR).

Solanum leucopogon Huber. Solanaceae. "Cubu nichi", "Intuto quiro". Leaf decoction as a topical antiinflammatory around Pucallpa (VDF).

♀ *Solanum mammosum* L. Solanaceae. "Vaca chucho", "Tinctona", "**Breast berry**". Used as an ornamental; fruit said to be POISONOUS. "Boras" use it to treat the sores of leishmaniasis, a worm infection (DAT). "Chocó" (JAD) and the "Chami" use the fruit to kill cockroaches (CAA). "Cuna" use fruit macerated in hot water for growths on the breast (doctrine of signatures?). In Tolima and Santander seeds are used as insecticides (FOR). Guatemalans use fruits as medicine and ornament during pilgrimages. In Costa Rica, the leaf decoction is used for kidney and bladder infections. The decoction of the fruit with all its juice is used for asthma; plant also used for sinusitis, arthritis and rheumatism (POV). "Kofán" use as a pacifier for small children (SAR).

Solanum sessiliflorum Dun. Solanaceae. "Cocona", "Topiro". Cultivated. Fruit edible and makes good juice, often served at Explorama. Juice used as a scabicide; also recommended after snakebite (RVM). "Waorani" rub juice on scalp to cleanse and gloss the hair (SAR). Boiled plant rubbed on spiderbites to heal necrotic tissue (SAR). Following scorpion sting, juice is drunk to prevent vomiting (SAR).

Solanum sisimbrifolium Lam. Solanaceae. "Misqui corrota", "Ocote mullaca". Leaves regarded as febrifugal around Pucallpa (VDF).

Sorghum bicolor (L.) Moench. Poaceae. "Sorgo", "Escoba", "Maiz guineo". Cultivated. As a forage, grain or as raw material for industry.

Sparattanthelium sp. Hernandiaceae. "Chundu huasca". Peruvians use the stem decoction for cough, diarrhea, headache and stomachache (SAR). "Tikuna" assert that powdered leaves of *S. glabrum* applied to the lip in oil for two weeks is depilatory (paralleling an activity reported for *Hernandia* in Oceania) (SAR).

♀ *Spartium junceum* L. Fabaceae. "Retamita", "Talhui", **"Spanish broom"**. Cultivated ornamental. Roasted flowers used for rheumatic pains and migraine headaches. They rub the fresh flowers on to clear freckles; also used as a diuretic, antirrheumatic, and for jaundice (RVM). Dried flowers smoked for asthma (SAR). Ecuadorians take the root infusion as abortifacient (SAR). Analgesic, diuretic, purgative (RAR). For albuminuria (RAR). Contains cytisine, methyl-cytisine, and sparteine.

Spathicalyx xanthophylla (DC.) A. Gentry. Bignoniaceae. "Huapan", "Quillo-panga huasca". Around Pucallpa, leaf maceration used in washes for fever and headache (VDF). "Tikuna" from Colombia treat conjunctivitis with leaf infusion (AYA).

Spathodea campanulata Beauv. Bignoniaceae. "Espatodea", **"African tulip tree"**. Cultivated ornamental. Elsewhere considered laxative, POISON, and stomachic; used for backache, dermatoses, dysentery, edema, enteritis, gastritis, gonorrhea, guineaworm, nephritis, stomach problems, ulcers, urethritis, and wounds (DAW).

Spigelia anthelmia L. Loganiaceae. "Pega pinto", **"Worm grass"**. Roots used as anthelmintic, POISONOUS when fresh (SOU). "Chocó" use it as a dangerous purge (JAD). "Cuna" respect this plant, using it in small doses to eliminate parasites; if used in bigger doses, it can kill a person (JAD). "Cuna" cook the roots with leaves of *Piper peltata*. The liquid from the decoction is used for stomachache, taking a small cup a day. Considered POISONOUS by the "Créoles" who use it as a vermifuge (GMJ). Root infusion used to bathe children as sedative and tranquilizer. Given, perhaps dangerously, as a soothing and cooling drink (SAR). (Fig. 208)

Spigelia humboldtiana Cham.& Schl. Loganiaceae. "Yape". Fruit reportedly mixed with *Bixa orellana* as face paint (RVM).

Spilanthes acmella L. Asteraceae. "Botoncillo". Brazilians boil the flowering tops for the lungs, specifically tuberculosis (BDS). (Fig. 209)

Spondias cytherea Sonn. Anacardiaceae. "Taperiba", "Tapisho", **"Otaheite apple"**. Cultivated fruit tree. Ripe fruits edible, pickled. Leaves eaten or used to tenderize meat (JFM). Tramil mentions antifertility effects (TRA).

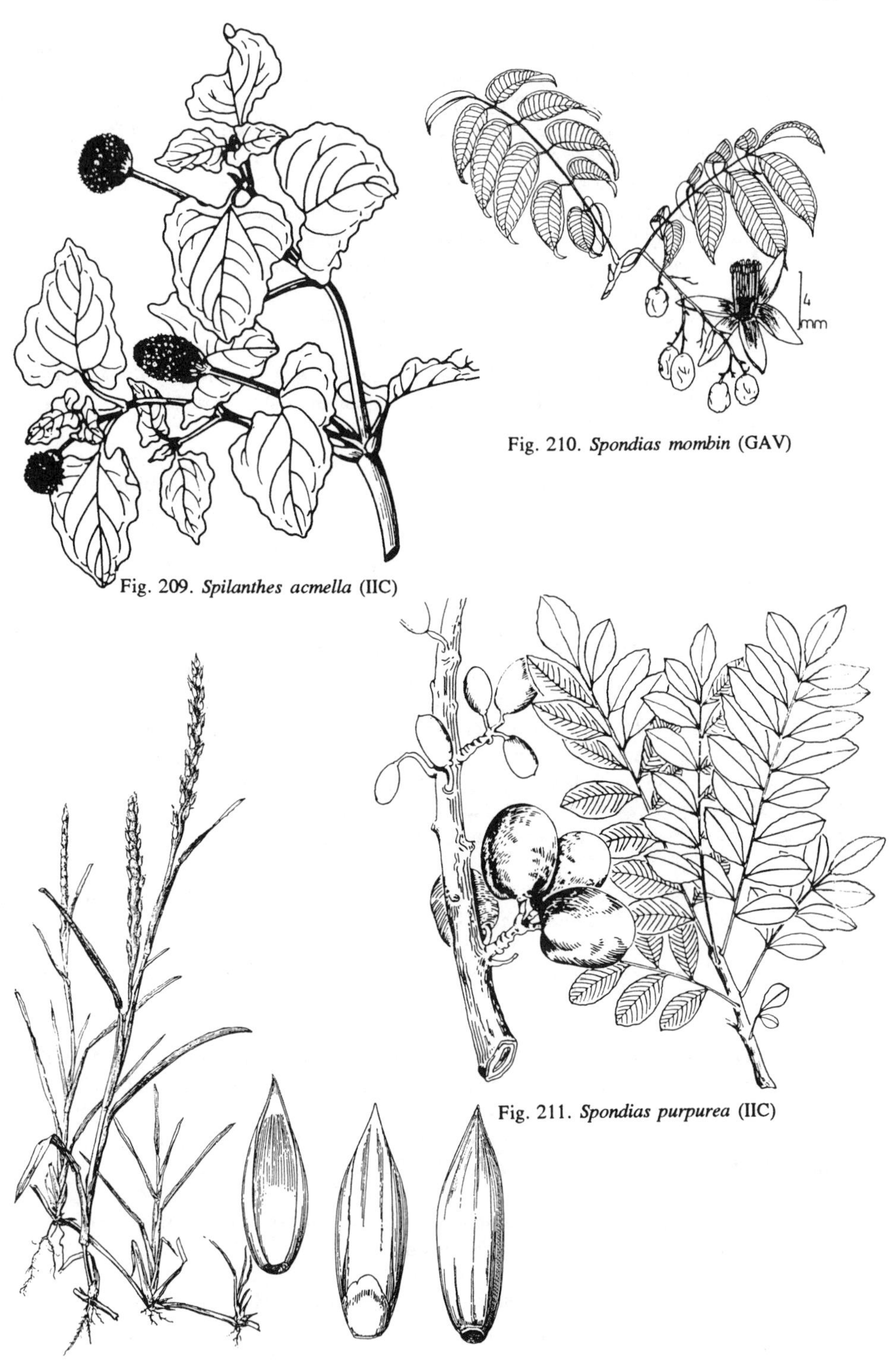

Fig. 209. *Spilanthes acmella* (IIC)

Fig. 210. *Spondias mombin* (GAV)

Fig. 211. *Spondias purpurea* (IIC)

Fig. 212. *Stenotaphrum secundatum* (HAC)

♀ *Spondias mombin* L. Anacardiaceae. "Ciruela", "Hubo", "Ubos, "Ushun", "**Hog Plum**". Fruit edible. Wood for lumber and veneer. Root decoction used for diarrhea, and for mothers after giving birth, taking small doses for two consecutive months. Itaya residents use it for tuberculosis, as an adjuvant with antibiotics. Docoction used for vaginal baths to treat infections and hemorrhoids (AYA). "Campas" use it to lure tapirs (RVM). "Créoles" use the bark for diarrhea and upset stomach (GMJ). "Tikuna" use bark decoction as anodyne and hemostat in diarrhea, metrorrhagia and stomachache (SAR). A single cup, given each day during the menses, is believed contraceptive; drunk one day after delivery, it is believed to lead to permanent sterility (SAR). Tramil mentions antiviral, myorelaxant and uterotonic activities (TRA). In Brazil, used in ice creams and liqueurs (MJP). (Fig. 210)

Spondias purpurea L. Anacardiaceae. "Ciruela" "**Spanish plum**". Cultivated fruit tree. Leaves used for amebiasis, diarrhea, fever, flu, sprains and trauma, bark for diarrhea and ulcers (TRA). (Fig. 211)

Spondias radlkoferi Donn. Smith. Anacardiaceae. "Ushum", "Tapisho sacha". Wood for lumber, plywood; fruit edible.

Spondias venulosa Mart. ex Engl. Anacardiaceae. "Ubos", "Ushum". Wood for lumber; fruit edible. Bark decoction for diarrhea.

Stachytarpheta cayennensis (Rich). Vahl. Verbenaceae. "Ocollucuy sacha", "Sacha verbena". The stems and leaves are soaked in some water, squeezed and mixed, the greenish extract drunk, one glass a day, for three consecutive months for diabetes (AYA). UHV natives use the plant in medicine for their dogs (RAF). "Créoles" use the leaf tea as a cholagogue purgative for dysentery. "Wayãpi" and "Palikur" use the plant decoction in baths to relieve colds and headaches (GMJ). Venezuelans have used it for tumors, Dominicans as a panacea, and Trinidadians as a collyrium and depurative in chest colds, dysentery, fever, heart attacks, ophthalmia and worms (DAW).

♀ *Stenomesson variegatum* R.&P. Amaryllidaceae. "Mayhua". Bulb decoction taken internally as abortifacient emmenagogue (FEO).

Stenosolen eggersii Markgf. Apocynaceae. "Papelillo", "Lobo sanango". Used for rheumatism (RVM).

Stenosolen van-huerchii (Muell.-Arg.) Mgf. Apocynaceae. "Huano sacha". Bark decoction used internally for rheumatism, leaf decoction a wash for headache (VDF).

Stenotaphrum secundatum (Walt.) Kuntze. Poaceae. "Grama inglesa", "**St. Augustine grass**". Cultivated lawn grass. (Fig. 212)

Sterigmapetalum obovatum Kuhlmann. Rhizophoraceae. "Quillobordón masha", "Mangle de altura". Wood for house construction.

Stigmaphyllum sp. Malpighiaceae. "Huasca sisa". The sap is used by the "Achuales" to extract decayed teeth (LAE).

Fig. 213. *Strychnos guianensis* (GAV)

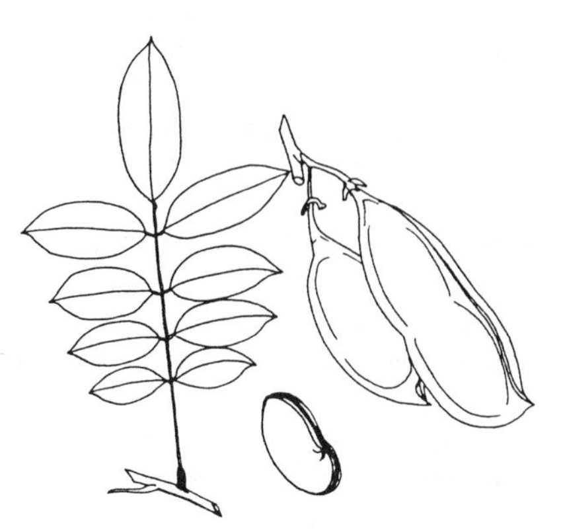

Fig. 214. *Swartzia polyphylla* (GAV)

Fig. 215. *Swietenia macrophylla* (GAV)

Fig. 216. *Symbolanthus calygonus* (GAV)

Fig. 217. *Symphonia globulifera* (GAV)

Fig. 218. *Syzygium cumini* (IIC)

Strychnos guianensis (Aubl.) Mart. Loganiaceae. "Comida del venado", "Anzuelo casha". Stems used to make "curaré"; recommended as an aphrodisiac. Mixed with *Uncaria guianensis*, the decoction is used in genital baths for venereal diseases (RVM). Contains brucine, eritocurarine, guaiacurarines, guaiacurine, c-guaianine, and strychnine (JAD). (Fig. 213)

Strychnos rondeletioides Spruce. Loganiaceae, "Anzuelo yacu". Similar to the previous species.

Strychnos sp. Loganiaceae, "Vona muca". Branches chewed for toothache (VDF).

Stryphnodendron polystachyum (Miq.) Klein. Fabaceae. "Pashaco", "Huamanzamana pashaco". Wood for lumber.

Swartzia laevicarpa Amshoff. Fabaceae. "Acero shimbillo", "Palo sangre dominante". Wood used for construction, beams, decks, tool handles, and dormers.

♀ *Swartzia polyphylla* A.DC. Fabaceae. "Añushi-remo caspi", "Cumaseba". Wood used for house construction, beams, decks, dormers. Alcoholic maceration of the duramen is used to hasten healing of dislocations, and to speed healing after childbirth. (Fig. 214)

Swartzia simplex (Sw.) Spreng. Fabaceae. "Porotillo". "Cuna" put the fruit and leaves in cold water, and use it for sudden cerebral pains, by washing their heads continuously with this liquid (FOR). "Kubeo" rub boiled crushed leaves on the abdomen four times a day for hepatitis (SAR).

Swietenia macrophylla G. King. Meliaceae. "Aguano", "Caoba", "**Mahogany**". Wood for lumber, decorative plaques, furniture and handicrafts. Considered one of the best quality woods in Amazonian Peru. (Fig. 215)

Syagrus tessmannii Burret. Arecaceae. "Bella vista", "Inchaui". Wood used for construction and lances. Larvae living in fallen stems may be edible (RVM).

Symbolanthus calygonus (R.&P.) Griseb. Gentianaceae. "Flor de mariposa", "**Butterfly flower**". As an ornamental. (Fig. 216)

Symphonia globulifera L.f. Clusiaceae. "Azufre caspi", "Navidad caspi", "Chullachaqui", "**Buckwax**". Wood used for house construction, canoes, paddles, keel plates, flooring, carpentry, tool handles, etc. It is good quality for construction, carpentry, and firewood (RVM). Latex used to caulk boats (RVM). "Créoles" use the latex for dermatosis, and to reinforce the binding of the arrows (RVM). Indians apply the bark ash to wounds and indolent ulcers (SAR). Brazilians use the seed oil for dermatoses (SAR). (Fig. 217)

Syngonium vellozianum Schott. Araceae. "Bonan rao". Aerial parts applied topically to relieve the itch of bugbites around Pucallpa (VDF).

Syzygium cumini (L.) Skeels. Myrtaceae. "Aceituna dulce", "**Java plum**", "**Sweet olive**". Cultivated ornamental tree; fruit edible. (Fig. 218)

Syzygium jambos (L.) Alst. Myrtaceae. "Pomarrosa", "**Rose apple**". Cultivated for the edible fruit. Leaves, containing limonene and pinene, used for conjunctivitis, fever, rheumatism, root for epilepsy (VAG).

Syzygium malaccensis (L.) Merr. & Perry. Myrtaceae. "Pomarrosa", "Mamey", "**Malay apple**". Cultivated ornamental; fruit edible. (Fig. 219)

Fig. 219. *Syzygium malaccensis* (IIC)

- T -

Tabebuia chrysanta (Jacq.) Nichols. Bignoniaceae. "Tahuarí negro", "Paliperro". Wood for lumber, posts, poles, handicrafts, parquets. "Yaguas" use the trunk to make jungle drums. Over-exported to the US as "tahebo" or "pao-d'arco", bark tea marketed for candidiasis, cancer, and malignant tumors (JAD).

Tabebuia incana A.Gentry. Bignoniaceae. "Tahuarí", "Tarota". Same applications as the previous species; bark decoction, or tincture used for tumors (RVM). "Urarina" use one "tahuarí" for kidney and liver disorders (NIC). (Fig. 220)

♀ *Tabebuia obscura* (Bur & K. Schum). Sandw. Bignoniaceae. "Purma tahuarí". Planted as an ornamental (RVM). "Taiwano" put dried flowers in food for dysmenorrhea, "Bora" believe the bark antirheumatic (SAR).

Tabebuia ochracea (Chan) Sandw. ssp. *heteropoda* (DC.) Gentry. Bignoniaceae. "Tahuarí colorado". Wood for lumber; parquet, posts, forked poles, dormers, support for sugar mills; the bark is used like *T. chrysantha.*

Tabebuia rosea (Bertol.) DC. Bignoniaceae. "Tahuarí", "Pali-perro", "**Trumpet tree**". Wood for posts, forked poles, dormers, and parquet.

Tabebuia serratifolia (Vahl) Nichol. Bignoniaceae. "Tahuarí", "**Surinam greenheart**". Wood and bark used as *T. chrysantha.* "Créoles" use the flower decoction mixed with sugar, as a pectoral syrup for colds, cough, and flu. "Palikur" use bark to poultice onto leishmaniasis sores. "Wayãpi" use the bark for fever (GMJ).

Tabernaemontana marcgrafiana Apocynaceae. "Sanango". Fruit edible (RVM).

Tabernaemontana maxima Markg. Apocynaceae. "Sanango ucho". Fruit edible (DAT).

Tabernaemontana muelleriana Mart. ex Muell.-Arg. Apocynaceae. "Sanango". Latex used by the "Boras" to POISON fish (DAT).

Tabernaemontana rimulosa Woods. Apocynaceae. "Lobo Sanango". Amazonian panacea (SAR). Around Iquitos regarded as antirheumatic, calmative, diuretic, emetic, febrifuge and vulnerary (SAR). Venezuelans boil a few leaves in milk as a sedative (SAR). (Fig. 221)

♀ *Tabernaemontana sananho* R.&P. Apocynaceae. "Sanango", "Lobo sanango", "Toomecocoriu". Much as *T. rimulosa.* The leaves, softened by fire, are applied to relieve rheumatic pains (RVM). In Pastaza, taken one week after delivery. "Pulp is used as a gargle for sore throat and colds" (SAR). "Tikuna" mix the latex with water for eye wounds (SAR). "Jivaro" apply the bark juice to toothache (SAR). Considered sudorific, tonic, used for colds, obesity, rheumatism, syphilis (RAR).

Tabernaemontana vanheurickii Muell.-Arg. Apocynaceae. "Ucho sanango". Similar to the previous species.

Tachigali formicarum Harms. Fabaceae. "Tangarana de altura". Wood for lumber; the small individuals are used for house construction, beams, decks, and columns.

Tachigali paniculata Aubl. Fabaceae. "Tangarana de altura", "Tangarana blanca". Wood used for house construction (RVM). "Makuna" rub the hot leaf infusion onto aching limbs (SAR). "Tikuna" take hot leaf tea as a stimulant. "Taiwano" rub leaf infusion over chest-ache, and believe the unripe pods aphrodisiac (SAR). "Kubeo" mix leaf ash with coca (SAR).

Tachigali polyphylla Poepp. & Endl. Fabaceae. "Tangarana amarilla". Wood used for house construction.

Tachigali tessmannii Harms. Fabaceae. "Tangarana". Wood for construction.

Tachigali sp. Fabaceae. "Uuapa". Pucallpa natives use the leaves for headache (VDF).

Tachigali sp. Fabaceae. "Tangarana". Wood for lumber for house construction.

Tagetes erecta L. Asteraceae. "Aya sisa", **"Aztec marigold"**. Cultivated ornamental (RVM). "Tikuna" use leaf decoction as an ophthalmic anodyne, bathing fever with the same decoction (SAR).

♀ *Tagetes minuta* L. Asteraceae. "Huacatay", **"Muster John Henry"**. Decoction considered antiabortive, cholagogue, digestive, gastric sedative, vermifuge (FEO).

Tagetes patula L. Asteraceae. "Flor de muerto", **"French marigold"**. Analgesic, antiasthmatic (RAR); leaf juice massage for fever in Tapajos (BDS).

Talisia cerasina (Benth.) Radlk. Sapindaceae. "Virote huayo". Fruit edible; supposed to make one good with the blowgun. Source of black stain; used for gonorrhea (RAR).

Talisia guianensis Aubl. Sapindaceae. "Virote huayo". Fruit edible (VAG).

Talisia reticulata Radlk. Sapindaceae. "Virote huayo". Fruit edible (VAG).

Tamarindus indica L. Fabaceae. "Tamarindo", **"Tamarind"**. Cultivated ornamental leguminous tree, the fruits used in making delicious beverages (JAD). Pulp diuretic and laxative, inhibiting gram-negative bacteria responsible for urinary infections. Aqueous leaf extracts exhibited antilipoperoxidative and hepatotropic activities while alcoholic extracts exhibited bactericidal, spasmolytic and vasodilator activities. Tramil approves use of the hepatotrophic leaves for jaundice and fruit pulp as antibacterial (TRA). In Amazonian Brazil, the bark tea is used for ameba (BDS). (Fig. 222)

Tanaecium nocturnum (Barb.-Rodr.) Bureau & K.Schum. Bignoniaceae. "Huangana huasca". Stem infusion used to eliminate lice and fleas from domestic animals. The leaves have high concentration of HCN. "Chocó" natives use it as an aphrodisiac. "Paunari" dry, toast, and pulverize the leaves, and mix with *Nicotiana tabacum* to inhale (RVM). "Créoles" believe it effective for pulmonary diseases. "Wayãpi" use the bark decoction to bathe cutaneous eruptions. The decoction of leaves and stems is used by the

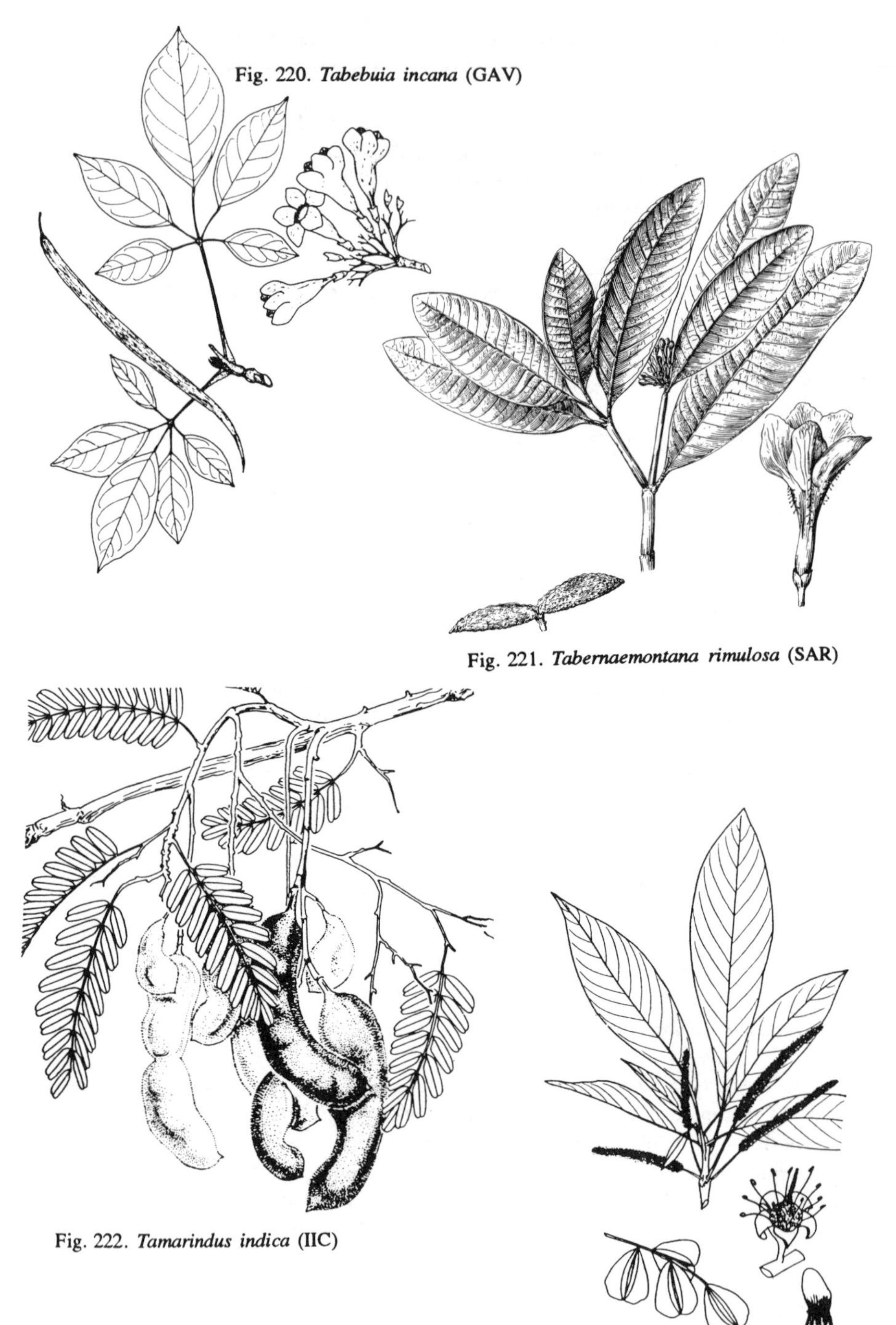

Fig. 220. *Tabebuia incana* (GAV)

Fig. 221. *Tabernaemontana rimulosa* (SAR)

Fig. 222. *Tamarindus indica* (IIC)

Fig. 223. *Terminalia dichotoma* (GAV)

"Palikur" in baths to treat migraine headaches (GMJ). "Paumari" make a hallucinogenic snuff from the vine (SAR).

Tapirira guianensis Aubl. Anacardiaceae. "Huira caspi", "Jemeco". Timber tree, "Wayãpi" use the bark to treat infants (RVM). "Taiwano" use flower tea for dysuria in the elderly (SAR). Fruits edible (RVM). Vesicant POISON (RAR).

Tapirira retusa Ducke. Anacardiaceae. "Huira caspi", "Wira caspi". Wood for lumber. Fruit edible (RVM).

Taralea oppositifolia Aubl. Fabaceae. "Cumaru-Rana", "Palo sangre dominante". Wood for lumber, poles, posts, dormers, and decorative plaques.

Tecoma stans (L.) Juss. Bignoniaceae. "Campanilla amarilla", "**Yellowbells**". Cultivated ornamental. In *SOME MEDICINAL FOREST PLANTS OF AFRICA AND LATIN AMERICA*, FAO (1986), note that the alkaloids tecomine and tecostanine lower the blood sugar in experimental animals. Leaf infusions lower the blood sugar in humans. In Mexico, its roots have shown antisyphilitic, diuretic and tonic properties (FAO).

Tectaria incisa Cav. Aspleniaceae. "Helecho". "Cuna" use cooked roots for stomachache and hepatoses (FOR).

Tectona grandis L. Verbenaceae. "Tectona", "**Teak**". Cultivated ornamental and timber tree.

Teliostachya lanceolata Nees. Acanthaceae. "Toé negro". Used as an additive to ahuayasca, but only in witchcraft (RVM). "Kokama" said to use it directly as an hallucinogen (SAR). "Secona" use for stomachache (SAR).

Tephrosia sinapou (Buch.) A. Chev. Fabaceae. "Cube", "Barbasco", "Kumu". Cultivated as a fish POISON.

Tephrosia toxicaria Sw. Pers. Fabaceae. "Barbasco", "Cube". Contains tephrosine, a fish POISON (SOU). Leaf decoction used by "Créoles" for snakebite and syphilis, by "Galibi" for blenorrhagia (GMJ). Used for the heart, considered POISON.

Terminalia amazonica (J.F.Gmel.) Exell. Combretaceae. "Yacushapana". Wood for lumber, heavy construction, parquet, decorative plaques, posts, and dormers (RVM).

Terminalia catappa L. Combretaceae. "Almendro", "Castañilla", "**Indian almond**". Cultivated ornamental (RVM). Kernel edible (JAD). Brazilians use the astringent bark for bilious fevers and dysentery (SAR) and the leaf tea for the liver (BDS). Tramil does not discourage the use of the leaf decoction for hypertension (TRA).

Terminalia dichotoma G.Meyer. Combretaceae. "Yacushapana". Wood for lumber, heavy construction, parquet, decorative plaques, posts, and dormers (RVM). (Fig. 223)

Terminalia ivorensis A.Chew. Combretaceae. "Almendrillo", "Terminalia". Cultivated ornamental.

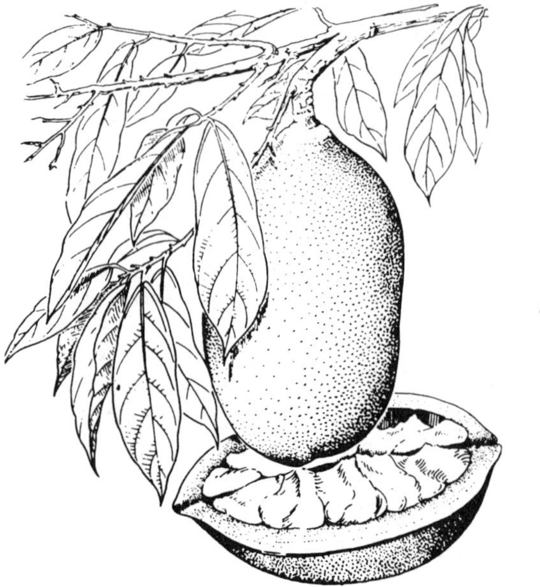

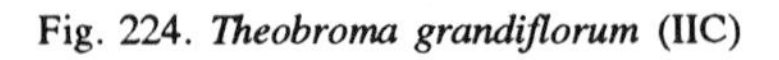

Fig. 224. *Theobroma grandiflorum* (IIC)

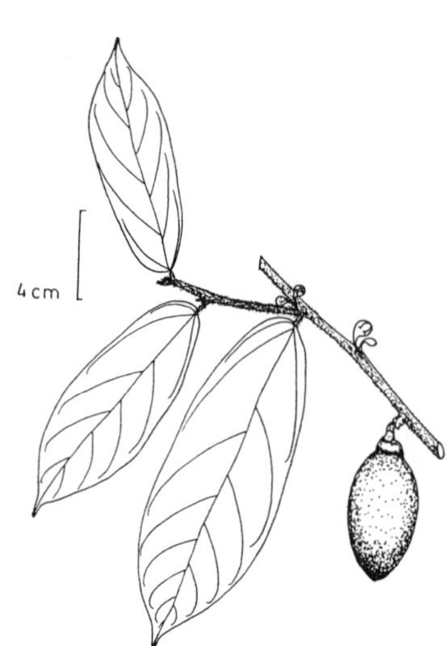

Fig. 225. *Theobroma obovatum* (RVM)

Fig. 226. *Thevetia peruviana* (GAV)

Fig. 227. *Tripsacum lanceolatum* (HAC)

Terminalia oblonga (R.&P.) Sten. Combretaceae. "Itauba", "Matsajcahe", "Yacushapana". Wood for lumber, heavy construction, parquet, decorative plaques, posts, and dormers (RVM).

Tessaria integrifolia R.&P. Asteraceae. "Huapariu", "Pajaro bobo", "Tseco". The wood is used to build provisional houses for the cultivators of rice; leaf infusion ingested for asthma (RVM). Leaf decoction ingested for gallstones and kidneystones (FEO). Around Pucallpa, leaf decoction used for fever, hepatitis, inflammation, and renitis (VDF).

Tetracera volubilis L. Dilleniaceae. "Paujil chaqui", **"Watervine"**. For drinkable water, cut a one-meter section of the stem and stand it vertically, the water will start flowing (RVM). Considered antisyphilitic, antitumor, diuretic, febrifuge and sudorific (DAW).

Tetragastris panamensis (Engl). O. Kuntze. Burseraceae. "Brea caspi", "Lacre", "Copal colorado". Wood for lumber; latex used to caulk boats.

Tetrameranthus pachycarpus Westra. Annonaceae. "Carahuasca". Wood used for house construction.

Tetrathylacium macrophyllum P.&E. Flacourtiaceae. "Amuu", "Anonilla". Wood used in construction (DAT).

Thelypteris opulenta (Klf). Fosberg. Thelypteridaceae. "Apáápalle". Used by the "Boras" to plug dental cavities (DAT).

Theobroma bicolor Humb. & Bonpl. Sterculiaceae. "Macambo". Cultivated. Fruit edible.

Theobroma cacao L. Sterculiaceae. "Cacao", **"Chocolate"**. Cultivated. The pulp of fruit edible. Food uses of chocolate, made from the seed, are well known (RVM). Not so well known is the fact that much cocoa butter ends up in suppositories. Leaf infusion widely used as cardiotonic and diuretic in Colombia (SAR). "Karijona" use toasted seed with manihot squeezings for a scalp condition like eczema. "Ingano" use the bark decoction as a wash for sarna (SAR). Theobromine and theophylline, like caffeine, all found in this plant, used in modern medicine as antiasthmatic (JAD). We are cooperating with one entrepreneur seeking a "lean green cacao bean" for renewable "organic low-fat rainforest chocolate".

Theobroma cacao L. ssp. *leiocarpum* (Bernoulli) Cuatr. Sterculiaceae. "Cacao amarillo". Fruit pulp edible.

Theobroma grandiflorum (Willd. ex Spreng) Schum. Sterculiaceae. "Copoasu", "Cupuasu". Cultivated. Fruit pulp edible. (Fig. 224)

Theobroma obovatum Kl. Sterculiaceae. "Cacahuillo", "Ushpa cacao". Fruit pulp edible. (Fig. 225)

Theobroma subincanum Mart. Sterculiaceae. "Cacahuillo", "Cacao macambillo"", "Macambillo", "Macambo sacha". Fruit pulp edible.Powdered inner bark (of pod) mixed with tobacco as an hallucinogen. "Tirio" value the bark as tinder for starting fires (MJP).

♀ *Thevetia peruviana* (Pers.) Schumann. Apocynaceae. "Árbol de Panama", "Bellaquillo", "Camalonga", "Flor amarilla", **"Lucky nut"**, **"Yellow oleander"**. Cultivated. As a POISONOUS ornamental (JAD). The latex is toxic. "Karaja" use the fruit for necklaces (RVM). Leaf juice used for toothache, the decongestant decoction for rheumatism; tincture of branches febrifugal and purgative (FEO). Elsewhere considered abortifacient, antidotal, anesthetic, cardiotonic, cathartic, emetic, insecticidal, piscicidal, and purgative; dangerously used for dropsy, fever, heart ailments, piles, rheumatism, toothache and tumors (DAW). (Fig. 226)

Thorococarpus bissectus (Vell.) Harl. Cyclanthaceae. "Puspo tamshi". Various species are used for temporary shelters.

Thunbergia grandiflora (Roxb. ex Rottl.) Roxb. Acanthaceae. "Soga de Cristo", **"Blue skyflower"**. Ornamental.

♀ *Tococa juruensis* Pilg. Melastomataceae. "Palo de hombre", "Tsaruwa", "Yojadataka". "Andoke" take the swollen leafbase during pregnancy, assuring the birth of a male child (SAR).

Tonina fluviatilis Aubl. Eriocaulaceae. "Yura pasto". Decoction used by "Palikur" in fortifying bath for infants (GMJ).

Tournefortia maculata Boraginaceae. "Boro shoanco rao". Around Pucallpa used for headache and hemicrania (VDF).

Tovomita sp. Clusiaceae. "Chullachaqui colorado". Wood sometimes used for house construction. Bark tincture antirrheumatic, added to ayahuasca.

Tradescantia elongata Mey. Commelinaceae. "Yona rao". Used in febrifugal baths around Pucallpa (VDF).

Tradescantia zebrina Hort. ex Bosse. Commelinaceae. "Bujeo", "Callicida", "Hiedra". Cultivated ornamental. Cockroach repellant. Suppurative cataplasm for infections.

Trattinickia aspera (Standl.) Sw. Burseraceae. "Copal", "Gallinazo copal". Wood for lumber.

Trattinickia peruviana Loesn. Burseraceae. "Caraña", "Copal caraña". Wood for lumber.

Trema micrantha (L.) Blume. Ulmaceae. "Atadijo". Bark used for cordage; stems used for fencing. The plant soaked in water makes an astringent liquid (PEA). "Cuna" use the bark as an antipyretic for infants (FOR).

Trichilia euneura. C.DC. Meliaceae. "Bola requia". Seed maceration mixed with aguardiente to poultice onto scabies.

Trichilia maynasiana C.DC. ssp. *maynasiana*. "Requia". Wood for lumber and decorative plaques.

Trichomanes elegans Rich. Hymenophyllaceae. "Helecho". Ornamental. "Waunana" rub the leaves on their hands for good luck when hunting peccary and boars (FOR). "Chocó" use it for snakebite (JAD). Leaf infusion drunk for colds (RVM).

Trichocereus pachanoi Britt. & Rose. Cactaceae. "Achuma", "San Pedro". Decoction applied topically to prevent alopecia (FEO).

Trichospermum galeottii (Turcz) Kost. Tiliaceae. "Atadijo blanco". Wood for lumber, beams, decks; bark for cordage.

Trigynaea duckei (R.E.Fries) R.E.Fries. Annonaceae. "Espintana blanca". Wood used for house construction, beams, and decks.

Triplaris peruviana Fisch. ex Meyer. Polygonaceae. "Tangarana sin madre". Wood for house construction and raft supports.

Triplaris poeppigiana Wedd. Polygonaceae. "Tangarana del bajo". Bites from ant inhabitants used for rheumatic pains (like those of bee and nettle stings? JAD). Around Pucallpa, the species called "tangarana" is used for diarrhea, enteritis, and fever (VDF).

Tripsacum lanceolatum Rupr. ex Fourn. Poaceae. "Grama amarga", "**Mexican gama grass**". Cultivated. Sometimes it escapes. As forage. (Fig. 227)

Trithrinax acanthocoma Drude. Arecaceae. "Palma brasilera", "**Brazilian palm**". Cultivated ornamental.

♀ *Triumfetta semitriloba* forma *althaeoides* (Lam.) Uittien. Tiliaceae. "Caballousa", "Pega-pega", "Carnaval huayo", "**Carnival fruit**". The fruits are used to molest people during holidays. Leaf infusion used for diarrhea and hernia (RVM). Used for uterine complaints around Iquitos (SAR). Astringent diuretic (RAR).

Trophis racemosa (L.) Urban. ssp. *meridionalis* (Bur.) W. Burger. Moraceae. "Urpay machinga", "Sachavaca micuna". Wood for dormers; this tree is a good one for hunting "motelos" (land tortoise). Leaves used for forage (RAR).

♀ *Tynnanthus panurensis* (Bur.) Sandw. Bignoniaceae. "Clavo huasca", "Inejkeu", "**Clove vine**". The pieces of roots and stems are macerated in aguardiente to make a stimulant liqueur, good for rheumatism (RVM). Resin used for fevers (DAT). Some explorama visitors have used it, effectivly, for toothache, being as effective as, and probably chemically similar to clove oil (JAD). Some visitors believe, others disbelieve, that the rays of the cross, steeped in aguardiente, are aphrodisiac, some for females, some for males, some for both. We have no incontrovertible empirical evidence, one way or the other.

- U -

♀ *Uncaria guianensis* (Aubl.) Gmel. Rubiaceae. "Uña de gato", "**Cat's claw**", "Paraguayo", "Garabato", "Uña de gavilán", "**Hawk's claw**". In Piura, the bark decoction, considered antiinflammatory, antirheumatic, and contraceptive, is used in treating gastric ulcers and tumors (FEO). Considered a remedy for cancer of the female's urinary tract; also used for gastritis, rheumatism and cirrhosis. The "Boras" use it for gonorrhea (RVM). Colombian and Guianan Indians use it for dysentery (SAR). Nicole Maxwell culimates her latest edition with an illustrated anecdote about this plant, now exported by the tons to Europe, for various cancers. Nicole even states that it turns grey hair black, including some of her own (NIC). See following entry. (Fig. 228)

Uncaria tomentosa (Aubl.) Gmel. Rubiaceae. "Uña de gato", "**Cat's claw**", "Paraguayo", "Garabato", "Uña de gavilán", "**Hawk's claw**". Widely used in Peru for antiinflammatory, contraceptive, and cytostatic activities, the plant has yielded an antiinflammatory antiedemic glycoside (JNP54{2}:453. 1991). In Piura, the bark decoction, considered antiinflammatory, antirheumatic, and contraceptive, is used in treating gastric ulcers and tumors (FEO). In her latest edition, Nicole Maxwell (1990) has added much information which may reflect the potential of the cat's claw. She informs us that Sidney McDaniel submitted samples to the NIH cancer screen.

Unonopsis elegantissima R.E.Fries. Annonaceae. "Espintanilla". Wood for house construction, beams, and decks.

Unonopsis floribunda Diels. Annonaceae. "Icoja". Wood for construction, beams, and decks. Alcoholic bark maceration used for arthritis, rheumatism, and diarrhea.

Unonopsis spectabilis Diels. Annonaceae. "Icoja". Bark used for arthritis, bronchitis, diarrhea, lung disorders, malaria and rheumatism (RVM).

♀ *Unonopsis veneficiorum* (Mart.) R.E. Fries. Annonaceae. "Icoja". Bark used like *U. floribunda* (RVM), also in curaré (SAR). "Maku" use in antifertility potions (SAR). (Fig. 229)

Urena lobata L. var. *reticulata* (Cav.) Gurke. Malvaceae. "Malva roja", "Yute", "**Jute**". Cultivated. Bark provides a hemp-like fiber; anthelmintic, sedative (RAR). (Fig. 230)

♀ *Urera baccifera* (L.) Gaud. Urticaceae. "Ishanga Moe", "Mara mara","**Stinging nettle**". The stinging hairs on the leaves are used to relieve rheumatic pains. "Chami" cook and eat the leaves and stems after removing the thorns (CAA). Around Pucallpa, applied to the body for persistent fever (VDF). Elsewhere considered diuretic, rubefacient and vesicant; used for amenorrhea, arthritis, chills, fever, gonorrhea, leucorrhea, malaria, rheumatism and venereal diseases (DAW). One M.D. speculated that the acetylcholine, choline and histamine injected with the stings, would stimulate the production of mast cells which might in turn result in antiinflammatory (and antiarthritic) activity, away from the sting.

♀ *Urera caracasana* (Jacq.) Griseb. Urticaceae. "Cunshi ishanga", "Ishanga macho". Fruit used as fishbait. "Wayãpi" use it in ritual at the first menstrual period. They rub leaves on their bodies to relieve fever (RVM). Vaupes Indians use an infusion for erysipelas, the roots for hemorrhage (SAR). Mexicans use it for poison ivy and syphilis, Panamanians for brain cancer (DAW).

Urera laciniata (Gaud.) Wedd. Urticaceae. "Mara mara". Same as the *U. baccifera* (RVM). "Siona" use externally for myalgia and to discipline children (SAR).

Urospatha sagittifolia (Rodach.) Schott. Araceae. "Jergón sacha". Believed to help snakebites. Some people sting their feet and legs to repel snakes (RVM).

Urtica magellanica Pois. Urticaceae. "Ishanga", "Quisa". Leaf infusion for rheumatism and sciatica (FEO).

Utricularia foliosa L. Utriculariaceae. "Maíz del tuqui tuqui". Aquate ornamental.

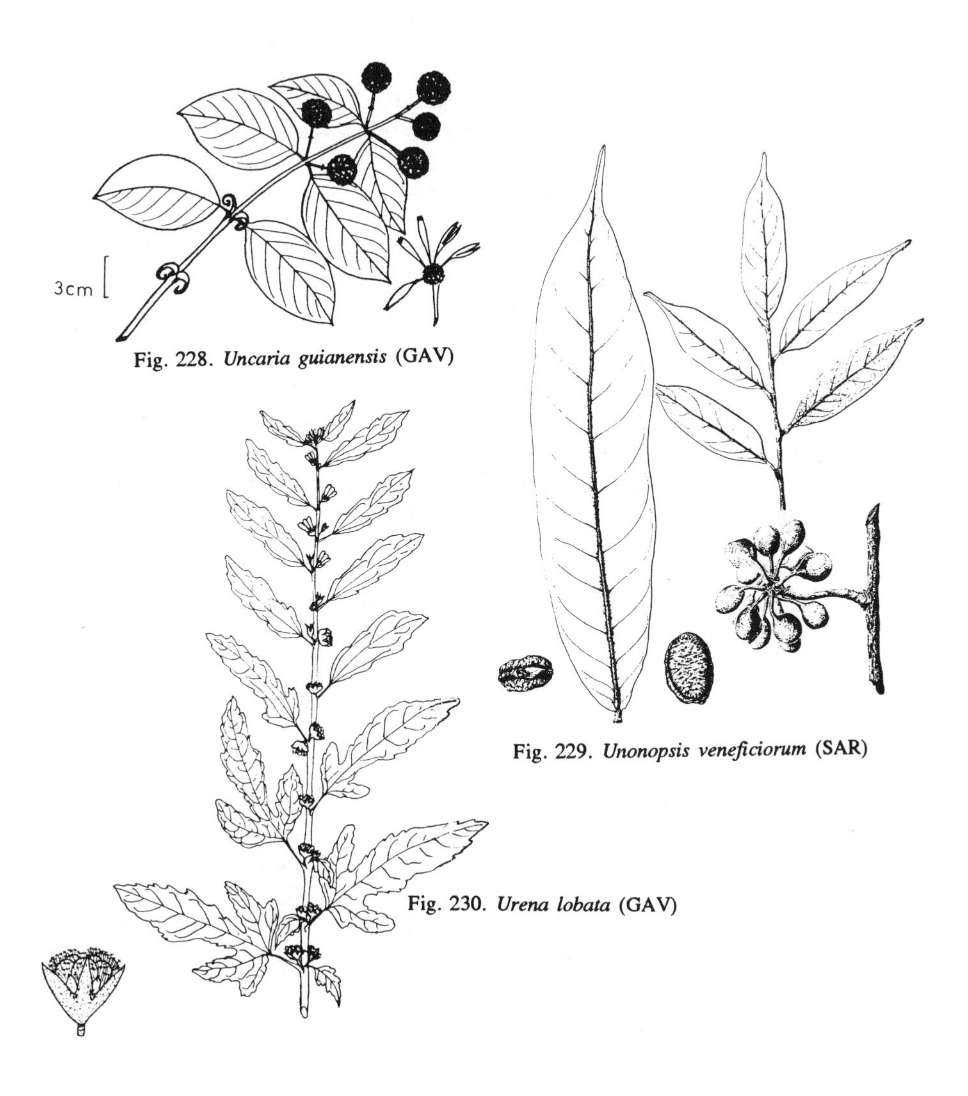

Fig. 228. *Uncaria guianensis* (GAV)

Fig. 229. *Unonopsis veneficiorum* (SAR)

Fig. 230. *Urena lobata* (GAV)

- V -

Vanilla sp. Orchidaceae. "Vanilla". As an aromatic; the commercial species, *V. planifolia* and *V. pompona*, are cultivated in Central America, and Madagascar. (Fig. 231)

Vantanea guianensis Aubl. Humiriaceae. "Manchari caspi", "Yerno prueba". Wood used for bridges, dormers, posts, poles, possibly for parquets (RVM).

Vantanea peruviana Macb. Humiriaceae. "Manchari caspi", "Yerno prueba". Wood is used as jam posts for bridges, dormers, posts, forked poles, and has good posibilities for parquets (RVM).

Vantanea tuberculata Ducke. Humiriaceae. "Manchari caspi", "Parinari sapo", "Loro shungo". Wood is used as jam posts for bridges, dormers, posts, forked poles, and has good posibilities for parquets (RVM).

Vatairea guianensis Aubl. Fabaceae. "Mari mari del bajo". Wood for posts and forked poles. The "Créoles" and "Palikur" use leaves and seeds in ointments to treat skin diseases (GMJ). Fresh seeds are poulticed onto mange, herpes, and other cutaneous eruptions; adding bark chips makes the treatment more effective (GMJ, RAR).

♀ *Verbena littoralis* H.B.K. Verbenaceae. "Verbena", "Yapo", "**Verbena**". Considered abortifacient around Napo (JAD), also antitussive, emetic, febrifuge, and vermifuge (RAR). Leaves used in antitussive febrifuges (VDF).

Vernonia baccharoides H.B.K. Asteraceae. "Ocuero", "Perma caspi", "Ticsa micuna", "Zui". Huallaga natives use the sap for conjunctivitis (RAF).

Victoria amazonica (Poepp.) Sowerby. Nymphaeaceae. "Victoria regia", "Sábana del lagarto". Aquatic ornamental. Mixed with andiroba oil for rheumatism, inflammation and hemorrhoids (RVM). Root, seed and stem edible (RAR).

Vigna unguiculata (L.) Walp. Fabaceae. "Chiclayo", "Caupi", "**Cowpea**". Cultivated. Forage; seeds edible.

Virola albidiflora Ducke. Myristicaceae. "Aguano cumala". Wood for interior decoration, furniture, and plywood veneer (RVM). "Kuneo" and "Tukano" use the resin on sores (SAR).

Virola calophylla Warb. Myristicaceae. "Cumala blanca". Wood for lumber. Some natives (e.g. "Bora" and "Huitoto"), use *Virola* as a powerful hallucinogen, taking it orally and nasally. They grate, dry, and toast the inner bark slowly until it becomes powder so they can inhale it. They also grate the cambium, boil it in water, mixing continuously until it forms a thick syrup; after it dries, they make pills and swallow them. The alkaloids found are mostly derivatives of tryptamine: DMT, MMT, 5-Me0-DMT, 5-Me0-MMT, and the derivatives of beta-carboline: 6-Me0-DMTHC; the percentage of such compounds vary according to the species, as well as their environment (RVM). Widely used for fungal diseases and scabies (SAR). Amazonian Peruvians use for bladder and stomach ailments (SAR). "Maku" use the bark tea for malaria (SAR). (Fig. 232)

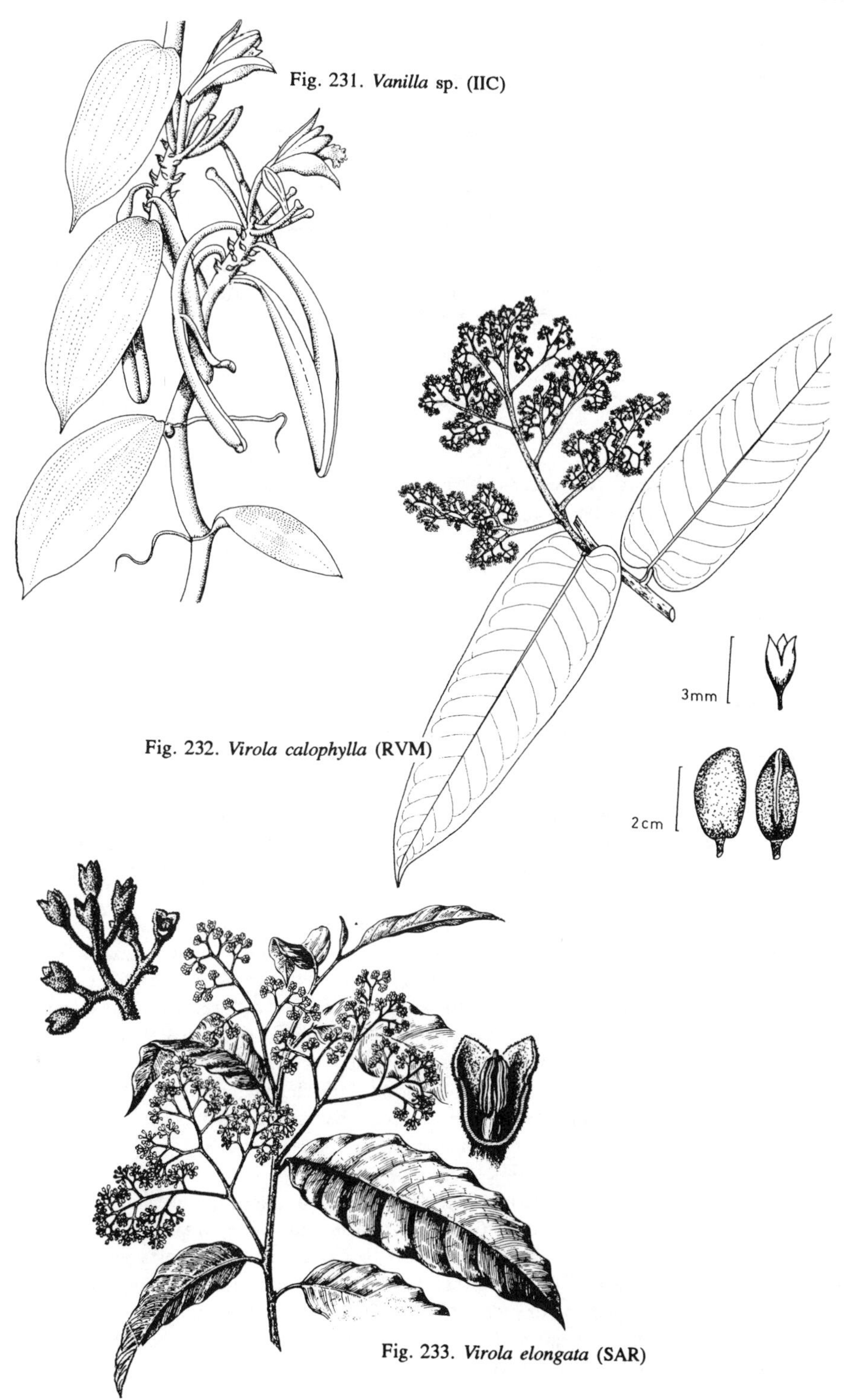

Fig. 231. *Vanilla* sp. (IIC)

Fig. 232. *Virola calophylla* (RVM)

Fig. 233. *Virola elongata* (SAR)

Virola decorticans Ducke. Myristicaceae. "Cumala negra". Wood for lumber (RVM). "Jivaro" rub the leaf juice on the gums of children cutting teeth (SAR).

Virola divergens Ducke. Myristicaceae. "Cumala negra". Wood for lumber.

Virola elongata (Benth.) Warb. Myristicaceae. "Cumala blanca". Wood for lumber (RVM). "Bora" and "Huitoto" use the cambium as a hallucinogenic snuff, but this use is more common in the native communities of the Orinoco, Rio Negro, and Amazonian Colombia (RVM). "Barasana" apply the leaf/twig decoction to arthritic swellings (SAR). Resin widely known for its fungicidal activity SAR. Bark contains sesartemin and yangambin, which reduces aggressiveness (JBH). (Fig. 233)

Virola flexuosa A.C.Smith. Myristicaceae. "Caupuri de altura", "Cumala blanca". Wood for lumber, plywood (RVM). "Taiwano" use powdered leaves as insect repellent (SAR).

Virola loretensis A.C.Smith. Myristicaceae. "Cumala blanca". "Boras" and "Huitotos" use as an hallucinogen.

Virola multinervia Ducke. Myristicaceae. "Cumala negra". Wood for lumber.

Virola pavonis (A.DC.) A.C.Smith. Myristicaceae. "Caupuri del bajo", "Aguano cumala". Wood for lumber, plywood. "Bora" and "Huitoto" use as hallucinogenic.

Virola peruviana (A.DC.) A.C.Smith. Myristicaceae. "Cumala blanca". Bark used as an hallucinogen (RVM); resin used to stop bleeding (SAR). "Tikuna" use the resin for mycoses (SAR). "Waorani" use the sap for mites and dermatoses (SAR).

Virola sebifera Aubl. Myristicaceae. "Cumala blanca". Wood for lumber; the leaves for tea; the sap, bark decoction, and aril for dyspepsia and intestinal colic, applied directly for erysipelas, also for cleaning and healing wounds and inflammations (RVM).

Virola surinamensis (Rol.) Warb. Myristicaceae. "Cumala blanca hoja parda". Wood for lumber, plywood. "Bora" and "Huitoto" use the cambium as a hallucinogen. The decoction of the aerial rootlets that appear on the base of the trunk is used for cough. "Palikur" prepare a bark emollient used for swellings and erysipelas; used as an oral antiseptic to treat canker sores and abscesses. For swelling, it is mixed with bark of *Humiria balsamifera,* the decoction used for external baths (GMJ). Tea of leaves, sap, and bark, mixed with *Physalis angulata*, is used for upset stomach, intestinal colic, erysipelas, and inflammations (RVM). Leaves contain the antitubercular compound galbacin, the antiaggregant veraguensin, and the antischistosomal surinamensis (JBH).

Vismia angusta Miq. Hypericaceae. "Pichirina hoja grande". The wood is used for rural construction; the decoction of the latex from the buds, mixed with the latex of *Euphorbia cotinifolia,* is used to treat ringworm or "caracha" (dermatosis caused by fungus) (RVM). Amazonian Colombians use the latex for infected sores and wounds. "Tikuna" use to treat herpes and mycoses (SAR). The latex of one *Vismia* is slated for studies by a California pharmaceutical company; preliminary tests suggest it to be effective (MJP). Both Segundo and JAD suffered long-lasting rashes as a result of the latex (JAD).

Vismia ferruginea HBK. Hypericaceae. "Pichirina". The yellowish resin is applied like iodine to wounds and dermatoses (SAR).

Vismia lateriflora Ducke. Hypericaceae. "Pichirina colorada". Wood used for construction.

Vismia macrophylla HBK. Hypericaceae. "Pichirina". Wood used for construction.

Vismia minutiflora Ewan. Hypericaceae. "Pichirina hoja menuda". Wood used for construction. The latex is fungicidal.

Vitex pseudolea Rusby. Verbenaceae. "Cormiñon", Paliperro". Seed edible.

Vitex triflora Vahl. Verbenaceae. "Paliperro". Seed edible.

Vitis vinifera L. Vitaceae. "Uva", "Parra", "**Grape**". Cultivated ornamental fruit vine. The tannins and anthocyanins in red wine have been shown to have antiherpetic activity (JAD).

Vochysia densiflora Spruce ex W. Vochysiaceae. "Quillo sisa". Wood used for panels, ceilings, and light drawers (RVM).

♀ *Vochysia lomatophylla* Standl. Vochysiaceae. "Quillo sisa", "Capiron" "Sacha alfaro", "Sacha casho". Wood used for panels, ceilings, and light drawers (RVM). "Barasana" give bark and pulverized leaves with warm chicha as an abortifacient (SAR).

Vochysia venulosa Warm. Vochysiaceae. "Mauba". Wood used for panels, ceilings, and light drawers (RVM).

Vochysia vismiifolia Spruce ex Warm. Vochysiaceae. "Quillo sisa". Wood used for panels, ceilings, light drawers and piling (RVM).

- W -

Warscewiczia coccinea (Vahl) Klotzch. Rubiaceae. "Puca sisa", "Bandera caspi", "**Flag tree**". Ornamental. "Chocó" wear the root on their ear as a perfume; they believe it is an aphrodisiac (JAD). In Colombia the bark is used as a hemostat (JAD). "Cuna" drink regularly a portion of the cooked root for nosebleed (FOR). (Fig. 234)

♀ *Wedelia trilobata* (L.) Hitch. Asteraceae. "Manzanilla cimarrona", "Botón de oro", "**Gold button**". Sometimes cultivated as an ornamental. For toothache the flowers are chewed and the extracted juice held in the mouth (RVM). In Trinidad used for amenorrhea, dysentery, fever, foot problems, kidney ailments, sores and stomach problems (DAW).

Wettinia maynensis Spruce. Arecaceae. "Ponilla", "Pona coto-shupa". The stems cut into strips are used for walls, room dividers etc.

Witheringia solanacea L'Her. Solanaceae. "Urpa coconilla". Used as fishbait. "Cuna" cook the roots for a long time, and drink the juice to relieve stomachache (FOR).

Witta amazonica K. Schum. Cactaceae. "Lagarto shupa". As an ornamental.

Wulffia baccata (L.f.) Kuntze. Asteraceae. "Manzanilla sacha", "Chirapa sacha". Rutter says fruit is edible (RAR). "Créoles" use flower tea to treat flu. Leaf decoction used as an antidiabetic; used by the "Wayãpi" as refreshing baths for fever. The decoction of the aerial parts is recommended for nausea (GMJ).

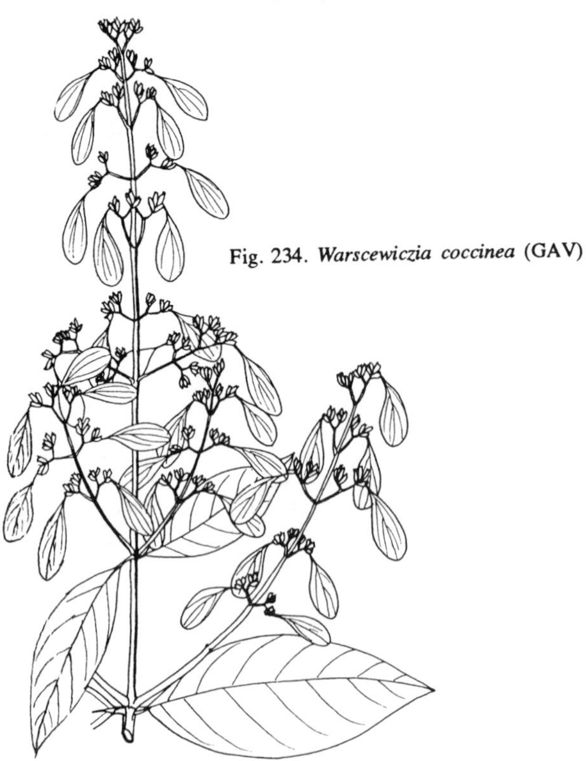

Fig. 234. *Warscewiczia coccinea* (GAV)

- X -

Xanthosoma helleborifolium (Jacq.) Schott. Araceae. "Mano abierta". Cultivated. As an ornamental; corms edible. Used elsewhere for snakebite (DAW).

Xanthosoma mafaffa Schott. Araceae. "Mafaffa". Corms and leaves edible.

Xanthosoma purpuratum Krause. Araceae. "Oreja de elefante", **"Elephant's ear"**. Cultivated. As an ornamental.

Xanthosoma saggitifolium Schott. Araceae. "Huitina". Cultivated. Leaves and corms edible (RVM). Elsewhere used for burns, cancers, polyps, sores and tumors (DAW). (Fig. 235)

Xanthosoma violaceum Schott. Araceae. "Huitina", "Uncucha". Cultivated. Corms edible.

♀ *Xiphidium caeruleum* Aubl. Haemodoraceae. "Sacha orquidia", **"Wild orchid"**. "Chocó" use for snakebites; "Cuna" use it to treat weakness in women (JAD). Flowers and leaves are left in cold water overnight by the "Cuna" and taken to hasten childbirth. The thick juice from the macerated leaves in cold water is applied to relieve pain (FOR). "Wayãpi" use the decoction of the entire plant for children who cry too much (GMJ).

Xylopia aromatica (Lam.) Mart. Annonaceae. "Espintana". Wood used for house construction, beams, and decks. Toasted seeds and stem bark are mashed and used as a carminative, stimulant, and aphrodisiac (RVM). (Fig. 236)

Xylopia barbata Mart. Annonaceae. "Espintana". Wood used for house construction, beams, and decks.

Xylopia benthamii R.E. Fries. Annonaceae. "Pinsha callo". Wood used for house construction, beams, and decks. Fruit tea used for stomachache (RVM).

Xylopia frutescens Aubl. Annonaceae. "Espintana". Wood used for house construction. Toasted seeds, and the aromatic bark, used as carminatives, aphrodisiac, stimulants for the bladder, and as a tonic for rheumatism (RVM).

Xylopia micans R.E.Fries. Annonaceae. "Espintana". Wood for lumber.

Xylopia parvifolia Spruce. Annonaceae. "Espintana del varillal". Wood used for house construction, lumber; the stems cut into strips are used for ceilings (RVM). Elsewhere used for sores (DAW).

Xylopia poeppiggii R.E.Fries. Annonaceae. "Yahuarachi caspi". Wood is used for house construction.

Xylopia sericea A.St.Hil. Annonaceae. "Espintana". Wood used for house construction, beams, decks, and columns. Used as a spice in Brazil (DAW).

Fig. 235. *Xanthosoma saggitifolium* (IIC)

Fig. 236. *Xylopia aromatica* (SAR)

Fig. 238. *Zingiber officinale* (IIC)

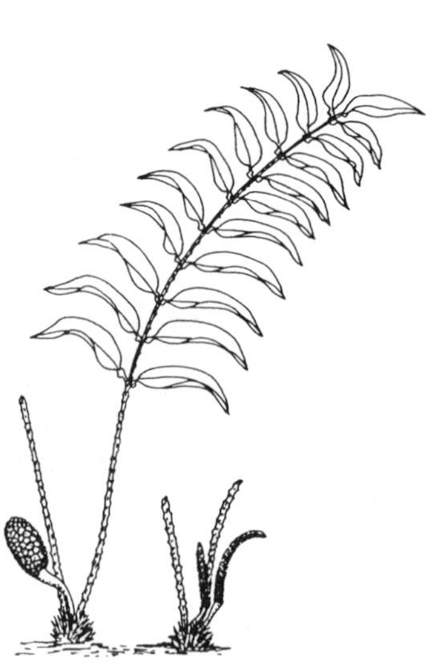

Fig. 237. *Zamia* sp. (GAV)

- Z -

Zamia ulei Dammer. Cycadaceae. "Palma de goma". Cultivated ornamental. (Fig. 237)

Zanthoxylum juniperinum. Poepp. & Endl. Rutaceae. "Hualaja", "Shapilleja". Wood for lumber.

Zanthoxylum cf *rigidum* Humb. ex Willd. Rutaceae. "Hualaja". Wood used for house construction.

Zanthoxylum sprucei (Engl.) Engl. Rutaceae. "Hualaja". Wood used for house construction.

Zanthoxylum valens Macb. Rutaceae. "Hualaja". Wood for lumber.

Zea mays L. Poaceae. "Maíz" (Corn). Cultivated. The hybrid commonly cultivated is yellow hard corn; when green, it is eaten as "sweet corn"; when ripe it is used for animal feed; in small amounts, it is used for flour for baking breads, and beverages such as "chicha". "Maiz polvo sara" is used for flour; other cvs are of lesser importance. For medicinal purposes, corn is soaked in aguardiente, and used in poultices for fever (RVM). In Piura, the silk decoction is considered antiblennorrhagic, diuretic, sedative, and tonic. The grain tincture is used for alopecia and rheumatism. Tramil recommends the use of cornsilk for edema and kidney pains (TRA).

♀ *Zingiber officinale* Roscoe. Zingiberaceae. "Jengibre", "Kión". Cultivated. Macerated rhizomes in aguardiente for arthritis and rheumatism; believed to invigorate males. Rhizome decoction used for diarrhea, and, with a pinch of cinnamon, stomachaches. Also used as an antiflatulent and spice. "Palikur" poultice the rhizomes onto migraine headaches (GMJ). Used also for bronchitis and rheumatic pains (RVM). Tramil reports that oral doses of 50-100 mg/kg of the alcoholic extract have antiinflammatory activity comparable to aspirin, and not so promising analgesic activity. The extract is active against gram negative and positive bacteria. Gingerol and shogoal show molluscicidal activity (TRA). Furanogermenone, at oral doses of 500 mg/kg helps prevent gastric ulcer. Shogoal is intensely antitussive, compared to dihydrocodeine (TRA). One gram of powdered ginger can prevent seasickness (JAD). Tramil all but recommends it for colds, coughs, flu, stomachache and vomiting (TRA). Rio Tapajos women drink the tea while in labor, giving the "baby the strength to come out" (BDS). They also take the tea for colic, menstrual cramps, sore throat. (Fig. 238)

INDEX OF COMMON NAMES

Apacas: *Phytolacca*
Apacharma: *Licania*
Apacharama blanca: *Licania*
Apiranga: *Mouriri*
Arazá: *Eugenia*
Árbol de Panama: *Thevetia*
Árbol del pan: *Artocarpus*
Árbol del tambor: *Cavanillesia*
Aretillo: *Lindernia*
Aripari: *Macrolobium*
Aripay: *Gordonia*
Aromuhe: *Pentagonia*
Arrowroot: *Maranta*
Arroz: *Oryza*
Arroz bravo: *Oryza*
Asahí: *Euterpe*
Ashipa: *Pachyrrhizus*
Asna huayo: *Siparuna*
Asna panga: *Cyphomandra*
Assacu: *Erythrina*
Assasu-rana: *Erythrina*
Ataco: *Amaranthus*
Atadijo: *Trema, Guazuma*
Atadijo blanco: *Trichospermum*
Avocado: *Persea*
Aya albaca: *Lantana*
Ayahuasca: *Banisteriopsis*
Ayahuasca sacha: *Banisteriopsis*
Ayahuma: *Couroupita*
Ayamurillo: *Hibiscus*
Ayaporoto: *Cassia*
Aya sisa: *Tagetes*
Azafran: *Curcuma*
Azar quiro: *Isertia*
Aztec marigold: *Tagetes*
Azúcar-huaillo: *Cynometra*
Azúcar huayo: *Hymenaea*
Azufre caspi: *Moronobea, Symphonia*
Azul: *Juanulloa*

Bacaba: *Oenocarpus*
Bahia grass: *Paspalum*
Bala huayo: *Gnetum*
Balata: *Eclinusa, Micropholis*
Balata rosada: *Micropholis*
Balata sapotina: *Chrysophylum*
Balatilla: *Micropholis*
Balatillo: *Chrysophyllum, Haploclathra*
Balsa: *Ochroma*
Balsa huasca: *Arrabidaea*
Balsam apple: *Momordica*
Balsamina: *Momordica*
Bálsamo: *Myroxylon*
Balsam of Peru: *Myroxylon*
Balsam pear: *Momordica*
Banana: *Musa*
Banano: *Musa*
Bandera caspi: *Warscewiczia*
Barba de chivo: *Eucharis*
Barbados cherry: *Malpighia*
Barbasco: *Clibadium*
Barbasco: *Clibadium, Lonchocarpus, Schoenobiblus, Tephrosia*
Barbasco negro: *Dictyoloma*
Barbusho: *Cordyline*
Barrigón: *Iriarte*
Basil: *Ocimum*
Bastón del emperador: *Nicolasia*
Bean: *Phaseolus*
Bedoca: *Passiflora*
Beggar-lice: *Desmodium*
Begonia: *Begonia*
Beguefide: *Phoradendron*
Bellaco caspi: *Himatanthus*
Bellaquillo: *Thevetia*
Bella vista: *Syagrus*
Bijao: *Calathea, Ichnosiphon*
Bijao chancaquero: *Calathea*
Bird's nest fern: *Asplenium*
Black pepper: *Piper*
Black physic nut: *Jatropha*
Bleeding heart: *Caladium*
Bloodflower: *Asclepias*
Blue skyflower: *Thunbergia*
Boa: *Monstera*
Boa caspi: *Haploclathra*
Boa sacha: *Philodendron*
Boca pintada: *Psychotria*
Boens: *Mansoa*
Bolaina: *Guazuma, Muntigia*
Bolaina blanca: *Guazuma*
Bola requia: *Guarea, Trichilia*
Bolsa quihua: *Priva*
Bolsa mullaca: *Physalis*
Bombonaje: *Carludovica*
Bombonaje sacha: *Ichnosiphon*
Botón caspi: *Anthodiscus*
Botón de oro: *Wedelia*
Botoneillo: *Spilanthes*
Bouquet de novia: *Ixora*
Bowstring hemp: *Sansevieria*
Brazilian mahogany: *Carapa*

Brazilian palm: *Trithrinax*
Brea caspi: *Caraipa, Protium, Tetragastris*
Breast berry: *Solanum*
Bride's bouquet: *Ixora*
Brinco de dama: *Clerodendron*
Bromilia: *Aechmea, Pitcairnea*
Bubinsana: *Calliandra*
Buckwax: *Symphonia*
Bujeo: *Zebrina*
Buriti: *Mauritia*
Butterfly lily: *Hedychium*
Butterfly's flower: *Symbolanthus*
Butterfly pea: *Clitoria*

Caballo sanango: *Faramea*
Caballo shupa: *Cespedesia*
Caballusa: *Triumfetta*
Cabalonga: *Fevillea*
Cabeza de negro: *Phytelephas*
Cacao: *Theobroma*
Cacao amarillo: *Theobroma*
Cacao macambillo: *Theobroma*
Cacahuillo: *Carpotroche, Erisma, Herrania, Theobroma*
Cachimbo: *Cariniana, Lecythis*
Cacto: *Opuntia*
Cadillo: *Bidens*
Cadillo cabezon: *Hyptis*
Café: *Coffea*
Cafecillo: *Rinorea*
Caferi: *Coffea*
Cafeto: *Coffea*
Cahuara micuna: *Crateva*
Caigua: *Cyclanthera*
Caimitillo: *Chrysophyllum, Diospiros, Micropholis, Pouteria*
Caimiitillo hoja grande: *Chrysophyllum*
Caimito: *Chrysophyllum, Pouteria*
Caimito brasilero: *Lucuma, Manilkara*
Calabaza: *Lagenaria*
Calabazo: *Cucurbita*
Calabur: *Muntingia*
Calaguala: *Polypodium*
Calzón panga: *Cyclanthus*
Camalonga: *Thevetia*
Camaroncillo: *Beloperone*
Camé: *Clusia*
Camona: *Iriartea*
Camote: *Ipomea*
Camote caspi: *Ipomoea*
Campanilla: *Hibiscus*
Campanilla amarilla: *Tecoma*
Campanilla de oro: *Allamanda*
Campanilla morada: *Ipomoea*
Campanita del campo: *Iribachia*
Camaroncillo: *Justicia*
Camu-camu: *Myrciaria*
Camu-camu arbol: *Myrciaria*
Camu-camu negro: *Myrciaria*
Canaria: *Allamanda*
Canela moena: *Licaria, Ocotea*
Caña agria: *Costus, Dimerocostus*
Caña brava: *Gynerium*
Caña de azucar: *Saccharum*
Cañagre: *Costus*
Caña isana: *Gynerium*
Canalete: *Aspidosperma*
Cañamazo: *Paspalum*
Canirca: *Lantana*
Canna lily: *Canna*
Cantaloupe: *Cucumis*
Caoba: *Swietenia*
Cape gooseberry: *Physalis*
Capinurí: *Maquira*
Capinurí de altura: *Clarisia, Naucleopsis*
Capinurí del bajo: *Maquira*
Capiron: *Vochysia*
Capirona: *Calycophyllum*
Capirona de altura: *Capirona*
Capulí cimarrón: *Physalis*
Capushi: *Microtea*
Caracha quinilla: *Pouteria*
Carahuasca: *Anaxagorea, Guatteria, Tetrameranthus*
Carahuasca-millua: *Guatteria*
Carambola: *Averrhoa*
Carana: *Bursera*
Carnaval huayo: *Triumfetta*
Carnival fruit: *Triumfetta*
Carpunya: *Piper*
Carricillo: *Arthrostylidium, Olyra*
Carrot: *Daucus*
Cascarilla: *Ladenbergia*
Cascarilla verde: Ladenbergia
Casha huayo: *Lindackeria*
Casha piña: *Ananas*
Cashapona: *Socratea*
Casharana: *Licania*
Casho: *Anacardium*
Cassabanana: *Sicana*

Coco: *Cocos*
Coco enano: *Cocos*
Cocona: *Solanum*
Coconilla colorada: *Solanum*
Coconilla con espiñas: *Solanum*
Coconut: *Cocos*
Cocotero: *Cocos*
Coffee: *Coffea*
Coffee senna: *Cassia*
Cola de caballo: *Andropogon, Axonopus*
Colchón quihua: *Homolepis, Imperata*
Comida delvenado: *Strychnos*
Congona: *Peperomia*
Conoco: *Micrandra*
Conta: *Attalea*
Copaiba: *Copaifera*
Copal: *Dacryodes, Protium, Trattinickia*
Copal: *Copaifera*
Copal blanco: *Crepidospermum*
Copal carana: *Protium, Trattinickia*
Copal colorado: *Protium, Tetragastris*
Copoasú: *Theobroma*
Coral hibiscus: *Hibiscus*
Coralito: *Capsicum*
Corazón de Jesús: *Caladium*
Corazón sangriento: *Caladium, Clerodendron*
Cordoncillo: *Piper*
Coriander: *Coriandrum*
Cormiñon: *Vitex*
Corn plant: *Dracaena*
Corona de neron: *Stomatophyllum, Philodendron*
Corpus sacha: *Mucuna*
Correa: *Plukenetia*
Cortadera: *Scleria*
Cosho: *Iriartea*
Costada sacha: *Passiflora*
Costilla de Adan: *Monstera*

Coto huayo: *Diclidanthera, Moutabea, Orthomene*
Cotton: *Gossypium*
Cotton rose: *Hibiscus*
Cow hoof: *Bauhinia*
Cowpea: *Vigna*
Cow's tongue: *Asplenium, Elephantopus*
Crepe myrtle: *Lagerstroemia*
Cucarda: *Hibiscus*
Cucarda caspi: *Malvaviscus*
Cubé: *Lonchocarpus, Tephrosia*
Cuchara caspi: *Ambelania*
Cuchara panga: *Limnocharis*
Cucumber: *Cucumis*
Culantrilla: *Lindsaea*
Culantrillo: *Adiantum, Lindsaea*
Culantro del pais: *Coriandrum*
Cumaseba: *Swartzia*
Cumala blanca: *Virola, Osteophloeum*
Cumala blanca hoja parda: *Virola*
Cumala colorada: *Iryanthera, Otoba*
Cumala negra: *Virola*
Cumaru-rana: *Taralea*
Cunshi ishanga: *Urera*
Cupana: *Paullinia*
Cupuasú: *Theobroma*
Curaré: *Chondrodendron*
Curarina: *Potalia*
Curihjau: *Sabicea*
Cushqui-huasca: *Dalbergia*
Custard apple: *Annona*
Cutana pashaco: *Parkia*
Cutgrass: *Scleria*
Cúwarahííba: *Bactris*
Cuya cuya: *Juanulloa*

Dale *dale: Calathea*
Delia: *Drymonia, Eucharis*
Devil's ear: *Psychotria*
Dictamo real: *Ichthyothere*
Doctor ojé: *Ficus*
Dog's ear: *Caladium*
Dormidera: *Cassia*
Dried love: *Bidens*
Drimonia: *Drymonia*

Elder: *Sambucus*
Elderberry: *Sambucus*
Elephant grass: *Pennisetum*
Elephant's ear: *Xanthosoma*
Enrredadera: *Ipomoea*
Entrada al baile: *Coleus*
Escoba: *Sorghum*
Escalera de mono: *Bauhinia*
Espatodea: *Spathodea*
Espintana: *Anaxagorea, Diclinanona, Guatteria, Malmea, Oxandra, Pseudoxandra, Ruizodendron, Xylopia*

Espintana blanca: *Trigynaea*
Espintana del varillal: *Xylopia*
Espintanilla: *Unonopsis*
Esponjilla: *Luffa*
Esponja huayo: *Ryania*
Espuma huayo: *Ryania*
Estoraque: *Myroxylon*

Fall panicum: *Panicum*
False balsa: *Croton*
Faveira: *Macrolobium*
Favorito: *Osteophloeum*
Fer-de-lance: *Dracontium*
Fiber palm: *Astrocaryum*
Fierro caspi: *Minquartia*
Fig: *Ficus*
Flag tree: *Warscewiczia*
Flor amarilla: *Thevetia*
Flor de agua: *Nymphaea*
Flor de cementerio: *Dracaena*
Flor de fuego: *Salvia*
Flor de mariposa: *Symbolanthus*
Flor de mediodia: *Portulaca*
Flor de muerto: *Asclepias, Tagetes*
Flor de once: *Portulaca*
Flor de sapo: *Isotoma*
Flor de seda: *Asclepias*
Flor variable: *Hibiscus*
Floripondio: *Brugmansia*
Fountain palm: *Livistona*
Four o'clock: *Mirabilis*
Frejol: *Phaseolus*
Frejol del tunchi: *Crotalaria*
French marigold: *Tagetes*
Frijol: *Phaseolus*
Frog's flower: *Isotoma*

Galicosa: *Ichthyothere*
Gallinazo copal: *Trattinickia*
Gallinazo panga: *Cyphomandra*
Gallito: *Erythrina, Securidaca*
Gallo cresta: *Celosia*
Gallo-cresta-rango: *Centropogon*
Garabato: *Uncaria*
Garland flower: *Hedychium*
Garlic: *Allium*
Garrafon piña: *Ananas*
Garza muena: *Rhodostemonodaphe*
Gebero piña: *Ananas*
Genip: *Melicocca*
Gherkin: *Cucumis*
Giant cane: *Gynerium*
Giant granadilla: *Passiflora*
Ginger lily: *Hedychium*
Girasol: *Helianthus*
Globe amaranth: *Gomphrena*
Gold button: *Wedelia*
Golden bell: *Allamanda*
Goma huayo: *Parkia*
Goma pashco: *Parkia*
Good-luck plant: *Cordyline*
Good Mary: *Pterocarpus*
Gordura: *Melinis*
Grama: *Acroceras, Axonopus, Brachiaria, Eleusine, Eragrostis, Leptochloa, Luziola, Panicum, Paspalum*
Grama amarga: *Tripsacum*
Grama chilco: *Setaria*
Grama dulce: *Imperata, Leptochloa*
Grama grass: *Acroceras*
Grama inglesa: *Stenotaphrum*
Grama nudilla: *Panicum*
Grama peineta: *Pennisetum*
Grama playa: *Hemarthria, Oryza*
Gramalote: *Hymenachne, Panicum, Paspalum*
Gramalote capo: *Echinochloa*
Gramalote negro: Hymenachne, Paspalum
Gramilla: *Digitaria*
Granadilla: *Passiflora*
Granadilla agria: *Passiflora*
Granadilla caspi: *Dilkea*
Granadilla venenosa: *Passiflora*
Grape: *Vitis*
Grape tree: *Pourouma*
Green bean: *Phaseolus*
Green sprangle top: *Leptochloa*
Gru-gru: *Acrocomia*
Guaba: *Inga*
Guaco: *Mikania*
Guanábana: *Annona*
Guanábana sacha: *Annona*
Guaraná: *Paullinia*
Guariuba: *Clarisia*
Guava: *Psidium*
Guayaba brasilera: *Eugenia*
Guayabilla: *Myrcia*
Guayabo: *Psidium*
Guayabo blanco: *Psidium*
Guayabilla: *Mouriri*

Guayusa: *Ilex*
Guinda: *Prunus*
Guinea grass: *Panicum*
Guineo: *Musa*
Guisador: *Curcuma*
Gutapercha: *Sapium*

Habilla: *Fevillea*
Hackberry: *Celosia*
Hamaca huayo: *Couepia*
Hambre huayo: *Gnetum*
Hawk's claw: *Uncaria*
Heart palm: *Euterpe*
Hedionda: *Cassia*
Helecho: *Lycopodium, Selaginella, Tectaria, Trichomanes*
Helecho arborescente: *Cyathea, Nephelea*
Hiedra: *Zebrina*
Hierba de la maestranza: *Lantana*
Hierba de la virgen: *Cestrum*
Hierba del conejo: *Pavonia*
Hierba del soldato: *Piper*
Hierba de San Martin: *Sauvagesia*
Hierba santa: *Cestrum*
Higo: *Ficus*
Higuera: *Ficus*
Higuerilla: *Ricinus*
Hispingo: *Ocotea*
Hoja del aire: *Kalanchoe*
Hog plum: *Spondias*
Holy weed: *Cestrum*
Hormiga caspi: *Duroia*
Horquetilla: *Paspalum*
Horse's tail: *Andropogon*
Hot pepper: *Capsicum*
Huacamayo caspi: *Simira*
Huacamayp piña: *Ananas*
Huacapú: *Minquartia*
Huacapurana: *Campsiandra*
Huacapusillo: *Lindackeria*
Huacatay: *Tagetes*
Huaco: *Clibadium*
Huacrapona: *Iriartea*
Hualaja: *Zanthoxylum*
Huama: *Pistia*
Huamanzamana: *Dictyoloma, Jacaranda*
Huamanzamana del varillal: *Jacaranda*
Huamanzamana pashaco: *Stryphnodendron*
Huambé: *Philodendron*
Huanarpo macho: *Jatropha*
Huangana huasca: *Tanaecium*
Huanguilla: *Eclipta*
Huarmi yoco: *Paullinia*
Huasahí: *Euterpe*
Huasahpi del varillal: *Euterpe*
Huasca bijao: *Ichnosiphon*
Huasca caimito: *Moutabea*
Huasca maronilla: *Olyra*
Huasca mullaca: *Sabicea*
Huasca sisa: *Stigmaphyllum*
Huayhuashi shupa: *Polypodium*
Huayra caspi: *Brosimum, Cedrelinga*
Huayra papa: *Dioscorea*
Huayruro: *Erythrina, Ormosia*
Huayruro amasisa: *Erythrina*
Huayruro colorado: *Batesia, Ormosia*
Hubo: *Spondias*
Huicungo: *Astrocaryum*
Huimba: *Ceiba*
Huingo: *Crescentia*
Huingo sacha: *Crescentia*
Huira caspi: *Nealchornea*
Huira palta: *Persea*
Huiririma: *Astrocaryum*
Huiririmi: *Astrocaryum*
Huitillo: *Alibertia, Palicourea, Posoqueria, Randia*
Huitillo del supay: *Duroia*
Huitina: *Xanthosoma*
Huito: *Genipa*
Huitol: *Genipa*
Humarí: *Poraqueiba*
Hungurahui: *Jessenia*

Ice cream bean: *Inga*
Icoja: *Unonopsis*
Iluicho caspi: *Lindackeria*
Imchic masha: *Cissampelos*
Imicayo: *Axonopus*
Inayuga: *Maximiliana*
Inchahui: *Syagrus*
Inchi: *Caryodendron*
Indano: *Bunchosia, Byrsonima*
Indano colorado: *Byrsonima*
Indian almond: *Terminalia*
Indian rubber tree: *Ficus*
Indigo: *Indigofera*
Inejkeu: *Tynnanthus*
Insira: *Chlorophora*

Insira amarilla: *Chlorophora*
Intimiracu: *Sauvagesia*
Intuto caspi: *Capparis*
Ipadú: *Erythroxylum*
Iporuro: *Alchornea*
Ipururo: *Alchornea*
Ipurosa: *Alchornea*
Irapay: *Lepidocaryum*
Irgapirina sacha: *Polygala*
Ironwood: *Caesalpinia, Minquartia*
Isabelita: *Cataranthus, Mirabilis*
Ishanga: *Urtica*
Ishanga blanca: *Laportea*
Ishanga macho: *Urera*
Ishpingo: *Amburana*
Ishtapi: *Jacaranda*
Isula huayo: *Siparuna*
Itauba: *Mezilaurus, Terminalia*
Itauba amarilla: *Pseudolmedia*
Itininga: *Philodendron*
Itininga sacha: *Philodendron*
Ivory palm: *Phytelephas*
Iwajyu: *Costus*

Jaboncillo: *Phytolacca*
Jacaranda: *Jacaranda*
Jambo piña: *Annanas*
Japanese lantern: *Hibiscus*
Jarabe huayo: *Macoubea*
Jarilla: *Ichthyothere*
Jasmin: *Plumeria*
Java plum: *Syzygium*
Jefe débil fino: *Hevea*
Jemeco: *Tapirira*
Jengibre: *Zingiber*
Jergón quiro: *Anthurium*
Jergón sacha: *Dracontium, Urospatha*
Jesus's heart: *Caladium*
Jimsonweed: *Datura*
Job's tear: *Coix*
Jocuchuchupa: *Sida*
Juanache: *Eugenia*
Jungle flame: *Ixora*
Jute: *Urena*
Juumyba: *Duguetia*

Kapok: *Ceiba*
Katio: *Piper*
Kéécui: *Aciotis*
Kííwahe: *Aciotis*
Kión: *Zingiber*
Knotroot-bristle-grass: *Setaria*
Kudtzú: *Pueraria*
Kudzu: *Calopogonium, Vigna*
Kumu: *Tephrosia*

Lablab bean: *Lablab*
Lacre: *Tetragastris*
Lagartillo: *Moronobea*
Lagartillo de altura: *Buchenavia*
Lagarto caspi: *Calophyllum*
Lagarto piña: *Ananas*
Lagarto shupa: *Epiphyllum, Witta*
Lágrima de Job: *Coix*
Lágrima de la virgen: *Coix*
Lancetilla: *Peperomia*
Lancetilla blanca: *Conmelina*
Lanche: *Myrcianthes*
Lanchi: *Myrcianthes*
Lanza caspi: *Mouriri*
Latapi: *Guarea*
Laurel: *Cordia*
Leche caspi: *Couma*
Leche huayo: *Couma*
Lechuga: *Lactuca*
Lechuga cimarrona: *Pistia*
Lechuga de agua: *Ceratopteris*
Lemon: *Citrus*
Lemon balm: *Melissa*
Lemongrass: *Cymbopogon*
Lengua de vaca: *Asplenium*
Lenteja de agua: *Salvinia*
Lettuce: *Lactuca*
Leucaena: *Leucaena*
Licopodio: *Lycopodium*
Lilea: *Eucharis*
Limón: *Citrus*
Limón ácido: *Citrus*
Limón cidra: *Citrus*
Limóncillo: *Rinorea*
Limón chino: *Averrhoa*
Limón sútil: *Citrus*
Lion's tail: *Acalypha*
Lirio: *Hedychium*
Llama plata: *Episcia, Lindernia*
Llanchama: *Naucleopsis, Poulsenia*
Llanchamillo: *Naucleopsis*
Llantén: *Plantago*
Llausaquiro: *Apeiba*
Lluichu lancetilla: *Justicia*
Lobo sanango: *Stenosolen, Tabernaemontana*

Locura: *Lagerstroemia*
Loro micuna: *Macoubea*
Loro shungo: *Duckesia, Humaria, Sacoglottis, Vantanea*
Loto azul: *Nymphaea*
Lucky nut: *Thevetia*
Lucma: *Lucuma, Pouteria*
Lupuna: *Ceiba, Chorisia*
Lupuna blanca: *Ceiba*
Lupuna bruja: *Cavanillesia*
Lucma sacha: *Clavija*

Maaihiiba: *Maranta*
Macambillo: *Theobroma*
Macambo: *Theobroma*
Macambo sacha: *Theobroma*
Macaquiño: *Dioscorea*
Machimango: *Eschweilera*
Machimango blanco: *Couratari, Eschweilera*
Machimango cachimbo: *Couratari*
Machimango colorado: *Eschweilera, Lecythis*
Machimango negro: *Eschweilera*
Machinga: *Brosimum*
Machinguilla: *Maprounea*
Madre selva: *Lonicera*
Mafaffa: *Xanthosoma*
Magaranduva: *Chrysophyllum*
Magua: *Mangifera*
Mahogany: *Swietenia*
Maicillo: *Axonopus*
Maidenhair fern: *Adiantum*
Mai tanpeshco: *Philodendron*
Maíz: *Zea*
Maíz del tuqui tuqui: *Utricularia*
Maíz guineo: *Sorghum*
Malay apple: *Syzygium*
Malva: *Malachra*
Malva roja: *Urena*
Malvavisco: *Malvaviscus*
Mamee apple: *Mammea*
Mamey: *Mammea, Syzygium*
Mamilla: *Otoba*
Manapeui: *Callichlamys*
Mancoa: *Grias*
Manchari: *Humiriastrum*
Manchari blanco: *Humiriastrum*
Manchari caspi: *Duckesia, Vantanea*
Mandarina: *Citrus*
Mandioca: *Manihot*
Maní: *Arachis*
Maní del monte: *Plukenetia*
Mangle de altura: *Sterigmapetalum*
Mango: *Mangifera*
Mango chico-rico: *Mangifera*
Mango inherto: *Mangifera*
Mangua dulce: *Mangifera*
Mano abierta: *Xanthosoma*
Manto de Cristo: *Gomphrena*
Manzanilla cimarrona: *Wedelia*
Manzanilla sacha: *Wulffia*
Maquizapa ñaccha: *Apeiba*
Maquizapa ñaccha blanco: *Apeiba*
Maracuyá: *Passiflora*
Mara mara: *Urera*
Marañón: *Anacardium*
Marco: *Ambrosia*
Margarita: *Desmodium*
María buena: *Clitoria, Pterocarpus*
María Luisa: *Cymbopogon*
Maricahua: *Brugmansia*
Marihuana: *Cannabis*
Mari Mari: *Hymenolobium*
Mari Mari del bajo: *Vatairea*
Marona: *Bambusa, Guadua*
Marupá: *Simarouba*
Masaranduvilla: *Chrysophyllum*
Mashishe: *Cucumis*
Mashonaste: *Batocarpus, Clarisia*
Mashushingo: *Pavonia*
Matapalo: *Clusia, Ficus*
Mataro: *Cassia*
Mataro chico: *Cassia*
Mataro grande: *Cassia*
Mataro huasca: *Cassia*
Matico: *Piper*
Matsajcahe: *Terminalia*
Mauba: *Erisma, Vochysia*
Mayhua: *Stenomesson*
Medicinal rose: *Rosa*
Melina: *Gmelina*
Melon: *Cucumis*
Meloncito blanco: *Celtis*
Meneco: *Jacaranda*
Meralla: *Peperomia*
Mesque: *Neea*
Meto huayo: *Caryodendron*
Mexican gama grass: *Tripsacum*
Micura: *Petiveria*
Milkbush: *Euphorbia*
Milk tree: *Couma*

Millua situlli: *Heliconia*
Misho chaqui: *Brosimum, Helicostylis, Pseudolmedia*
Misho quiro: *Alseis*
Misho runto: *Rauwolfia*
Mishquina: *Eugenia*
Mishquipanga: *Renealmia*
Mishquipanguilla: *Geogenanthus*
Mishuisma: *Hibiscus*
Mistletoe: *Phoradendron*
Mocosa: *Petiveria*
Moena: *Caryodaphnopsis, Cinnamomum, Endlicheria, Licaria, Nectandra, Ocotea, Phoebe*
Moena alcanfor: *Ocotea*
Moena amarilla: *Aniba, Nectandra, Ocotea*
Moena negra: *Aniba, Ocotea*
Moena sin olor: *Didymopanax, Qualea, Ruitzernania*
Mojarra caspi: *Alchornea, Hieronima*
Molasses grass: *Melinis*
Molle: *Schinus*
Monkey-cap palm: *Manicaria*
Monkey comb: *Apeiba*
Monkey ladder: *Bauhinia*
Moriche: *Mauritia*
Moriche palm: *Mauritia*
Motelilla: *Calathea*
Motelilla enana: *Calathea*
Motelillo: *Fittonia*
Motelo chaqui: *Naucleopsis, Pseudolmedia*
Motelo huasca: *Bauhinia*
Motelo sanangri: *Abuta*
Mountain soursop: *Annona*
Mucuna: *Mucuna*
Mucura: *Petiveria*
Muena: *Licaria, Rhodostemonodaphe*
Muena rifarillo: *Coupia*
Mullaca: *Clidemia, Physalis*
Mullaca morada: *Clidemia*
Mullo huayo: *Coix*
Musk: *Abelmoschus*
Muster John Henry: *Tagetes*
Mútsújkeu: *Blepharodon*
Mututi: *Pterocarpus*

Najahe: *Batocarpus*
Ñame: *Dioscorea*
Naparo cimarrón: *Eclipta*
Napier grass: *Pennisetum*
Naranja: *Citrus* (**Orange**)
Naranja agria: *Citrus* (**sour orange**)
Naranja dulce: *Citrus* (**sweet orange**)
Naranjilla: *Murraya*
Naranjo: *Aspidosperma*
Naranjo podrido: *Parahancornia*
Nashum: *Calatola*
Ñati-papa: *Dioscorea*
Navidad caspi: *Symphonia*
Ñeja: *Bactris*
Ñejilla: *Bactris*
Nescafé: *Canavalia, Mucuna*
Nescao: *Mucuna*
Nia boens: *Mansoa*
Nina caspi: *Leonia, Licania*
Nipirihe: *Phytolacca*
Nishi bata: *Pleonotoma*
Niu weoko: *Piper*
Ñorbo cimarrón: *Passiflora*
Ñucñu pichana: *Scoparia*
Nudillo: *Brachiaria, Ichnanthus, Leptochloa, Axonopus*
Nuez moscada: *Myristica*
Numiallamihe: *Quararibea*
Nupe: *Pachyrrhizus*
Nutmeg: *Myristica*

Ocajiníímune: *Orthoclada*
Ocollucuy sacha: *Stachytarpheta*
Ocuera negra: *Pollalesta*
Ocuero: *Vernonia*
Ojé: *Ficus*
Ojé macho: *Ficus*
Ojé de hoja menuda: *Ficus*
Ojé del cauchero: *Ficus*
Ojo de pollo: *Alternanthera*
Oleander: *Nerium*
Onion: *Allium*
Ooniyatso: *Picramnia*
Opuntia: *Opuntia*
Orange: *Citrus*
Orange jasmine: *Murraya*
Oreja de elefante: *Xanthosoma*
Oreja del diablo: *Psychotria*
Oreja sacha: *Hydrocotyle*
Orquidia: *Cattleya*
Otaheite apple: *Spondias*

Pablo manchana: *Chimarrhis*

Pacay: *Inga*
Pacunga: *Bidens*
Paddle tree: *Aspidosperma*
Padojcohe: *Remijia*
Paichecara: *Kalanchoe*
Paico: *Chenopodium*
Pairajo de altura: *Inga*
Pájaro bobo: *Tessaria*
Palillo: *Campomanesia, Curcuma, Selaginella*
Paliperro: *Tabebuia, Vitex*
Palisangre: *Brosimum*
Palisangrillo: *Haploclathra*
Palma aceitera: *Elaeis*
Palma africana: *Elaeis* (**African palm**)
Palma brasilera: *Cycas, Trithrinax* (**Brazilian palm**)
Palma de goma: *Zamia*
Palmicha: *Geonoma*
Palmito: *Euterpe*
Palo brujo: *Brosimum*
Palo cruz: *Brownea*
Palo de ajo: *Gallesia*
Palo de balsa: *Ochroma*
Palo de cebolla: *Gallesia*
Palo de hombre: *Tococa*
Palo de rosa: *Aniba*
Palo de sangre: *Hieronima*
Palo de vaca: *Alseis*
Palometa huayo: *Alchornea, Neea*
Palo perro: *Chimarrhis*
Palo sangre: *Dialium*
Palo sangre dominante: *Swartzia, Taralea*
Palo santo: *Bursera*
Palta: *Persea*
Pamarrosa: *Syzygium*
Pampa moena: *Endlicheria*
Pampa orégano: *Lippia*
Pampa remocaspi: *Chimarrhis, Duroia*
Panama hat palm: *Carludovica*
Pan del árbol: *Artocarpus*
Pandisho: *Artocarpus*
Papa china: *Colocasia*
Papailla: *Carica, Momordica*
Papaya: *Carica*
Papaya caspi: *Jacaratia*
Papaya del venado: *Jacaratia*
Papelillo: *Bougainvillea, Cariniana, Couratari, Stenosolen*
Papelillo caspi: *Cariniana*
Paraguayo: *Uncaria*
Paraguita: *Schizaea*
Paraiso: *Melia*
Parinari: *Couepia, Humiria, Licania*
Parinari colorado: *Licania*
Parinari sacha: *Schistostemon*
Parinari sapo: *Vantanea*
Parra: *Vitis*
Pashaca (o): *Macrolobium, Parkia, Pentaclethra, Piptadenia, Schizolobium, Stryphnodendron*
Pashaco colorado: *Macrolobium*
Pashaco curtidor: *Parkia*
Pashaco oreja de negro: *Enterolobium*
Pashaquilla: *Jacqueshuberia, Leucaena, Macrolobium*
Pashna huachuna: *Croton*
Pasto de cuaresma: *Digitaria*
Pasto elefante: *Pennisetum*
Pasto estrada: *Eleusine*
Pasto guinea: *Panicum*
Pasto yaragua: *Hyparrhenia*
Pata de gallina: *Digitaria, Eleusine* (**hen's leg**)
Pata de vaca: *Bauhinia, Cynometra* (**cow's leg**)
Pate: *Crescentia*
Patiquina: *Dieffenbachia*
Paujil chaqui: *Davilla, Pharus, Tetracera*
Paujil huasca: *Pinzona*
Paujil ruro: *Gnetum, Guarea*
Pava chaqui: *Begonia*
Pawpaw: *Carica*
Peach grass: *Hemarthria*
Peach palm: *Bactris*
Peanuts: *Arachis*
Péécojúhe: *Casearia*
Pega josa: *Boerhavia*
Pega pega: *Boerhavia, Desmodium, Pavonia, Triumfetta*
Pega perro: *Adenaria*
Pega pinto: *Spigelia*
Pejibaye: *Bactris*
Pelo ponto: *Cycas, Elaeis, Phytelephas*
Penca: *Fourcroya*
Pepino: *Cucumis*
Pepper hibiscus: *Malvaviscus*
Perma caspi: *Vernonia*
Peruvian onions: *Allium*

Peruvian ragweed: ***Ambrosia***
Philippine waxflower: ***Nicloaia***
Physic nut: ***Jatropha***
Piasabas: *Phytelephas*
Pichana: *Sida*
Pichana albaca: *Ocimum*
Pichirina: *Vismia*
Pichirina colorada: *Vismia*
Pichirina hoja grande: *Vismia*
Pichirina hoja menuda: *Vismia*
Picho e muela: *Psychotria*
Picho huayo: *Siparuna*
Picsho: *Codonanthe*
Picurillo: *Alternanthera*
Pigeon pea: ***Cajanus***
Pijuayo: *Bactris*
Pimienta negro: *Piper*
Pimiento: *Capsicum*
Piña: *Ananas* (**Pineapple**)
Piña amarilla: *Ananas*
Piña negra: *Ananas*
Piñaraño: *Floscopa*
Piñaquiro colorado: *Hieronima*
Pineapple: ***Anacardium***
Pink primrose: ***Oenothera***
Piñón: *Jatropha*
Piñón blanco: *Jatropha*
Piñón ceqeati: *Malvaviscus arboreus*
Piñón negro: *Jatropha*
Pino regional: *Alseis, Podocarpus* (**Regional pine**)
Pinsha callo: *Xylopia*
Pinsha caspi: *Aspidosperma*
Pió: *Calatola*
Pióó: *Piper*
Piosha: *Neea*
Piri piri: *Cyperus*
Piri piri de víbora: *Cyperus*
Pishco huichi: *Dichorisandra*
Pishco isma colorado: *Juanulloa*
Pishco isma: *Oryctanthus, Phoradendron, Phthirusa*
Pitanga: *Eugenia*
Pitomba: *Melicocca*
Pituca: *Colocasia*
Planta china: *Dieffenbachia*
Planta navideña: *Euphorbia* (**Christmas plant**)
Planta plástica: *Pedilanthus* (**Plastic plant**)
Plantain: ***Musa, Plantago***
Plantilla pashaco: *Enterolobium*
Plata pashaco: *Macrolobium*
Platanillo: *Phenakospermum*
Plátano: *Musa* (**Plantain**)
Playa huasca: *Mikania*
Plumilla: *Nephrolepis*
Podocarp: ***Podocarpus***
Poe-hoe: *Phytolacca*
Poeinseta: *Euphorbia*
Poinsettia: ***Euphorbia***
Pokeberry: ***Phytolacca***
Pomarrosa: *Syzygium*
Pomelo: *Citrus*
Pona: *Socratea*
Pona coto-shupa: *Wettinia*
Ponilla: *Catoblastus, Iriartella, Wettinia*
Porotillo: *Swartzia, Vigna*
Poroto: *Phaseolus*
Poroto huango: *Geophila*
Poroto shimbillo: *Inga*
Poto-pate: *Lagenaria*
Pretino: *Cavanillesia*
Pretino-punga: *Pseudobombax*
Prickly pear: ***Opuntia***
Pride of Barbados: ***Caesalpinia***
Provision tree: ***Pachira***
Puca cabacina: *Gasteranthus*
Puca huasca: *Doliocarpus*
Puca huayo: *Neea*
Puca lupuna: *Cavanillesia*
Puca panga: *Arrabidaea*
Puca quiro: *Simira*
Puca sisa: *Besleria, Warscewiczia*
Puca varilla: *Adenaria*
Pucuna caspi: *Iryanthera*
Pucunucho: *Capsicum*
Puma barba: *Pharus*
Puma caspi: *Roucheria*
Puma chaqui: *Naucleopsis*
Puma yarina: *Elaeis*
Pumpkin: ***Cucurbita***
Punga: *Pachira, Pseudobombax*
Punga de altura: *Pachira*
Pungar muena: *Pleurethyrium*
Punguilla: *Rodongnaphalopsis*
Punguilla del varillal: *Rhodognaphalopsis*
Purma cetico: *Cecropia*
Purma sisa: *Palicourea*
Purma tahuarí: *Tabebuia*

Purple granadilla: *Passiflora*
Purslane: *Portulaca*
Puru pagic sacha: *Polygala*
Puspo moena: *Ocotea*
Puspo poroto: *Cajanus*
Puspo quihua: *Priva*
Puspo tamshi: *Asplundia, Torococarpus*
Putu putu: *Eichhornia*
Puzanga caspi: *Sloanea*

Quillobordón: *Aspidosperma*
Quillobordón masha: *Sterigmapetalum*
Quillo-panga huasca: *Spathicalyx*
Quillo sisa: *Erisma, Qualea, Ruitzerania, Vochysia*
Quinilla: *Chrysophyllum, Manilkara, Micropholis*
Quinilla blanca: *Pouteria*
Quinilla blanca del bajo: *Pouteria*
Quinilla caimitillo: *Pouteria*
Quinilla colorada: *Chrysophyllum*
Quinilla negra: *Humiria, Micropholis, Pouteria*
Quisa: *Urtica*

Rabo de zorro: *Andropogon*
Raya balsa: *Montrichardia*
Raya caspi: *Banara*
Rayan: *Sambucus*
Red granadilla: *Paspalum*
Red sprangle top: *Leptochloa*
Red spurge: *Euphorbia*
Remo caspi: *Aspidosperma*
Remolina: *Paspalum*
Renaco: *Ficus*
Renaquilla: *Clusia*
Requia: *Carapa, Guarea, Trichilia*
Retama: *Cassia*
Retamita: *Spartium*
Rice: *Oryza*
Rifari: *Miconia*
Ringworm senna: *Cassia*
Rosa: *Rosa*
Rosa Alejandrina: *Rosa*
Rosa china: *Hibiscus*
Rosa del monte: *Brownea*
Rosa del remedio: *Rosa*
Rosca shimbillo: *Inga*
Rose: *Rosa*
Rose apple: *Syzygium*
Rosella: *Hibiscus*
Roselle: *Hibiscus*
Rosewood: *Aniba*
Rotenone: *Lonchocarpus*
Rubber: *Hevea*
Ruda: *Ruta*
Ruichao: *Sabicea*
Rumo sacha: *Psychotria*
Rupiña: *Myrcia*

Sábana del lagarto: *Victoria*
Sacha ajo: *Petiveria*
Sacha alfaro: *Vochysia*
Sacha ampihuasca: *Curarea*
Sacha anona: *Rollinia*
Sacha bombonaje: *Chelyocarpus*
Sacha café: *Picrolemma*
Sacha caimito: *Chrysophyllum*
Sacha casho: *Anacardium, Vochysia*
Sacha culantro: *Eryngium*
Sacha chope: *Gustavia*
Sacha foster: *Phyllanthus*
Sacha guayaba: *Lacunaria, Eugenia*
Sacha huito: *Palicourea, Posoqueria*
Sacha inchi: *Plukenetia*
Sacha mangua: *Grias, Potalia*
Sacha níspero: *Bellucia*
Sacha nisperillo: *Loreya*
Sacha ojé: *Ficus*
Sacha oje del cauchero: *Ficus*
Sacha orquidia: *Xyphidium*
Sacha papa: *Dioscorea*
Sacha papa morada: *Dioscorea*
Sacha pichana: *Croton*
Sacha piña: *Aechmea*
Sacha pandisho: *Pachira*
Sacha punga: *Cochlospermum*
Sacha rifari: *Banara*
Sacha tulpay: *Batocarpus*
Sacha uvilla: *Didymopanax, Pourouma*
Sachavaca micuna: *Trophis*
Sacha verbena: *Stachytarpheta*
Sacha yuyu: *Peperomia*
Sage: *Salvia*
Salvia: *Salvia*
Sameruca: *Pyschotria*
Sam Pange: *Picramnia*
Sanango: *Bonafousia, Faramea, Rauwolfia, Tabernamontana*
Sanango ucho: *Tabernamontana*
Sananguillo: *Psychotria*
Sandia: *Citrullus*

Tahuarí colorado: *Tabebuia*
Tahuarí negro: *Tabebuia*
Talhuí: *Spartium*
Talla: *Caesalpinia*
Tamamuri: *Brosimum*
Tamara: *Leonia*
Tamara blanca: *Crateva, Leonia*
Tamarillo: *Gleospermum*
Tamarind: *Tamarindus*
Tamarindo: *Tamarindus*
Tamshi: *Heteropsis*
Tamshi canastero: *Heteropsis*
Tamshi delgado: *Heteropsis*
Tanasharina: *Citrus*
Tangarana: *Tachigalia*
Tangarana amarilla: *Tachigalia*
Tangarana blanca: *Tachigalia*
Tangarana de altura: *Sclerolobium, Tachigalia*
Tangarana de hoja menuda: *Sclerolobium*
Tangarana del bajo: *Triplaris*
Tangarana sin madre: *Triplaris*
Tansharina: *Citrus*
Taótaco: *Catablatus*
Taperiba: *Spondias*
Tapisho: *Spondias*
Tara: *Caesalpinia*
Tarota: *Tabebuia*
Teak: *Tectona*
Tectona: *Tectona*
Terminalia: *Terminalia*
Téwatacaá: *Mucuna*
Texas palmetto: *Sabal*
Thatch palm: *Lepidocaryum*
Ti plant: *Cordyline*
Ticsa micuna: *Vernonia*
Timbó: *Paullinia*
Tinaja caspi: *Licania*
Tinctona: *Solanum*
Toad vine: *Cissus, Omphalea*
Toa-toé: *Brugmansia*
Tobacco: *Nicotiana*
Toé: *Brugmansia*
Toé negro: *Teliostachya*
Tocino caimito: *Pouteria*
Tomate: *Lycopersicum*
Tomato: *Lycopersicum*
Tonipulmon: *Maytenus*
Toomecocoriu: *Tabernaemontana*
Topa: *Ochroma*
Topillo: *Croton*
Topiro: *Solanum*
Torch ginger: *Nicloaia*
Tornillo: *Cedrelinga*
Toronja: *Citrus* (**Grapefruit**)
Toronjil: *Melissa*
Tortuga blanca: *Diclinanona* (**White tortoise**)
Tortuga caspi: *Duguetia*
Torurco: *Digitaria, Homolepis, Panicum, Paspalum*
Tree cotton: *Gossypium*
Tree papaya: *Jacaratia*
Tree tomato: *Cyphomandra*
Trigo: *Coix* (**Wheat**)
Trompetero sacha: *Abuta*
Trompo huayo: *Lacistemma*
Tropical milkweed: *Asclepias*
Trueno shimbillo: *Brownea*
Trujillo: *Impatiens*
Trujillo amarillo: *Impatiens*
Trumpet tree: *Tabebuia*
Tsaruwa: *Tococa*
Tscaahe: *Senefeldera*
Tsutsihe: *Philodendron*
Tuaruubia: *Cleome*
Tuberose: *Polianthes*
Tulpay: *Batocarpus, Clarisia*
Tumbo: *Passiflora*
Tuna: *Opuntia*
Tunchi albaca: *Lantana*
Tupamaqui: *Neea, Psychotria*
Turmeric: *Curcuma*
Tutumo: *Crescentia*

Ubos: *Spondias*
Ucsha-coconilla: *Solanum*
Ucsha gramalote: *Leptochloa*
Ucho caspi: *Casearia*
Ucho mullaca: *Humiriastrum*
Ucho sanango: *Tabornamontana*
Ucumi-micuna: *Psychotria*
Ukshaquiro: *Aparisthmium*
Ullucuy chuchuashi: *Rhacoma*
Umarí: *Pouraqueiba*
Umari del monte: *Paypayrola*
Uncucha: *Xanyhosoma*
Ungushurato: *Piper*
Uña de gato: *Macfadenya, Uncaria*
Uña de gavilán: *Uncaria*
Uña de tigre: *Chelonanthus*

Zapallito: *Gurania*
Zapallo: *Cucurbita*
Zapatito de Jesús: *Pedilanthus*
Zapayo: *Cucurbita*
Zarandeja: *Lablab*
Ziu: *Vernonia*
Zorro caspi: *Couratari, Guatteria*

MEDICINAL INDEX TO VASQUEZ' and SCHULTES' AMAZONIAN COMPENDIA

Jim Duke

(Note: This index does not cover the extra-Amazonian entries which are computerized in Duke and Wain, Medicinal Plants of the World. 1981. The index is fairly complete for Rodolfo's entries.)

abortifacient: *Ananas, Brunfelsia, Caesalpinia, Carica, Citrus, Crescentia, Cyperus, Eleuterine, Genipa, Gossypium, Mangifera, Melia, Passiflora, Persea, Rhacoma, Ruta, Scoparia, Siparuna, Verbena, Vouacoupa*

abscess: *Brosimum, Leonia, Piper, Potalia, Virola*

acaricide: *Iryanthera, Virola*

adenopathy: *Davilla*

albuminuria: *Eclipta, Lindernia*

alopecia: *Entada, Guazuma, Mabea, Phyllanthus*

altitude sickness: *Erythroxylum*

ameba: *Couma, Eleuterine, Humiria, Persea*

amenorrhea: *Gossypium, Pityrogramma*

analgesic (anodyne): *Abuta, Anadenathera, Asclepias, Bonafousia, Brugmansia, Cissampelos, Cissus, Crotalaria, Croton, Duguetia, Erisma, Erythrina, Erythroxylum, Gossypium, Goupia, Hedychium, Hibiscus, Ilex, Iribachia, Jatropha, Mansoa, Mayna, Odontocarya, Paullinia, Pilea, Piper, Plantago, Potalia, Psychotria, Sabicea, Scoparia, Securidaca, Siparuna, Spartium, Spondias, Swartzia, Tabernaemontana, Tachigalia, Tagetes, Urera, Warszewiczia, Wedelia, Xiphidium, Zingiber*

anemia: *Abuta, Lantana, Persea, Phoradendron*

anesthetic: *Piper, Wedelia*

angina: *Bidens*

anorexic: *Paullinia*

anthelmintic: *Ananas*

antiabortive: *Pavonia*

antibechic: *Lantana, Pityogramma*

antibilious: *Gurania, Iribachia, Paullinia, Solanum, Terminalia* antidote: *Bixa* (HCN), *Jatropha, Manihot, Maranta, Mollia, Mucuna, Phytolacca* (capsaicin), *Piper, Potalia, Solanum*

antiflatulent: *Melissa*

antiseptic: *Aspidosperma, Chlorophora, Clidemia, Hibiscus, Lantana, Mangifera, Physalis, Phytolacca, Piper, Plantago, Schoenobiblus, Scoparia, Virola*

antispasmodic: *Annona, Brugmansia, Clidemia, Hibiscus, Indigofera, Mayna, Melissa, Muntingia, Petiveria, Phyllanthus, Piper, Siparuna*

antitussive: *Erythrina, Hibiscus, Macfadenya, Pavonia, Petiveria, Psychotria, Tabebuia, Virola*

antivenereal: *Brunfelsia, Cassia, Chlorophora, Copaifera, Hamelia, Ilex, Jatropha, Lycopodium, Melia, Mucuna, Potalia, Strychnos, Uncaria*

anxiety: *Allamanda,*

apertif: *Panicum*

aphidicide: *Quassia*

aphrodisiac: *Abuta, Bixa, Caraipa, Davilla, Ilex, Maytenus, Mimosa, Nymphaea, Persea, Siparuna, Strychnos, Tachigalia, Tanaecium, Tynnanthus, Warscewiczia, Xylopia, Zingiber*

aphthae: *Bidens, Piper*

arteriosclerosis: *Caryodendron*

arthritis: *Alchornea, Brugmansia, Carapa, Lycopodium, Mansoa, Maytenus, Pentagonia, Solanum, Unonopsis, Virola, Zingiber*

asthma: *Brosimum, Couma, Crescentia, Eclipta, Gynerium, Hura, Jessenia, Lantana, Osteophloeum, Physalis, Solanum, Tessaria*

astringent: *Chlorophora, Pityogramma, Poeppigia, Poraqueiba, Sabicea, Trema*

atherosclerosis: *Paullinia*

athlete's foot: *Hymenaea, Peperomia*

atticide: *Euphorbia, Lonchocarpus*

backache: *Warszewiczia*

bactericide: *Eleuterine, Kalanchoe, Phyllanthus, Poraqueiba*

balm: *Erisma*

bat-repellent: *Scleria*

bladder: *Xylopia*

blenorrhagia: *Chlorophora, Piper, Tephrosia*

blenorrhea: *Costus, Dimerocostus,*

boils: *Croton, Dioscorea, Hancornia, Kalanchoe*

bot-fly: *Anacardium*

breasts: *Miconia, Solanum*

bronchitis (osis): *Adiantum, Anadenanthera, Brosimum, Cajanus, Caryodendron, Cordia, Costus, Crescentia, Dimerocostus, Genipa, Hibiscus, Jessenia, Kalanchoe, Lantana, Macoubea, Myroxylum, Nicotiana, Petiveria, Piper, Pityrogramma, Plantago, Psychotria, Smilax, Tanaecium, Unonopsis, Zingiber*

bruises: *Abuta, Brosimum, Carludovica, Curcuma, Dipteryx, Eleusine, Jatropha, Kalanchoe, Microtea, Nicotiana, Passiflora*

bugbite (sting): *Adiantum, Carapa, Cyclanthus, Indigofera, Lycopodium, Mammea, Omphalea, Otoba, Potalia, Urera*

burns: *Allium, Microtea*

callus: *Luffa*

calmant: *Annona, Brugmansia, Potalia, Tabernaemontana*

cancer: *Cariniana, Miconia, Tabebuia, Uncaria*

candidiasis: *Tabebuia*

canker: *Virola*

cardiopathy: *Cecropia, Chondrodendron*

cardiotonic: *Annona, Carica, Paullinia, Theobroma*

caries: *Anacardium, Calatola, Dendropanax, Duroia, Ficus, Manettia, Neea, Petiveria, Phenakospermum, Piper, Potalia, Simira, Stigmaphyllum, Thelypteris, Virola*

carminative: *Ananas, Capsicum, Lycopodium, Melissa, Piper, Salvia, Siparuna, Zingiber*

cataracts: *Goupia, Pityogramma*

catarrh: *Eleusine*

cathartic: *Caesalpinia, Genipa*

cellulitis: *Luffa*

chickenpox: *Cicer, Lantana*

chills: *Bidens, Brugmansia, Brunfelsia, Manihot, Nicotiana*

cholagogue: *Mikania, Nicotiana, Stachytarpheta*

cholecystosis: *Virola*

cholera: *Chenopodium, Hamelia*

choleretic: *Crescentia, Pilea*

cicatrizant: *Andira, Brosimum, Copaifera, Coussapoa, Cyathea, Jacaranda, Jacaratia, Jatropha, Kalanchoe, Maprounea, Myroxylum, Parahancornia, Piper, Potalia, Socratea, Virola*

cirrhosis: *Uncaria*

CNS-depressant: *Petiveria*

coffee: *Canavalia, Mucuna, Pourouma*

colds: *Alchornea, Brugmansia, Chenopodium, Cordia, Crotalaria, Eleusine, Eryngium, Faramea, Jacaranda, Lantana, Macfadyena, Malachra, Maximiliana, Melia, Myroxylum, Osteophloeum, Piper, Solanum, Stachytarpheta, Tabebuia, Tabernaemontana, Trichomanes*

colic: Abuta, Chenopodium, Eleuterine, Iribachia, Momordica, Ocimum, Physalis, Sabicea, Salvia, Sambucus, Virola

collyrium: Nicotiana, Ocimum, Passiflora, Pityogramma

conjunctivitis: *Abuta, Arrabidaea, Asclepias, Kalanchoe, Lepidocaryum, Martinella, Nicotiana, Ocimum, Paspalum, Passiflora, Plantago, Spathicalyx, Syzygium, Tabernaemontana, Zingiber*

contraceptive (male): Curaréa,

contraceptive: Anacardium, Chenopodium, Citrus, Curaréa, Cyperus, Desmodium, Guatteria, Mangifera, Persea, Poraqueiba, Pourouma, Priva, Scoparia, Spondias, Unonopsis, Zingiber

convulsions: Monstera, Sabicea, Siparuna

cornea: Caraipa,

cosmetic: Goupia, Hibiscus, Jessenia, Mabea, Palicourea, Solanum, Xylopia

cough: Costus, Dimerocostus, Ficus, Hibiscus, Lantana, Macfadenya, Tabebuia, Melia, Osteophloeum, Pavonia, Piper, Polypodium, Solanum, Sparattanthelium, Tabebuia

cystitis: Chlorophora, Costus, Dimerocostus, Eleusine, Kalanchoe debility: Abuta, Brosimum, Canna, Cyphomandra, Gnetum, Persea, Phoradendron, Piper, Xiphidium

depilatory: Sparattanthelium

depurative: Ampelozizyphus, Jatropha, Pentagonia, Persea, Pityogramma

dermatosis: Arrabidaea, Artocarpus, Bixa, Caraipa, Calycophyllum, Caryodendron, Cassia, Cestrum, Chenopodium, Couma, Curaréa, Dioscorea, Eclipta, Eleusine, Geophila, Gossypium, Hamelia, Lindernia, Manihot, Maprounea, Mikania, Moronobea, Olyra, Philodendron, Physalis, Phytolacca, Picramna, Picrolemma, Scizaea, Securidaca, Simira, Solanum, Symphonia, Tanaecium, Theobroma, Vatairea, Virola, Vismia, Warszewiczia

diabetes: Alternanthera, Annona, Calophyllum, Calycophyllum, Coutarea, Momordica, Persea, Portulaca, Stachytarpheta, Wullfia

diaphoretic: Brunfelsia, Isertia, Lantana

diarrhea: *Alchornea, Anacardium, Asclepias, Calophyllum, Casearia, Cedrela, Couma, Crescentia, Cyperus, Dalbergia, Eleuterine, Eleusine, Eryngium, Erythroxylum, Ficus, Genipa, Hamelia, Helosis, Hibiscus, Hymenaea, Iryanthera, Jessenia, Lippia, Machaerium, Manihot, Maprounea, Maytenus, Musa, Paullinia, Persea, Pharus, Posoqueria, Psidium, Scoparia, Sparattanthelium, Spondias, Triumfetta, Unonopsis*

digestive: Bixa, Cymbopogon, Piper, Portulaca

disinfectant: *Copaifera*

dislocation: Maquira, Oryctanthus, Phoradendron, Phthirusa, Swartzia

diuretic: *Abuta, Andropogon, Bauhinia, Bidens, Brunfelsia, Cajanus, Canna, Cassia, Chlorophora, Chondrodendron, Copaifera, Dioscorea, Eleusine, Gynerium, Imperata, Indigofera, Laportea, Lycopodium, Microtea, Mirabilis, Mucuna, Phyllanthus, Piper, Tabernaemontana, Pothomorphe, Sauvagesia, Sida, Spartium, Theobroma, Urera*

divination: Banisteriopsis, Brugmansia,

dropsy: Bidens, Canna, Chondrodendron, Luffa,

dye: Anacardium, Arrabidaea, Bellucia, Chlorophora, Faramea, Genipa, Guatteria, Neea, Oenocarpus, Orthoclada, Palicourea, Picramnia, Picrolemma, Renealmia, Simira, Warszewiczia

dysentery: Anacardium, Bidens, Bixa, Bunchosia, Cajanus, Celtis, Clidemia, Eleusine, Grias, Guazuma, Hamelia, Helosis, Humiria, Lycopodium, Maytenus, Persea, Poraqueiba, Sabicea, Simarouba, Stachytarpheta, Terminalia, Uncaria

dysmenorrhea: Abuta, Brownea, Caryocar, Clidemia, Cymbopogon, Faramea, Gossypium, Gurania, Isertia, Lantana, Matisia, Maytenus, Persea, Piper, Pityogramma, Psidium, Quararibea, Spondias, Tabebuia dyspepsia: Ananas, Carica, Cymbopogon, Erythroxylum, Lippa, Maranta, Microtea, Siparuna, Virola

dyspnea: Capsicum

dysuria: Cedrela, Indigofera, Kalanchoe, Microtea, Sambucus, Uncaria

earache: Cassia, Gossypium, Juanulloa, Kalanchoe, Musa, Petiveria, Physalis, Psychotria

eczema: Amaranthus, Caraipa

edema: *Zea*

emetic: Abuta, Adiantum, Asclepias, Banisteriopsis, Brownea, Faramea, Grias, Guarea, Ilex, Jatropha, Mansoa, Melia,, Momordica, Palicourea, Pothomorphe, Scoparia, Tabernaemontana, Tachigalia

emmenagogue: *Ambrosia, Bidens, Chenopodium, Chondrodendron, Gossypium, Hibiscus,Lantana, Momordica, Persea, Sicana*

emollient: Carapa, Lantana, Piper, Portulaca, Virola

enteritis (osis): Crescentia, Croton, Erythroxylum, Kalanchoe, Physalis, Piper, Virola

epilepsy: *Cissus, Eleuterine, Indigofera, Syzygium*

epistaxis: Warscewiczia

erysipelas: Brugmansia, Physalis, Solanum, Urera, Virola

evil-eye: Brosimum,

excitant: Piper, Xylopia

expectorant: Bixa, Croton, Lantana

fever: *Abuta, Aciotis, Allamanda, Ambelania, Ambrosia, Aspidosperma, Bixa, Brosimum, Caesalpinia, Canna, Carapa, Cassia, Ceiba, Chenopodium, Chondrodendron, Cissampelos, Citrus, Cochlospermum, Coix, Costus, Cymbopogon, Cyperus, Dimerocostus, Dipteryx, Eryngium, Erythrina, Faramea, Hamelia, Hibiscus, Hyptis, Indigofera, Iribachia, Isertia, Jatropha, Justicia, Kalanchoe, Ladenbergia, Lindernia, Ludwigia, Lycopodium, Macfadenya, Malchra, Manihot, Mansoa, Martinella, Melia, Mikania, Momordica, Myroxylum, Nectandra, Ocimum, Petiveria, Phyllanthus, Piper, Plantago, Polypodium, Portulaca, Potalia, Pothomorphe, Psychotria, Ricinus, Sambucus, Sauvagesia, Schizolobium, Scoparia, Sida, Simarouba, Siparuna, Syzygium, Tabebuia, Tabernaemontana, Tagetes, Terminalia, Trema, Tynanthus, Urera, Wulffia, Zea*

filariasis: Caryocar, Gossypium

flu: Anacardium, Chenopodium, Costus, Croton, Cymbopogon, Cyperus, Dimerocostus, Eleusine, Eryngium, Erythrina, Hibiscus, Macfadenya, Ocimum, Panicum, Pityogramma, Sambucus, Stachytarpheta, Tabebuia, Wulffia

food-poisoning (started lated): Solanum

fracture: Kalanchoe, Passiflora

freckles: Musa, Spartium

fright: Cyperus

fumitory: Brugmansia,

fungicide: Aspidosperma, Calycophyllum, Caraipa, Caryocar, Cassia, Chelonanthus, Chrysophyllum, Curaréa, Erythrina, Geophila, Helicostylis, Hymenaea, Iryanthera, Omphalea, Otoba, Siparuna, Solanum, Tabebuia, Virola, Vismia, Warszewiczia

gastritis (osis): Allium, Cleome, Croton, Cymbopogon, Erythroxylum, Himatanthus, Ilex, Jessenia, Lantana, Malachra, Maytenus, Microtea, Mollis, Ocimum, Piper, Uncaria, Virola

gingivitis: Chrysophyllum, Croton, Jatropha

gonorrhea: Canna, Copaifera, Lycopodium, Mucuna, Uncaria

gout: Lycopodium, Musa, Persea, Psidium

grippe: *Allium*

gums: *Cecropia*

hairdye: Casearia

hairloss: *Bertholletia*

hallucinogen: Banisteriopsis, Brosimum, Brugmansia, Capsicum, Erythroxylum, Gloeospermum, Helicostylis, Ocimum, Paullinia, Scoparia, Tanaecium, Teliostachya, Virola

hangover: Begonia, Ilex

headache: Bonafousia, Cassia, Cymbopogon, Dioclea, Erythrina, Erythroxylum, Faramea, Gnetum, Jatropha, Jessenia, Kalanchoe, Lippia, Mansoa, Pavonia, Petiveria, Pothomorphe, Scoparia, Sida, Sparattanthelium, Spartium, Stachytarpheta, Swartzia, Tanaecium, Zingiber

heart: *Lippia, Cecropia*

heartburn: Kalanchoe

hematochezia: Diplotropis, Erythroxylum

hematomas: Justicia

hemoptysis: Anacardium, Campomanesia

hemorrhoids: Jatropha, Leonia, Mucuna, Psychotria, Scoparia, Victoria, Virola

hemostat (styptic): Asclepias, Brownea, Catharanthus, Clidemia, Costus, Croton, Dimerocostus, Faramea, Helosis, Lantana, Manihot, Piper, Pityogramma, Spondias, Urera, Warscewiczia

hepatitis (osis): *Asplenium, Banara, Bertholettia, Bixa, Cassia, Curcuma, Cyclopeltis, Erythrina, Gossypium, Guarea, Momordica, Nicotiana, Phenakospermum, Phyllanthus, Physalis, Piper, Plantago, Socratea, Swartzia, Tectaria, Uncaria*

herbicide: Duroia

hernia: Artocarpus, Himatanthus, Maquira, Piper, Pothomorphe, Swartzia, Triumfetta

herpes: Caraipa, Cassia, Vismia

hex ("bad luck"): Gallesia, Jatropha, Lantana, Lycopodium, Mansoa, Petiveria, Ruta

hypertension: Cyclanthera, Lantana, Microtea, Pityogramma, Portulaca, Rauwolfia, Siparuna

hypoglycemic: *Chrysophyllum, Phyllanthus*

hypothermia: Lantana

hysteria: Ambrosia

ictericia: Bidens

impetigo: Caraipa,

inappetence: Gnetum, Grias

incontinence: Bidens,

infection: Brugmansia, Cassia, Codonanthe, Eleuterine, Erythrina, Ficus, Genipa, Geophila, Gossypium, Himatanthus, Indigofera, Iryanthera, Jatropha, Macrolobium, Mammea, Martinella, Momordica, Otoba, Phytolacca, Piper, Plantago, Potalia, Schoenobiblus, Simira, Spondias, Symphonia, Vismia, Warszewiczia

inflammation: *Amaranthus, Arrabidaea, Calophyllum, Carapa, Cissampelos, Coix, Costus, Croton, Cucumis, Dimerocostus, Dioscorea, Erythrina, Genipa, Hamelia, Hibiscus, Humiria, Kalanchoe, Malachra, Ocimum, Physalis, Phytolacca, Piper, Pothomorphe, Victoria, Virola*

insanity: Chondrodendron

insecticide: Anthodiscus, Cordia, Euphorbia, Himatanthus, Iryanthera, Lonchocarpus, Melia, Piper, Ryania, Solanum, Tanaecium, Vitex

insectifuge (repellent): Ambelania, Anthodiscus, Aspidosperma, Carapa, Cassia, Chelonanthus, Virola

insomnia: Eryngium, Mimosa, Orthomene

itch: Caraipa, Curaréa, Lantana, Mayna, Mikania, Phytolacca

jaundice: Spartium

kidney: *Bauhinia, Zea*

lactagogue: Alibertia, Bactris, Gossypium,

laryngitis: Chenopodium, Costus, Dimerocostus, Piper

laxative: *Amaranthus, Caryodendron, Copaifera, Dichorisandra, Hura, Jessenia, Mikania, Plukenetia, Ricinus, Sicana*

legache: Mayna, Odontocarya, Tachigalia

leishmaniasis: *Callichlamys, Jacaranda, Monstera, Solanum, Tabebuia*

leprosy: Aspidosperma

leucorrhea: Costus, Dimerocostus, Lycopodium, Urera

liniment: Jessenia, Ricinus

lumbago: Himatanthus, Maquira

lung ailments: *Cordia*

malaria: *Abarema, Ampelozizyphus, Campsiandra, Doliocarpus, Erythrina, Grias, Ilex, Isertia, Ladenbergia, Melia, Mikania, Paullinia, Piper, Pithecellobium, Roucheria, Sabicea, Simarouba, Unonopsis, Virola*

mange: Caraipa, Theobroma, Vatairea

measles: Citrus, Lantana, Momordica, Sambucus,

metrorrhagia: Brownea, Clidemia, Faramea, Hibiscus, Spondias

migraine: *Erythrina, Faramea, Kalanchoe, Mucuna, Nicotiana, Petiveria, Scoparia, Sida, Spartium, Tanaecium, Zingiber*

miscarriage: Cocos

myalgia: Alchornea, Brugmansia, Duguetia, Erisma, Mansoa, Mayna, Ricinus, Urera mycosis: Calycophyllum, Cassia, Chelonanthus, Chrysophyllum, Geophila, Helicostylis, Iryanthera, Kalanchoe, Lantana, Olyra, Omphalea, Otoba, Pistia, Simaruba, Solanum, Virola, Vismia, Warszewiczia

myorelaxant: Abuta, Chondrodendreon, Curaréa,

narcotic: Erythrina, Paullinia, Physalis

nausea: Salvia, Scoparia, Wulffia

nervousness: Abuta, Desmodium, Ilex, Mansoa, Mimosa

niguas: Cassia

ophthalmia: *Cecropia,* Goupia, Kalanchoe, Lepidocaryum, Martinella, Ocimum, Palicourea, Pityogramma, Plantago, Potalia, Spathicalyx, Tabernaemontana, Tagetes

otitis: Kalanchoe

@oxytocic: Cyperus, Ruta

parasiticide: Bothriospora, Dieffenbachia, Ficus, Helicostylis, Mammea, Spigelia, Stomatophyllum

parturition: Abuta, Arrabidaea, Brosimum, Chenopodium, Croton, Erythroxylum, Ficus, Grias,Hibiscus, Maprounea, Oryctanthus, Phoradendron, Phthirusa, Piper, Plantago, Socratea, Spondias, Swartzia, Tabernaemontana, Tococa, Xiphidium

pectoral: Brosimum, Cordia, Justicia, Lantana, Pityogramma, Tabebuia

pediculicide: Caraipa, Sida, Piper, Tanaecium

pertussis (whooping cough) Costus, Dimerocostus, Polypodium, Priva

pharyngitis: Chenopodium, Costus, Dimerocostus, Piper

philtre (love potion): Desmodium, Jacaranda, Maranta, Sloanea

phlegm: Leonia

pimples: Schizaea

piscicide: Anthodiscus, Caesalpinia, Caryocar, Clibadium, Dictyoloma, Diplotrophis, Euphorbia, Jatropha, Lonchocarpus, Manihot, Minquartia, Neoalchornea, Palicourea, Piper, Ryania, Schoenobiblus, Tabernaemontana, Tephrosia

pneumonia: Bougainvillea, Jacaranda, Justicia, Petiveria,

POISON: Abuta, Anomospermum, Asclepias, Bothriospora, Guarea, Guatteria, Hamelia, Helicostylis, Hura, Impatiens, Isotoma, Lonchocarpus, Malouetia, Maquira, Nerium,

Ormosia, Pachyrrhizus, Passiflora, Paullinia, Rauwolfia, Ryania, Socratea, Solanum, Spigelia, Tanaecium, Thevetia

polyuria: Psychotria

pruritis: Caraipa, Phytolacca

psoriasis: Copaifera

pulmonosis (see brcnchosis)

purgative: Anacardium, Bactris, Banisteriopsis, Bixa, Caesalpinia, Cassia, Crescentia, Erythrina, Fevillea, Ficus, Genipa, Grias, Gustavia, Hamelia, Hymenaea, Indigofera, Jatropha, Luffa, Paullinia, Ricinus, Spigelia, Stachytarpheta

rash: Lindernia, Phytolacca

renitis (osis): Bauhinia, Cassia, Malchra, Persea, Phyllanthus, Polypodium, Scleria, Scoparia, Solanum

resolvent: Abuta, Persea, Pothomorphe

rheumatism: *Abuta, Alchornea, Ambrosia, Anaxagorea, Brosimum, Brugmansia, Brunfelsia, Campsiandra, Canna, Caraipa, Clusia, Crateva, Dipteryx, Eleusine, Erythrina, Ficus, Hamelia, Hymenaea, Lantana, Laportea, Mansoa, Maytenus, Mikania, Piper, Pterocarpus, Salix, Solanum, Spartium, Stenosolen, Syzygium, Tabebuia, Tabernaemontana, Tovomita, Triplaris, Tynnanthus, Uncaria, Unonopsis, Urera, Victoria, Xylopia, Zingiber*

rickets: Annona,

rodenticide: Ryania

scabies: Iryanthera, Manihot, Otoba, Solanum, Theobroma, Trichilia, Virola

scar-preventive: Bixa,

scorpion stings: Crotalaria, Solanum

scrofula: Bougainvillea,

scurvy: Hamelia

sedative (soporific): Annona, Bactris, Brugmansia, Lantana, Lippia, Melissa. Mimosa, Orthomene, Spigelia, Xylopia

@shampoo: Solanum

shock: *Desmodium, Mansoa, Psidium (emotional)*

shyness: Desmodium

sinusitis: *Jatropha, Luffa, Petiveria, Solanum*

snakebite: Abuta, Adiantum, Andropogon, Brunfelsia, Cordia, Cyclanthus, Cyperus, Dipteryx, Dracontium, Hibiscus, Lindsaea, Mikania, Mucuna, Nicotiana, Persea, Potalia, Solanum, Tephrosia, Trichomanes, Xiphidium

snoring: Cyperus,

soap: Bertholettia, Carapa, Phytolacca

soporific: Hyptis, Lippia

sores (external ulcers): Canna, Cassia, Codonanthe, Copaifera, Croton, Ficus, Genipa, Indigofera, Kalanchoe, Machaerium, Macrolobium, Martinella, Melia, Miconia, Monstera, Muntingia, Omphalea, Orthomene, Picramnia, Piper, Poeppigia, Pothomorphe, Psychotria, Solanum, Symphonia, Tabebuia, Virola, Vismia

sorethroat: Capsicum, Erythroxylum, Olyra, Tabernaemontana

spice: Allium, Capsicum, Coriandrum, Eryngium, Ocimum

spiderbite: Lycopodium, Mucuna, Solanum

sprain: Anthodiscus, Nicotiana

sterility: Abuta, Guarea, Manihot, Spondias,

stimulant: *Copaifera, Anadenanthera, Dipteryx, Ficus, Maytenus, Murraya, Paullinia, Piper, Plumeria, Sambucus, Tachigalia, Tynnanthus, Xylopia*

stingray: Pentagonia, Potalia

stomachache: Adiantum, Cassia, Chelonanthus, Chenopodium, Costus, Cyclanthus, Cyclopeltis, Cymbopogon, Cyperus, Dimerocostus, Diospyros, Erythroxylum, Ficus, Herrania, Ilex, Jatropha, Jessenia, Justicia, Lantana, Lippia, Malachra, Maytenus, Microtea, Mollia, Ocimum, Physalis, Piper, Pityogramma, Priva, Sabicea, Sauvagesia, Selaginella, Sparattanthelium, Spigelia, Spondias, Tectaria, Teliostachya, Virola, Witheringia, Xylopia, Zingiber

stomachic: Crateva, Indigofera, Lantana, Mikania, Momordica, Persea, Sambucus

stomatosis: Banisteriospsis,

stones (calculus): Persea, Phyllanthus, Scleria

sudorific: Lantana, Petiveria, Pothomorphe, Salix, Sambucus

suppurative: Momordica

"susto": Desmodium, Xylopia

swellings: Brugmansia, Caperonia, Codonanthe, Eleusine, Geogenanthus, Jatropha, Microtea, Pentagonia, Portulaca, Potalia, Pothomorphe, Scoparia, Virola

syphilis: Brunfelsia, Hamelia, Indigofera, Ilex, Melia, Potalia, Tephrosia

tachycardia: Lippia

tuberculosis: Jessenia, Musa, Spondias

tetanus: Eleuterine,

tick-repellent: Carapa,

tonic (stimulant): Abuta, Ambrosia, Brosimum, Carapa, Crateva, Dipteryx, Hibiscus, Jessenia, Lantana, Mansoa, Mikania, Odontadenia, Paullinia, Persea, Piper, Remijia, Schizaea, Spigelia, Tachigalia, Tonina, Xylopia.

tonsilitis: Bixa, Costus, Dimerocostus,

toothache: Abuta, Asclepias, Capsicum, Chlorophora, Coix, Couroupita, Crescentia, Erythroxylum, Ficus, Fittonia, Humiria, Jatropha, Securidaca, Stigmaphyllum, Tabernaemontana, Virola, Wedelia, Zingiber

tooth-black: Arrabidaea, Calatola, Duroia, Lantana, Manettia, Neea, Piper

tooth-extraction: Asclepias, Chlorophora, Genipa, Stigmaphyllum

tranquilizer (entered late): Spigelia, Xylopia

trauma: Cassia, Plantago, Pothomorphe

tremors: Dracontium, Eleuterine,

tuberculosis: Bauhinia, Mansoa, Jessenia

tumors (malignant sores): *Anacardium*, Caperonia, Chenopodium, Cocos, Himatanthus, Leonia, Miconia, Muntingia, Solanum, Tabebuia

ulcer(internal): Croton, Genipa, Himatanthus, Indigofera

urethritis: Kalanchoe, Potalia, Sambucus

uricosuric: Persea

uterosis: Triumfetta

uterotonic: Lantana, Pavonia

vaginitis: *Costus, Desmodium, Dimerocostus, Genipa, Spondias*

NOTE: Not a complete index, just a hastily contrived summary index, taking many liberties. A rash might have been entered under both rash and dermatitis. Going through several translations, the references above may refer either to the translation or the original. Medicinal terms, like common names, have different meanings in different places.(I did not always index uses that are clearly external to Amazonia)

MAJOR REFERENCES

Ayala Flore, F. 1984. Notes on Some Medicinal and Poisonous Plants of Amazonian Peru. pp. 1-8 in *Advances in Economic Botany* 1: 1984. (Cited as AYA)

Balick, M.J. 1985. Useful Plants of Amazonia: A Resource of Global Importance. Chap. 19 in Prance, G.T. and Lovejoy, T.E., eds. *Amazonia*. Pergamon Press. 1985. (Cited as MJB)

Balick, M.J. and Gersgoff, S.N. 1990. A Nutritional Study of *Aiphanes caryotifolia* (Kunth) Wendl. (Palmae) Fruit: An Exceptional Source of Vitamin A and High Quality Protein from Tropical America. *Advances in Econ. Bot.* 8:35-40. (Cited as MJB)

Branch, L.C. and da Silva, I.M.F. 1983. Folk Medicine of Alter do Chao, Para, Brazil. *Acta Amazonica* 13(5/6):737-797. Manaus. (Cited as BDS)

Cayon, A.E. and Aristizabal, G.S. 1980. List of Plants Used by the Indigenous Chami of Riseralda (in Spanish). *Cespedesia* 9(33-4):5-115. (Cited as CAA)

CSIR (Council of Scientific and Industrial Research). 1948-1976. *The Wealth of India*. 11 vols. New Delhi. (Cited as WOI)

Denevan, W.M. and Treacy, J.M. 1988. Young Managed Fallows at Brillo Nuevo. pp. 8-46 in Denevan, W.M. and Padoch, C. Swidden-Fallow Agroforestry in the Peruvian Amazon. *Advances in Econ. Bot.* 5. New York Botanical Garden, NY. 107 pp. (Cited as DAT)

Duke J.A. 1985. *CRC Handbook of Medicinal Herbs*. CRC Press, Boca Raton, FL. 677 pp. (Cited as CRC)

Duke J.A. 1986a. *Handbook of Northeastern Indian Medicinal Plants*. Quarterman Publications, 212 pp., Lincoln, MA. (Cited as JAD)

Duke J.A. 1986b. *Isthmian Ethnobotanical Dictionary*. Third Edition, 325 pp, Scientific Publishers, Jodhpur, India. (Cited as JAD)

Duke J.A. 1989. *CRC Handbook of Nuts*. CRC Press, Inc., Boca Raton, FL. 343 pp. (Cited as CRC)

Duke J.A. 1992a. *CRC Handbook of Biologically Active Phytochemicals and their Bioactivities*. CRC Press, Boca Raton, FL. (Published both as hardcopy book and as WordPerfect Database). 183 pp. (Cited as CRC)

Duke J.A. 1992b. *CRC Handbook of Phytochemical Constituents of GRAS Herbs and Other Economic Plants*. CRC Press, Boca Raton, FL. (Published both as hardcopy book and as WordPerfect Database). 654 pp. (Cited as CRC)

Duke J.A. 1992c. *CRC Handbook of Edible Weeds*. CRC Press, Boca Raton, FL. 246 pp. (Cited as CRC)

Duke, P.K. sin date. Illsutrations published in some of J.A. Duke's books. (Cited as PKD)

Duke, J.A. and duCellier, J.L. 1993. *CRC Handbook of Alternative Cash Crops*. CRC Press, Inc., Boca Raton, FL, 536 pp. (Cited as DAD)

Duke J.A. and Wain, K.K. 1981. *Medicinal Plants of the World*. Computer index with more than 85,000 entries, 3 vols. (Cited as DAW)

Elisabetsky, E. and Posey, D.A. 1989. Use of Contraceptive and Related Plants by the Kayapo Indians (Brazil). *J. Ethnopharm.* 26:299-316. (Cited as EAP)

FAO. 1986. *Some medicinal forest plants of Africa and Latin America*. Food and Agriculture Organization of the United Nations. Rome, 1986. (Cited as FAO)

de Feo, V. 1992. Medicinal and magical plants in the northern Peruvian Andes. Fitoterapia 63:417-440. (Cited as FEO)

Ferreyra, R. 1970. *Flora Invasora de los Cultivos de Pucallpi y Tingo Maria*. (Cited as RAF)

Forero, P.L.E. 1980. Ethnobotany of the Cuna and Waunana Indigenous Communities, Choco (Colombia) (in Spanish). *Cespedesia* 9(33):115-302. (Cited as FOR)

Garcia-Barriga, H. 1974-5. *Flora Medicinal de Colombia, Botanica-Medica*. Inst. Cien. Nat. Bogota. 3 vols. (Cited as GAB)

Gentry, A.H. 1993. *A Field Guide to the Families and Genera of Woody Plants of Northwest South America (Colombia, Ecuador, Peru)*. Illustrations by R. Vasquez Martinez. Conservation International. Washington, DC. 895 pp. (Cited as GAV, source of most of the illustrations)

Grenand, P., Moretti, C. and Jacquemin, H. 1987. *Pharmacopées taditionnels en Guyane: Créoles, Palikur, Wayãpi*. Editorial l-ORSTROM, Coll. Mem. No. 108. Paris 569 pp. (Cited as GMJ)

Hartwell, J.L. 1982. *Plants used against cancer*. A survey. Reissued in one volume by Quarterman Publications, Inc. Lawrence, MA. 710 pp. (Cited as JLH)

Hitchcock, A.S. 1950. *Manual of the Grasses of the United States*. 2nd Ed. Revised by A. Chase. USGPO, Washington, DC. 1051 pp. (Cited as HAC)

Lamb, F.B. 1985. *Rio Tigre and Beyond, the Amazon Jungle Medicine of Manuel Cordova*. North Atlantic Books, Berkeley, 227 pp. (Cited as FBL)

Leon, J. 1987. *Botanica de los Cultivos Tropicales*. Instituto Interamericano de Cooperacion Para la Agricultura. San José, Costa Rica. 445 pp. (Cited as IIC)

Lewis, W. and Elvin-Lewis, M., 1977. *Medical Botany*. John Wiley & Sons, NY. 515 pp. (Cited as LAE)

List, P.H. and Hohammer, L., 1969-1979. *Hager's Handbuch der Pharmazeutischen Praxis*, Vols. 2-6, Springer-Verlag, Berlin. (Cited as HHB)

Little, E.L., Jr. and Wadsworth, F.H. 1964. *Common Trees of Puerto Rico and the Virgin Islands*. Ag. Handbook No. 249, USDA, Washington. 548 pp. (Cited as LAW)

MacBride, J.F. 1936-. *Flora of Perú*. Field Museum of Natural History, Botanical Services, Chicago. (Cited as MAC)

Maxwell, N. 1990. *Witch Doctor's Apprentice, Hunting for Medicinal Plants in the Amazonian*, 3rd Edition, Citadel Press, New York. 391 pp. (Cited as NIC)

Missouri Botanical Garden. 1993. *Florula de las Reservas Biologicas de Iquitos*. Computer Printout. (Cited as RBI)

Mors, W.B. and Rizzini, C.T., 1966. *Useful Plants of Brazil*, Holden-Day, Inc., San Francisco, Calif. (Cited as MAR)

Morton, J.F., 1981. *Atlas of Medicinal Plants of Middle America, Bahamas to Yucatan*, C.C. Thomas, Publisher, Springfield, IL. (Cited as JFM)

Perez-Arbelaez, E., 1956. *Plantas Utiles de Colombia*, Liberia Colombiana, Bogota. (Cited as PEA)

Plotkin, M.J. 1993. *Tales of a Shaman's Apprentice*. Viking Press, NY. 318 pp. (Cited as MJP)

Poveda, L.J. 1985-6. *Marvels of our Medicinal Flora* (in Spanish). Biocenosis Vols. 1 and 2. (Cited as POV)

Rutter, R.A. 1990. *Catalogo de Plantas Utiles de la Amazonia Peruana*. Instituto Linguistico de Verano. Yarinacocha, Peru. 349. (Cited as RAR)

Schultes, R.E. and Raffauf, R.F. 1990. *The Healing Forest*. Dioscorides Press, Portland, 484 pp. (Cited as SAR)

Soukup, J. 1970. *Vocabulary of the Common Names of the Peruvian Flora and Catalog of the Genera*. Editorial Salesiano, Lima. 436 pp. (Cited as SOU)

Robineau, L., Ed. 1991. *Towards a Caribbean pharmacopoeia*, TRAMIL-4 Workshop, UNAH, Enda Caribe, Santo Domingo. (Cited as TRA)

Valdizan, H. and Maldonado, A. 1982. *La Medicina Popular Peruana* (Documentos Ilustrativos). Imp. Torres Aguirre. Lima. 3 vols. (Cited as VAM)

Vasquez M., R. 1990. *Useful Plants of Amazonian Peru*. Spanish Typescript. Second Draft. Filed with USDA's National Agricultural Library. (Cited as RVM)

Vilmorin-Andrieux, M.M. 1976 (reprint of the 1885 English edition). *The Vegetable Garden. Illustrations, Descriptions, and Culture of the Garden Vegetables of Cold and Temperate Climates*. The Jeavons-Leler Press, Palo Alto, CA. 620 pp. (Cited as MVA)